PRINCIPLES OF SYSTEMS

Jay W. Forrester

Productivity Press
Portland, Oregon

First published in 1971 by Wright-Allen Press, Inc.

First reprinted by Productivity Press, Inc. in 1990.

Productivity Press
P.O. Box 13390
Portland, OR 97213-0390
United States of America
Telephone: (503) 235-0600
Telefax: (503) 235-0909

ISBN: 1-915299-87-9

Cover design by William Stanton
Printed and bound by Edwards Brothers in the United States of America

98 97 96 95 94 10 9 8 7 6 5 4 3 2

PRINCIPLES OF SYSTEMS

Jay W. Forrester

Table of Contents

Contents

CHAPTER 1

SYSTEMS

1.1 The Ubiquity of Systems

Man lives and works within social systems. His scientific research is exposing the structure of nature's systems. His technology has produced complex physical systems. But even so, the principles governing the behavior of systems are not widely understood.

As used here a "system" means a grouping of parts that operate together for a common purpose. An automobile is a system of components that work together to provide transportation. An autopilot and an airplane form a system for flying at a specified altitude. A warehouse and loading platform is a system for delivering goods into trucks.

A system may include people as well as physical parts. The stock clerk and office workers are part of the warehouse system. Management is a system of people for allocating resources and regulating the activity of a business. A family is a system for living and raising children.

If systems are so pervasive, why do not the concepts and principles of systems appear more clearly in our literature and in education? Is it because there has been no need for understanding the basic nature of systems? Or have systems seemed to possess no general theory and meaning? Or is it because the principles of systems, while sought after, have been so obscure that they have evaded detection? The answer seems to have been each of these three in turn.

In a primitive society, the existing systems were those arising in nature and their characteristics were accepted as divinely given and as being beyond man's comprehension or control. Man simply adjusted himself to the natural systems around him and to the family and tribal social systems which were created by gradual evolution rather than by design. Man adapted to systems without feeling compelled to understand them.

As industrial societies emerged, systems began to dominate life as they manifested themselves in economic cycles, political turmoil, recurring financial panics, fluctuating employment, and unstable prices. But these social systems suddenly became so complex and their behavior so confusing that no general theory seemed possible. A search for orderly structure, for cause and effect relationships, and for a theory to explain system behavior gave way at times to a belief in random, irrational causes.

Gradually over the last hundred years it has become clear that the barrier to understanding systems has been, not the absence of important general concepts, but only the difficulty in identifying and expressing the body of universal principles that explain the successes and failures of the systems of which we are a part. Economics has identified many basic relationships within our industrial system. Psychology and religion have described some of the interactions between systems of people. Medicine has treated biological systems. Political science has explored governmental and international systems. But most such analysis has been verbal and qualitative. Mere description has not been sufficient to expose the true nature of systems. Mathematics, which has been used to structure knowledge in science, has not been adequate for handling the essential realities of our important social systems. We have been overwhelmed by fragments of knowledge but have had no way to structure this knowledge.

(See Section W1.1 of the accompanying Workbook)

1.2 System Principles as the Structure of Knowledge

A structure (or theory) is essential if we are to effectively interrelate and interpret our observations in any field of knowledge. Without an integrating structure, information remains a hodge-podge of fragments. Without an organizing structure, knowledge is a mere collection of observations, practices and conflicting incidents.

Such a state of unrelated facts describes much of our knowledge about managerial and economic systems. Our separate and often conflicting impressions have not yet been brought into focus by being assembled into a unified structure. Without a structure to interrelate facts and observations, it is difficult to learn from experience, it is difficult to use the past to educate for the future.

The importance of structure in education is well argued by Jerome S. Bruner of Harvard.[1] He says, "Grasping the structure of a subject is understanding it in a way that permits many other things to be related to it meaningfully. To learn structure, in short, is to learn how things are related.....good teaching that emphasizes the structure of a subject is probably even more valuable for the less able student than for the gifted one, for it is the former rather than the latter who is most easily thrown off the track.....There are two ways in which learning serves the future. One is through its specific applicability to tasks that are highly similar to those originally learned.....A second way is through the transfer of principles or attitudes.....the continuity of learning that is produced by the second type of transfer, transfer of principles, is dependent upon mastery of the structure of the subject matter.....Inherent in the preceding discussions are at least four general claims that can be made for teaching the fundamental structure of a subject. The first is that understanding fundamentals makes a subject more comprehensible.....The second relates to human memory. Perhaps the most basic thing that can be said about human memory, after a century of intensive research, is that unless detail is placed into a structured pattern, it is rapidly forgotten.....Third, an understanding of fundamental principles and ideas, as noted earlier, appears to be the main road to adequate 'transfer of training.' To understand something as a specific instance of a more general case--which is what understanding a more fundamental principle or structure means--is to have learned not only a specific thing but also a model for understanding other things like it that one may encounter.....The fourth claim for emphasis on structure and principles in teaching is that by constantly reexamining material one is able to narrow the gap between 'advanced' knowledge and 'elementary' knowledge."

---

[1] Bruner, Jerome S., The Process of Education, Harvard University Press, 1960. A short book and well worth reading.

The laws of physics form a structure to interrelate our many observations about nature. This structure of physical knowledge is the foundation for today's technology.

But in management systems, such a basic structure of principles is only now being developed. Managers and educators have long searched for a structure to unify the diverse manifestations of psychological, industrial, and economic processes. Management education has been criticized as being merely descriptive without a unifying structure. Indeed, structure has long been pursued, even though the nature of a suitable structure was elusive.

But now the concepts of "feedback" systems seem to be emerging as the long-sought basis for structuring our observations of social systems. Over the last century the theory of systems has slowly been developed to apply to mechanical and electrical systems. However, physical systems are far simpler than social and biological systems and it is only in the last decade that the principles of dynamic interactions in systems have been developed far enough to become practical and useful in dealing with systems of people.

Around the system principles discussed in this book it should be possible to structure our confusing observations about political and business systems. When a structure and governing principles for systems have been accepted, they should go far to explain the contradictions, clarify the ambiguities, and resolve the controversies in the social sciences. A systems structure should give to education in human affairs the same impetus that the structure of physical laws has given to technology. The social sciences should become easier to teach if they can rest on a body of principles that are common to all systems, be they human systems or technical systems. In the concepts of systems we should find a common foundation that underlies and unites the "two cultures" of the sciences and humanities. Education in many areas should be accelerated. As Bruner says, "structure.....is able to narrow the gap between 'advanced' knowledge and 'elementary' knowledge."

This book deals with the structure and principles of systems giving special emphasis to systems in economics and industrial organization and to systems that combine people, finance, and technology.

(See Section W1.2 of the accompanying Workbook)

1.3 Systems--Open and Feedback

Systems can be classified as "open" systems or "feedback" systems.

An open system is one characterized by outputs that respond to inputs but where the outputs are isolated from and have no influence on the inputs. An open system is not aware of its own performance. In an open system, past action does not control future action. An open system does not observe and react to its own performance. An automobile is an open system which by itself is not governed by where it has gone in the past nor does it have a goal of where to go in the future. A watch, taken by itself, does not observe its own inaccuracy and adjust itself--it is an open system.

A feedback system, which is sometimes called a "closed" system, is influenced by its own past behavior. A feedback system has a closed loop structure that brings results from past action of the system back to control future action. One class of feedback system--negative feedback--seeks a goal and responds as a consequence of failing to achieve the goal. A second class of feedback system--positive feedback--generates growth processes wherein action builds a result that generates still greater action.

A feedback system controls action based on the results from previous action. The heating system of a house is controlled by a thermostat which responds to the heat previously produced by the furnace. Because the heat already produced by the system controls the forthcoming generation of heat, the heating system represents a negative feedback system that seeks the goal of proper temperature. A watch and its owner form a negative feedback system when the watch is compared with the correct time as a goal and is adjusted to eliminate errors. An engine with a governor senses its own speed and adjusts the throttle to achieve the preset speed goal--it is a negative feedback system. Bacteria multiply to produce more bacteria which increase the rate at which new bacteria are generated. In this positive feedback system the generation rate of new bacteria depends on the bacteria accumulated from past growth of bacteria.

Whether a system should be classified as an open system or a feedback system is not intrinsic to the particular assembly of parts but depends on the observer's viewpoint in defining the purpose of the system.

The way in which the purpose of the system determines whether it is an open or a feedback system can be illustrated by considering a gasoline engine in terms of a series of viewpoints.

1. The engine, operating without a governor, has no goal for speed. It is an open system in terms of speed regulation. Changing the throttle will change the speed but the speed has no effect on the throttle. Furthermore, changes in load will change the speed without causing a throttle adjustment.

2. Adding a governor produces a feedback system in terms of a constant-speed goal. Changes in load cause changes in speed which produce a compensating change in throttle setting as the governor tries to hold the speed for which it has been set.

3. But suppose the engine is part of a lawn mower and we change the goal from constant-speed operation to a goal of mowing the lawn. Now, from the broader purpose of cutting grass, the lawn mower is an open system because it has no awareness of what grass has been cut or where to cut next.

4. By adding the person operating the lawn mower, we again see a feedback system in terms of the goal of cutting a particular lawn. The operator and mower form a feedback system (that is, a goal seeking system) rather than an open system (that is, one not striving for an objective) because the guidance of the mower is in accordance with the pattern of grass already cut.

5. But if the viewpoint is broadened again to that of the owner of a lawn-care enterprise with a goal of meeting his customer demands, the operator and his lawn mower are properly considered a component of a larger management system. As such, the operator and his equipment represent an open system that is undirected in its sequence of separate tasks.

6. By adding the management function, instructions arising from customer requirements are introduced as a guide. In terms of the goal of properly scheduled work, the operator, equipment, and owner must be taken together to form a feedback system for the purpose of serving customer lawn-care needs.

A broad purpose may imply a feedback system having many components. But each component can itself be a feedback system in terms of some subordinate purpose. One must then recognize a hierarchy of feedback structures where the broadest purpose of interest determines the scope of the pertinent system.

This book is devoted to the theory, principles, and behavior of feedback systems. It is in the positive feedback form of system structure that one finds the forces of growth. It is in the negative feedback, or goal-seeking, structure of systems that one finds the causes of fluctuation and instability.

(See Section W1.3 of the accompanying Workbook)

1.4 The Feedback Loop

The basic structure of a feedback loop appears in Figure 1.4a. The feedback loop is a closed path connecting in sequence a decision that controls action, the level* of the system, and information about the level of the system, the latter returning to the decision-making point.

---

*The term "level" is used in this book to mean a state or condition of the system.

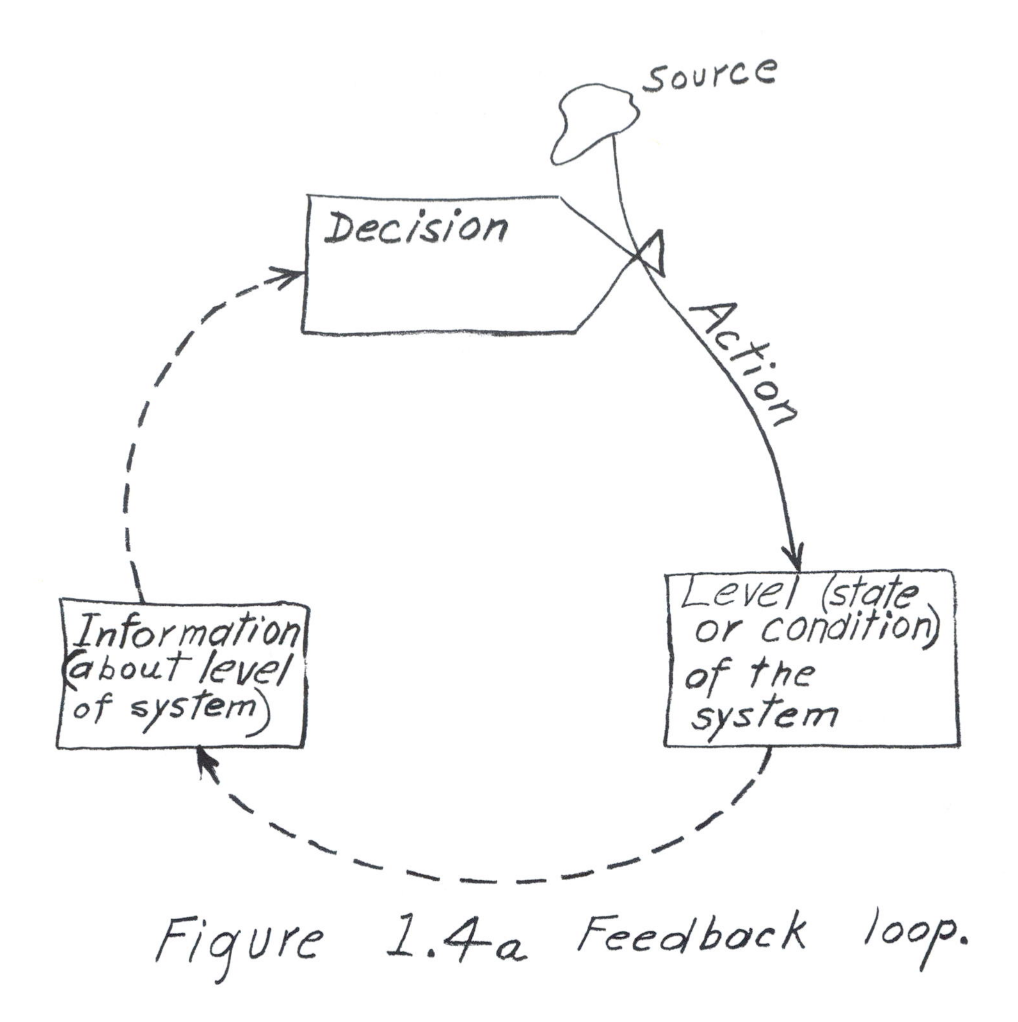

Figure 1.4a Feedback loop.

The available information, as it exists at any moment, is the basis for the current decision that controls the action stream. The action alters the level of the system. The level (true level) of the system is the generator of information about the system, but the information itself may be late or erroneous. The information is the apparent level of the system which may differ from the true level. It is the information (apparent level), not the true level, that is the basis for the decision process.

(Section 1.4)

The single-loop structure of Figure 1.4a is the simplest form of feedback system. There may be additional delays and distortions appearing sequentially in the loop. There may be many loops that interconnect.

Ordering replacement goods to maintain an inventory in a warehouse illustrates the circular cause-and-effect structure of the feedback loop as in Figure 1.4b.

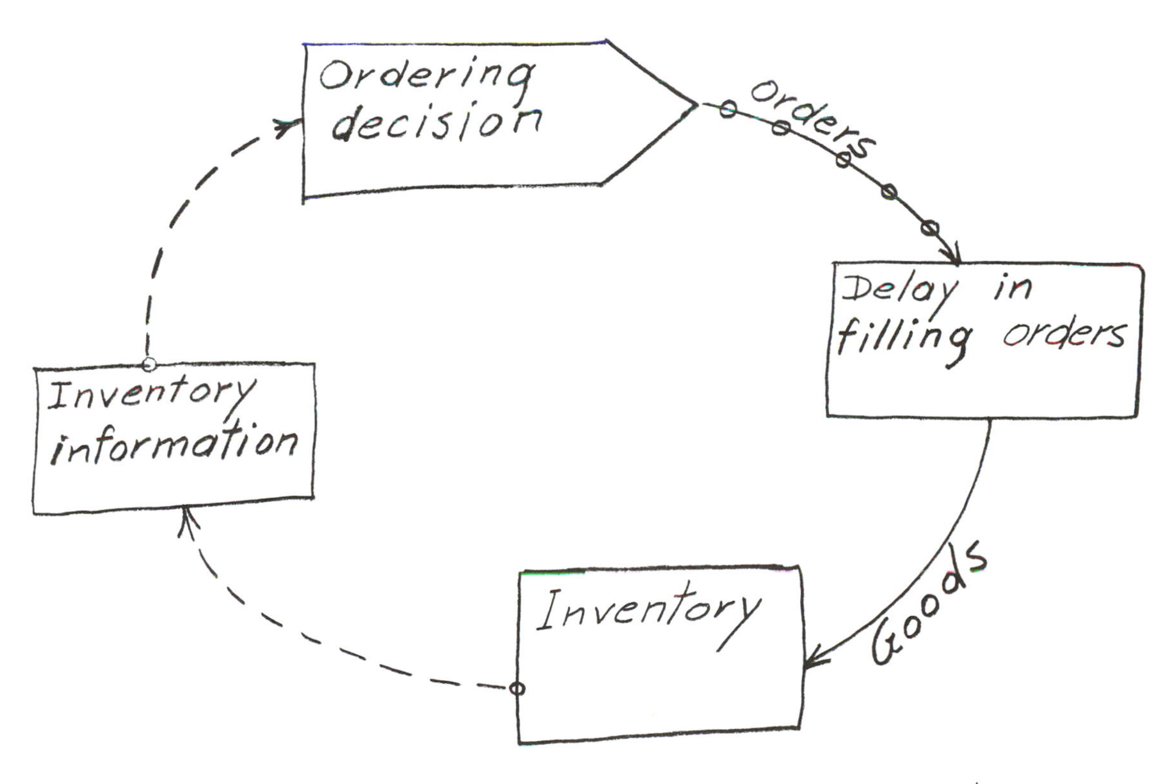

Figure 1.4b Inventory ordering loop.

Here the ordering decision generates a stream of orders to the supplier. The supplier, after a delay to ship or to manufacture, delivers the goods to the inventory. The inventory is the source of information about the inventory, but that information may contain errors and may be delayed so that it does not always reflect the true present level of the inventory. The information about inventory is the input on which the ordering decision is based. (In a more complete system there can, of course, be other inputs to the ordering decision.) The inventory-control loop is in continuous operation. Changes can be occurring at all times at each point around the loop.

The present action stream corresponds to the present decision that in turn depends on the present information. However, the present level of the system does not depend on the present action but is instead an accumulation from all past actions. For example, consider a tank that is being filled with water. The height of the water is the system level. The level depends on the accumulation produced by the past flow of water but the level is not determined by how fast water is being added at the present instant. A large stream into an empty tank does not imply a full tank, an already filled tank is not affected if the flow ceases entirely.

Information is itself one of the levels of the system (referred to earlier as apparent level). The information changes as it becomes evident that the information differs from the true variable that it is presumed to represent. Information is not determined by the present true condition, which is not instantaneously nor exactly available, but instead by the past conditions that have been observed, transmitted, analyzed and digested. The discrepancy between a true system level and the information level that governs decisions always exists in principle. As a practical matter, the information is sometimes good enough that no distinction is necessary between true and apparent level.

(See Section W1.4 of the accompanying Workbook)

CHAPTER 2

PREVIEW OF FEEDBACK DYNAMICS

2.1 Diversity of Behavior

The interplay of activity within "negative" feedback loops can range from smooth achievement of the goal that the loop is seeking, to wild fluctuation in search of the goal. "Positive" feedback loops show growth or decline. "Nonlinear" coupling can cause a shift of dominance from one system loop to another. As an introduction to the dynamic (that is, the time-varying) behavior of feedback loops, this chapter presents several feedback systems to illustrate the kinds of time responses sketched in Figure 2.1. These curves show the changing value of some system variable as time progresses from left to right.

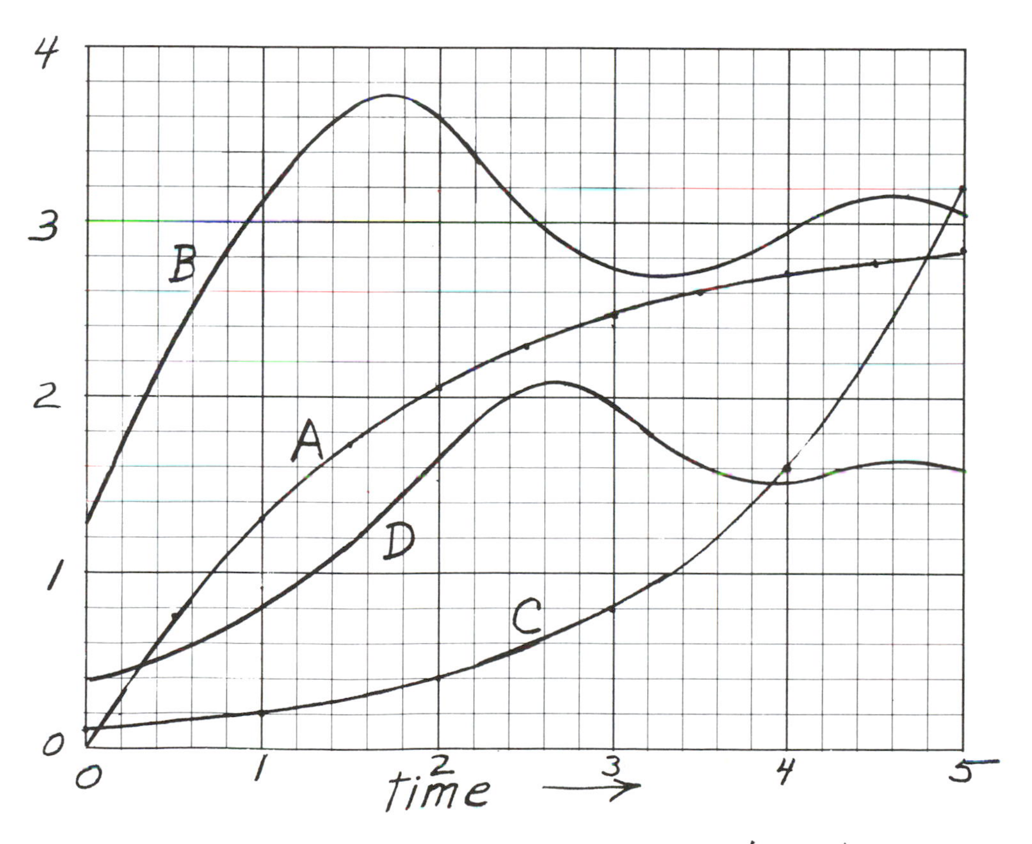

Figure 2.1 Dynamic behavior

Curve A is typical of the simplest kind of feedback system in which the variable rises at a decreasing rate toward a final value (here a value of 3). The value of the variable shown by Curve A here happens to be zero at the start, where the start is defined as time being zero. The value rises to 1.3 at time 1, to 2.05 at time 2, and continues to approach, without actually reaching, an ultimate value of 3. Such a simple approach to equilibrium might represent the increase in an employee group as hiring expands the group toward an authorized level. Or the curve might represent information that gives the apparent level (that is, the condition or state) of a system as understanding increases toward the true value. Or the curve might represent the way the water in a toilet tank approaches the final full level. In all these, the change toward the final value is more rapid at first and approaches more and more slowly as the discrepancy decreases between present and final value.

Curve B is a more complicated approach to the final value (here again the value of 3) where the system overshoots the final value, then falls below in trying to recover from the earlier overshoot. Such behavior can result from excessive time delays in the feedback loop or from too violent an effort to correct a discrepancy between apparent system level and the system goal. Such fluctuation can be observed in the unsteady speed as a defective governor tries to regulate an engine, in the rise and fall of industrial production as seen in economic cycles, in the fluctuation of commodity prices as supply and demand seek one another, and in a drunk trying to put a key in a keyhole.

Curve C shows growth where there is, in each succeeding time interval, the same fractional increase in the variable. In the illustration, the vertical value doubles in each unit of time. Such "exponential" growth is seen in cell division, in the sales growth of a product where salesmen produce sales to yield revenue to hire more salesmen, in the chain reaction of an atomic explosion, and in the multiplication of rabbits.

Curve D shows an initial section of exponential growth followed by a leveling out. Curve D is a composite of an early section similar to Curve C that yields to a later section having the characteristics of Curve B (the later section could be as in Curve A with no overshoot beyond the final value). Such behavior (without overshoot beyond the final value) is seen in the growth of an animal which at first is increasingly rapid and then slows in its approach to final size. This kind of growth that gives way to a continuing balance might also represent a rabbit population that rises rapidly to the point where the food supply is overtaxed and no more rabbits can be supported. Curve D is seen in the nuclear activity of an atomic power plant as fission rate rises to the operating level and is moderated by the control system. It represents the early growth of a product that stagnates because market demand has been satisfied, or because production capacity has been reached, or because quality may have declined.

In the following sections several systems will be examined to see how the foregoing behavior patterns can occur.

(See Workbook Section W2.1)

2.2 First-Order, Negative-Feedback Loop

The simplest structure to be found in feedback loops appears in Figure 2.2a. Here a single decision (order rate) controls the input to one system level (inventory). There is no delay or distortion in the information channel going from inventory I to the ordering decision OR, that is, the apparent system level is assumed identical to the actual system level.

The loop in Figure 2.2a is classified as a "first-order" system because there is only one level variable (the inventory).*

---

*In these flow diagrams, which will be given in more detail in Chapter 7, the rectangle always represents a level variable, the valve (order rate in Figure 2.2a) represents a rate variable.

The diagram illustrates an elementary inventory control system where there is no delay between the ordering of goods and their receipt into inventory. For the present, assume that the order rate OR can be either positive or negative, that is, goods can either be ordered into inventory or returned to the supplier. The goal of the system is to maintain the desired inventory DI which is shown on the diagram as a constant in the decision process.

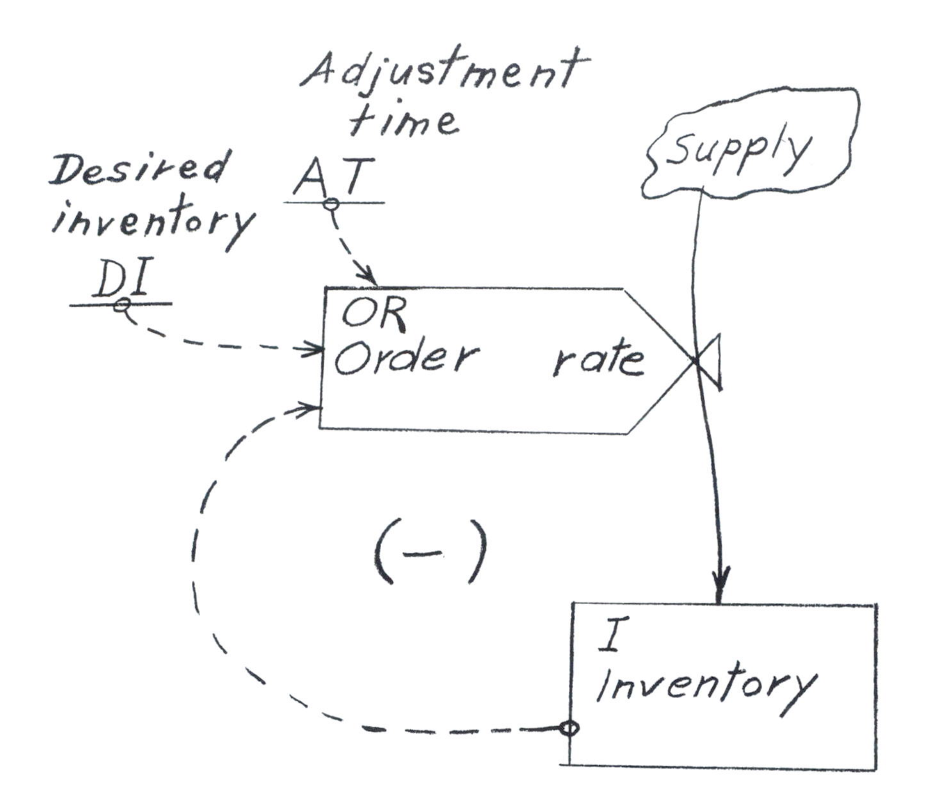

Figure 2.2a First-order negative feedback.

The order rate decision, if it is to bring actual inventory toward desired inventory, must increase the order rate as inventory falls below desired inventory. Conversely, as inventory rises toward the desired inventory, order rate should be reduced. If inventory becomes greater than desired inventory, the order rate should become increasingly negative, indicating that goods are being returned to the supply. As a very simple ordering policy, we could arbitrarily specify that order rate depends on the difference between desired inventory and actual inventory. This might at first seem to be accomplished by expressing the order rate as:

$$OR = DI - I$$

OR--Order rate	(units/week)
DI--Desired inventory	(units)
I--Inventory	(units)

But such an equation is not dimensionally correct. On the left is a term measured in "units/week"; on the right the terms are measured in "units." The preceding paragraph says "order rate depends on the difference between desired inventory and actual inventory," but what is the nature of that dependence? Does order rate rise rapidly or slowly as inventory declines? Is it proportional to the inventory discrepancy? Possible answers to these questions are illustrated graphically by curves showing relationships between inventory and order rate as in Figure 2.2b. When inventory equals desired inventory, the

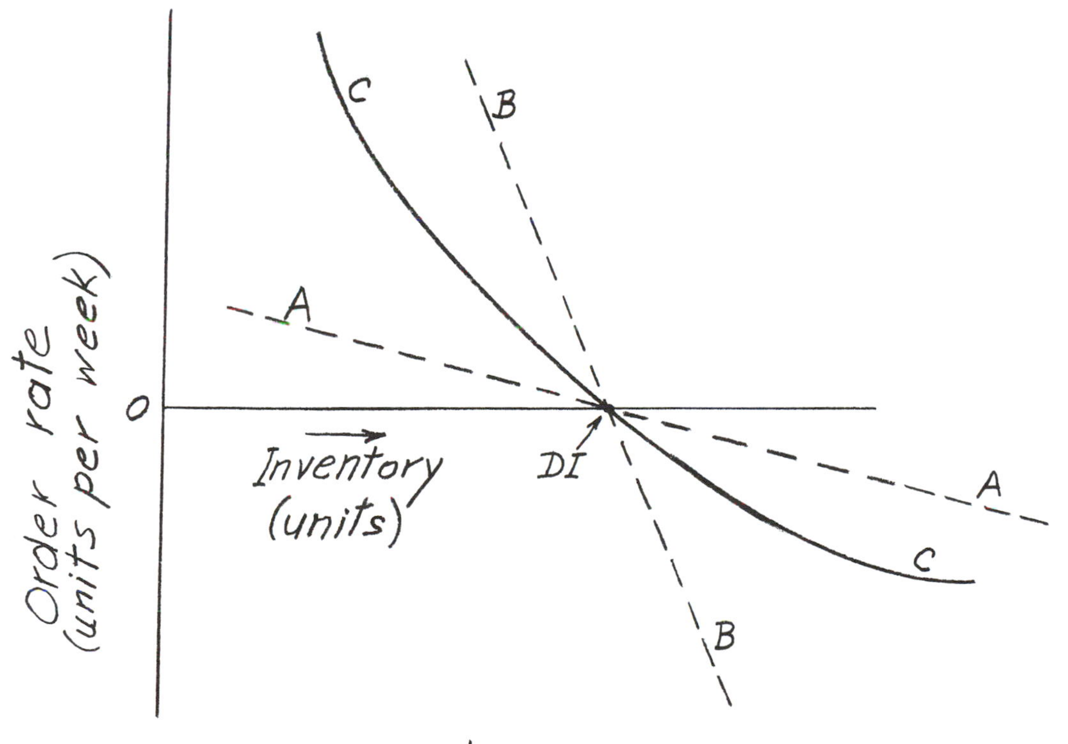

Figure 2.2b Order rate.

order rate should be zero to fit the foregoing description of intended ordering policy. But, as inventory declines below desired inventory DI, how fast should order rate rise? As in Curve A? Or faster as in Curve B? Or slowly at first and then more rapidly as in Curve C? (Curves A and B are "linear," that is, the order rate is proportional to the inventory discrepancy as seen by the straight-line relationship. Curve C is "nonlinear" as seen by the way the order rate line bends as inventory changes.)

So, to be complete, the ordering rate equation should specify how the order rate depends on changes of inventory. Suppose we wish to use a simple linear relationship like Lines A or B. Another term is required in the equation for order rate to determine which of the possible straight-line relationships is desired. That term will specify how rapidly the inventory discrepancy is to be corrected. It must also yield the dimensions "units per week" on the right side of the order rate equation. The term by itself will indicate the units per week of order rate for each unit of inventory discrepancy so it will have the measure

$$\frac{\text{units/week}}{\text{unit}} = \frac{1}{\text{week}}$$

It will determine the slope of the proper line in Figure 2.2b. Such is accomplished in the following:

$$OR = \frac{1}{AT}(DI-I) \qquad \text{Eq. 2.2-1}$$

OR--Order rate (units/week)
AT--Adjustment time (weeks)
DI--Desired inventory (units)
I--Inventory (units)

This is a dimensionally correct statement of a simple, arbitrary, ordering policy. It says that ordering is to be at the rate of one ATth per week of the difference between desired and actual inventory.

If given the constants AT and DI and the variable I, one can compute the order rate OR in Equation 2.2-1. The units of inventory error (that is, the discrepancy between desired and actual inventory) are divided by the time AT measured in weeks. The time AT is the time that would be required to correct the inventory if the order rate OR were to persist without changing (of course, as goods flow into inventory, the inventory changes and as a result the order rate changes).

Feedback systems are of interest because of the way they act through time. As a simple introduction to dynamic (that is, time-varying) behavior we can examine how the foregoing inventory-control loop would correct an inventory discrepancy.

Suppose the desired inventory DI is 6000 units. Further assume that AT is 5 weeks indicating the time that any current order rate would require to correct the inventory. The Equation 2.2-1 which was

$$OR = \frac{1}{AT} (DI-I)$$

then becomes

$$OR = \frac{1}{5} (6000-I) \qquad \text{Eq. 2.2-2}$$

OR--Order rate (units/week)
I--Inventory (units)

If the initial inventory is known, the initial order rate can be calculated. Assume the starting inventory is 1000 units. Then the order rate from Equation 2.2-2 is 1000 units per week. If that rate of 1000 units per week persists for two weeks before the order rate is again calculated, 2000 units will have been added to inventory which by then has become 3000 units. Using this new value of inventory in Equation 2.2-2 yields a new order rate of 600 units per week. If we again allow 600 units per week to flow into inventory for two weeks, the new inventory at the end of the fourth week will be 4200 units. In a continuing sequence one can calculate, step-by-step, the succeeding values of order rate and inventory as in Table 2.2. The first column gives the number of weeks since the start.

(1) (weeks)	(2) Change in inventory (units)	(3) Inventory (units)	(4) Inventory error (units)	(5) Order rate (units/week)
TIME	CI	I	IE	ØR
.000		1000.	5000.	1000.
2.000	2000.	3000.	3000.	600.
4.000	1200.	4200.	1800.	360.
6.000	720.	4920.	1080.	216.
8.000	432.	5352.	648.	130.
10.000	259.	5611.	389.	78.
12.000	156.	5767.	233.	47.
14.000	93.	5860.	140.	28.
16.000	56.	5916.	84.	17.
18.000	34.	5950.	50.	10.
20.000	20.	5970.	30.	6.
22.000	12.	5982.	18.	4.
24.000	7.	5989.	11.	2.

Table 2.2 Computation of inventory.

Note that the first value in the table in column three (1000 units of inventory) was given as an initial condition of the system. All system level variables, here only inventory, must be assigned initial values to give the level from which the system starts.

The fourth column is the subtraction term (6000-I) in Eq. 2.2-2. The last column gives order rate and is, according to Equation 2.2-2, one-fifth of the fourth column. The second column is the change in inventory occurring during the preceding two weeks and is two weeks multiplied by the order rate in units per week taken from the last column of the previous line. The new inventory in the third column is computed as the previous inventory plus the change given in the second column. The reader should calculate the successive values in the table until the procedure is clear.

The inventory values from Table 2.2 are plotted in Figure 2.2c. The rate at which inventory approaches the desired value of 6000 units

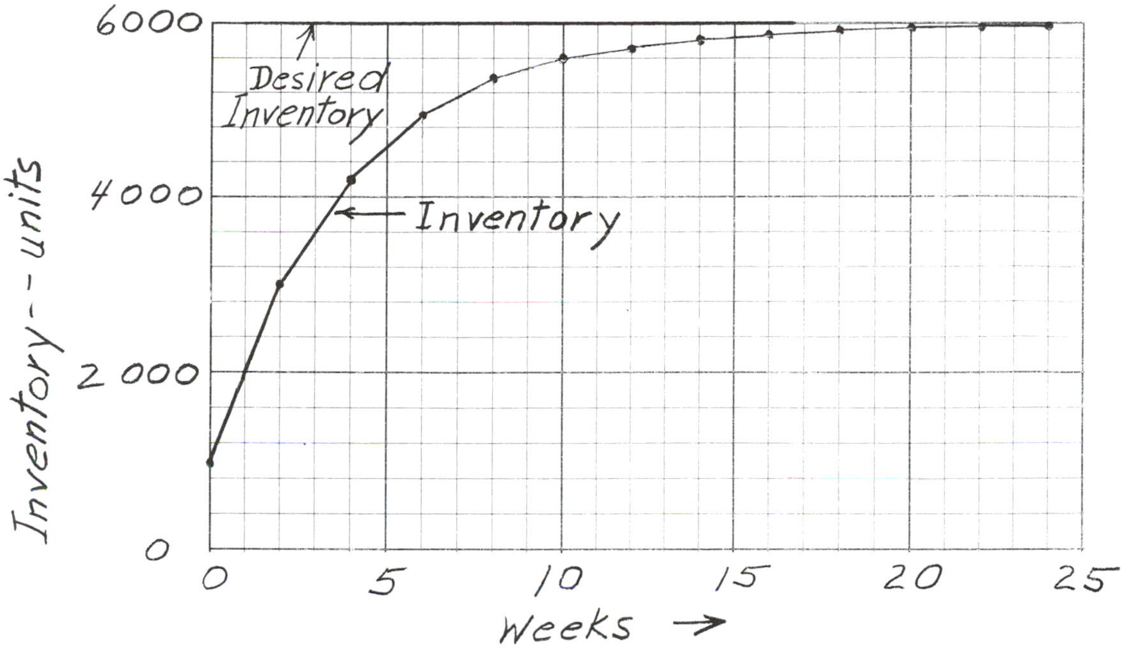

Figure 2.2c First-order system response.

is proportional to how far inventory is from 6000, so, the rate of approach becomes ever more gradual as the gap between inventory and desired inventory diminishes. In Table 2.2, order rate declines as inventory approaches 6000. The time behavior in Figure 2.2c is the type illustrated by Curve A of Figure 2.1 and shows the "exponential" response typical of first-order, negative-feedback loops to be discussed further in chapters yet to be written.

A negative-feedback loop is a loop in which the control decision attempts to adjust some system level to a value given by a goal introduced from outside the loop. In Figure 2.2a the goal is the constant DI which specifies the desired inventory. The desired inventory DI is not itself generated within the feedback loop. The term "negative" feedback implies an algebraic sign reversal in the decision process. This reversal is seen in Equation 2.2-1 where the negative sign associated with inventory indicates that the larger the inventory the smaller will be the order rate. The reversal of influence is also shown by the downward sloping lines in Figure 2.2b where a rising inventory produces a falling order rate.

(See Workbook Section W2.2)

2.3 Second-Order, Negative-Feedback Loop

A "second-order" system has two level variables. (Levels represent the condition or state of the system.) In the previous section a first-order system (with one level variable, the inventory) was seen to have a time response that simply approached its final value without overshoot or oscillation. Now, a second level variable will be introduced into the feedback loop to show how the mode of behavior can change to one of oscillation.

The previous feedback loop of Figure 2.2a will be extended to include a delay between the ordering of goods and receiving them into inventory. Figure 2.3a is like Figure 2.2a except that the pool of goods on order GO and the receiving rate RR have been added.

The combination of the level variable representing goods on order GO and the flow variable representing the receiving rate RR giving the flow from goods on order into inventory, when taken together, have the effect of introducing a time delay between order rate and receiving rate. An explanation of and justification for choosing such a combination to create a delay will be postponed until Chapter . In this section, we will merely observe in the system response that a delay is indeed created between order rate and receiving rate.

The goods on order GO has the same nature as the inventory. Both are level variables which are created by accumulating the effects of the rates flowing in and out. A new value for goods on order would be calculated by starting with the old value, adding the units that have come in from order rate, and subtracting the units that have gone out to receiving rate.

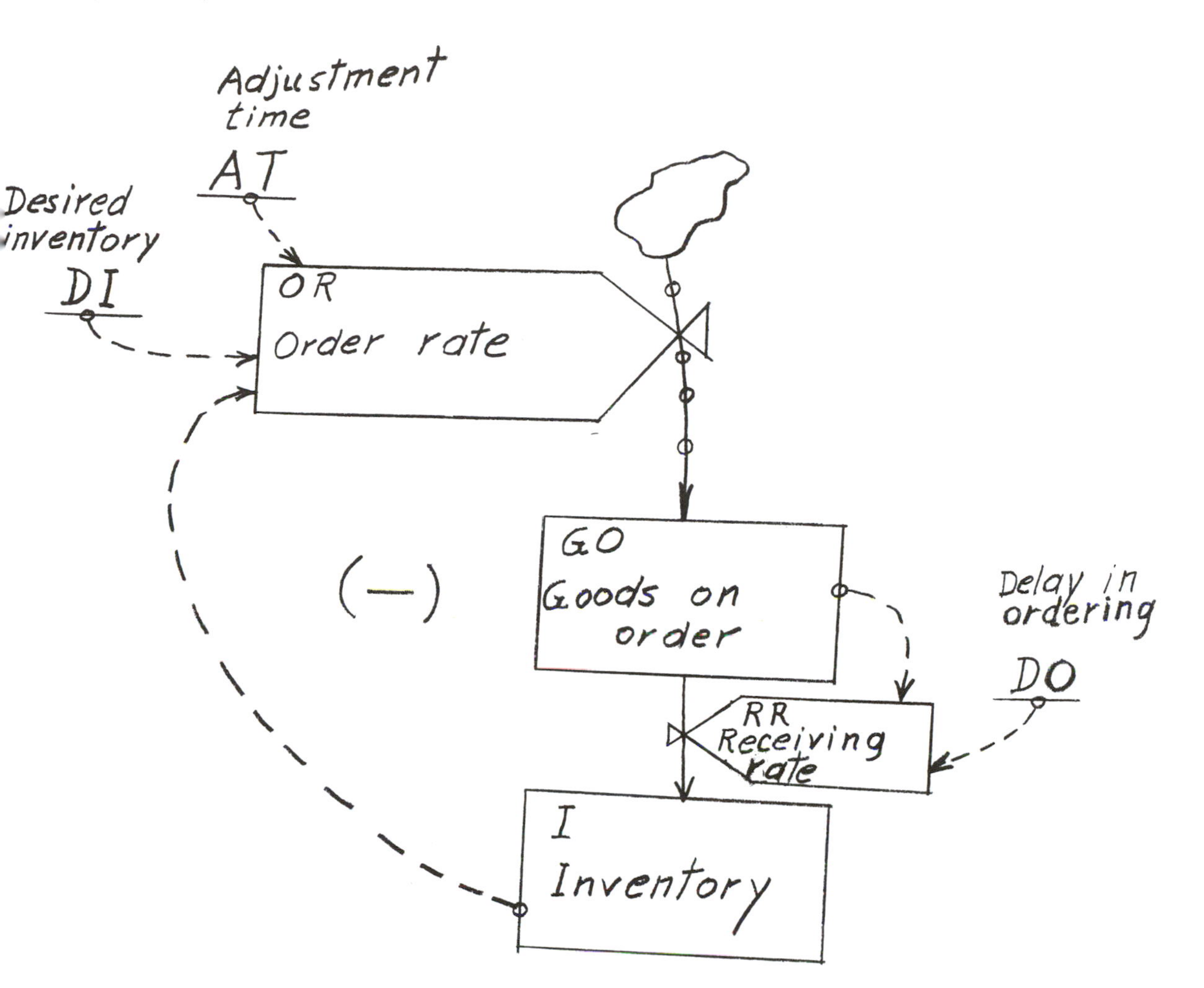

Figure 2.3a Second-order negative feedback.

To create a so-called "exponential" delay of DO weeks between order rate and receiving rate, the following equation can be used for receiving rate:

$$RR = \frac{GO}{DO} \qquad \text{Eq. 2.3-1}$$

RR--Receiving rate (units/week)
GO--Goods on order (units)
DO--Delay in ordering (weeks)

This equation says that one DOth of the goods on order are received each week. In operation, if the order rate increases, it causes the goods on order to build up and this leads to an increased receiving rate. Should the order rate stop, the receiving rate would deplete the goods on order toward zero which in turn would cause the receiving rate to decline. The equation is dimensionally correct--each side is measured in units per week.

For this example, delay in ordering DO will be taken as 10 weeks. As noted in the previous section, an initial value is necessary for the level variable, goods on order. This could be any arbitrary value but will here start with 10,000 units. By taking this initial value of goods on order, we will have the same initial flow of 1000 units per week into inventory as in the example of Section 2.2. All other numerical values will be the same as in Section 2.2. The two rate of flow equations are then:

Repeating Equation 2.2-2

$$OR = \frac{1}{5}(6000-I) \qquad \text{Eq. 2.3-2}$$

OR--Order rate (units/week)
I--Inventory (units)

Substituting the value of DO into Equation 2.2-1,

$$RR = \frac{GO}{10} \qquad \text{Eq. 2.3-3}$$

RR--Receiving rate (units/week)
GO--Goods on order (units)

These equations, along with the initial values of inventory and goods on order are sufficient to permit the calculations in Table 2.3. In the Table, the first values of inventory and goods on order are the initial values already selected arbitrarily. The sixth column is the subtraction term (6000-I) giving the "inventory error" in Equation 2.3-2 and can be calculated from the

(1)	(2)	(3)	(4)	(5)	(6)	(7)	(8)
	Change in Inventory	Inventory	Change in Goods on Order	Goods on Order	Inventory Error	Order Rate	Receiving Rate
TIME	CI	I	CG	GØ	IE	ØR	RR
.00		1000.		10000.	5000.	1000.	1000.
2.00	2000.	3000.	0.	10000.	3000.	600.	1000.
4.00	2000.	5000.	-800.	9200.	1000.	200.	920.
6.00	1840.	6840.	-1440.	7760.	-840.	-168.	776.
8.00	1552.	8392.	-1888.	5872.	-2392.	-478.	587.
10.00	1174.	9566.	-2131.	3741.	-3566.	-713.	374.
12.00	748.	10315.	-2175.	1566.	-4315.	-863.	157.
14.00	313.	10628.	-2039.	-473.	-4628.	-926.	-47.
16.00	-95.	10533.	-1757.	-2229.	-4533.	-907.	-223.
18.00	-446.	10087.	-1367.	-3597.	-4087.	-817.	-360.
20.00	-719.	9368.	-916.	-4512.	-3368.	-674.	-451.
22.00	-902.	8465.	-445.	-4957.	-2465.	-493.	-496.
24.00	-991.	7474.	5.	-4952.	-1474.	-295.	-495.
26.00	-990.	6484.	401.	-4551.	-484.	-97.	-455.
28.00	-910.	5573.	717.	-3834.	427.	85.	-383.
30.00	-767.	4807.	937.	-2897.	1193.	239.	-290.
32.00	-579.	4227.	1057.	-1840.	1773.	355.	-184.
34.00	-368.	3859.	1077.	-763.	2141.	428.	-76.
36.00	-153.	3707.	1009.	246.	2293.	459.	25.
38.00	49.	3756.	868.	1114.	2244.	449.	111.
40.00	223.	3979.	675.	1789.	2021.	404.	179.
42.00	358.	4336.	451.	2240.	1664.	333.	224.
44.00	448.	4784.	217.	2457.	1216.	243.	246.
46.00	491.	5276.	-5.	2452.	724.	145.	245.
48.00	490.	5766.	-201.	2251.	234.	47.	225.
50.00	450.	6216.	-357.	1895.	-216.	-43.	189.
52.00	379.	6595.	-466.	1429.	-595.	-119.	143.
54.00	286.	6881.	-524.	905.	-881.	-176.	91.
56.00	181.	7062.	-533.	372.	-1062.	-212.	37.
58.00	74.	7137.	-499.	-128.	-1137.	-227.	-13.
60.00	-26.	7111.	-429.	-557.	-1111.	-222.	-56.
62.00	-111.	7000.	-333.	-890.	-1000.	-200.	-89.
64.00	-178.	6822.	-222.	-1112.	-822.	-164.	-111.
66.00	-222.	6599.	-106.	-1218.	-599.	-120.	-122.
68.00	-244.	6356.	4.	-1214.	-356.	-71.	-121.
70.00	-243.	6113.	101.	-1114.	-113.	-23.	-111.
72.00	-223.	5890.	178.	-936.	110.	22.	-94.
74.00	-187.	5703.	231.	-705.	297.	59.	-70.
76.00	-141.	5562.	260.	-445.	438.	88.	-45.
78.00	-89.	5473.	264.	-181.	527.	105.	-18.
80.00	-36.	5437.	247.	66.	563.	113.	7.
82.00	13.	5450.	212.	278.	550.	110.	28.
84.00	56.	5506.	164.	443.	494.	99.	44.
86.00	89.	5594.	109.	552.	406.	81.	55.
88.00	110.	5704.	52.	604.	296.	59.	60.
90.00	121.	5825.	-3.	601.	175.	35.	60.
92.00	120.	5945.	-50.	551.	55.	11.	55.
94.00	110.	6056.	-88.	463.	-56.	-11.	46.
96.00	93.	6148.	-115.	348.	-148.	-30.	35.
98.00	70.	6218.	-129.	219.	-218.	-44.	22.
100.00	44.	6261.	-131.	88.	-261.	-52.	9.

Table 2.3 Second-Order Loop

value of inventory on the same line. Order rate is then calculated by taking one-fifth of column six. The last column giving receiving rate can be calculated as one-tenth of goods on order, using Equation 2.3-3. This completes the first row.

In the second row, change in inventory in column two is the receiving rate from the preceding line multiplied by the two week interval between computations. The new inventory in column three is the old inventory plus the change. The change in goods on order in column four is two weeks multiplied by the difference, order rate minus receiving rate, between inflow and outflow (on line two the difference is zero because OR and RR are equal on line one). Goods on order GO is calculated as the preceding value plus the change. In a similar manner, the entire table can be calculated one step at a time. The reader should trace the calculation until the procedure is evident.

A table of numbers does not present a clear image of how the variables are related. If a picture is worth a thousand words, a picture should be worth ten thousand numbers. To show better what is happening in the system of Figure 2.3a, the variables from Table 2.3 have been plotted for 100 weeks in Figure 2.3b. The curves belong to the same class as Curve B in Figure 2.1. Inventory no longer gradually approaches its final value as in Figure 2.2c. Instead, inventory rises higher than desired inventory by the 4th week with a peak at the 18th week in Figure 2.3b. Goods are returned to the supplier, as shown by the order rate falling below the zero axis which is in the center of the figure. The order rate is negative from the 4th to the 27th week. But too much is returned so that inventory again falls below desired inventory. Fluctuation continues at a decreasing amplitude. The curve is like that of a pendulum swinging in a tub of oil that gradually brings the pendulum to rest. In fact, such a pendulum could be represented by equations and computation nearly identical to those that here represent inventory. The two level variables for the pendulum would be the two forms of energy--kinetic energy represented by the pendulum

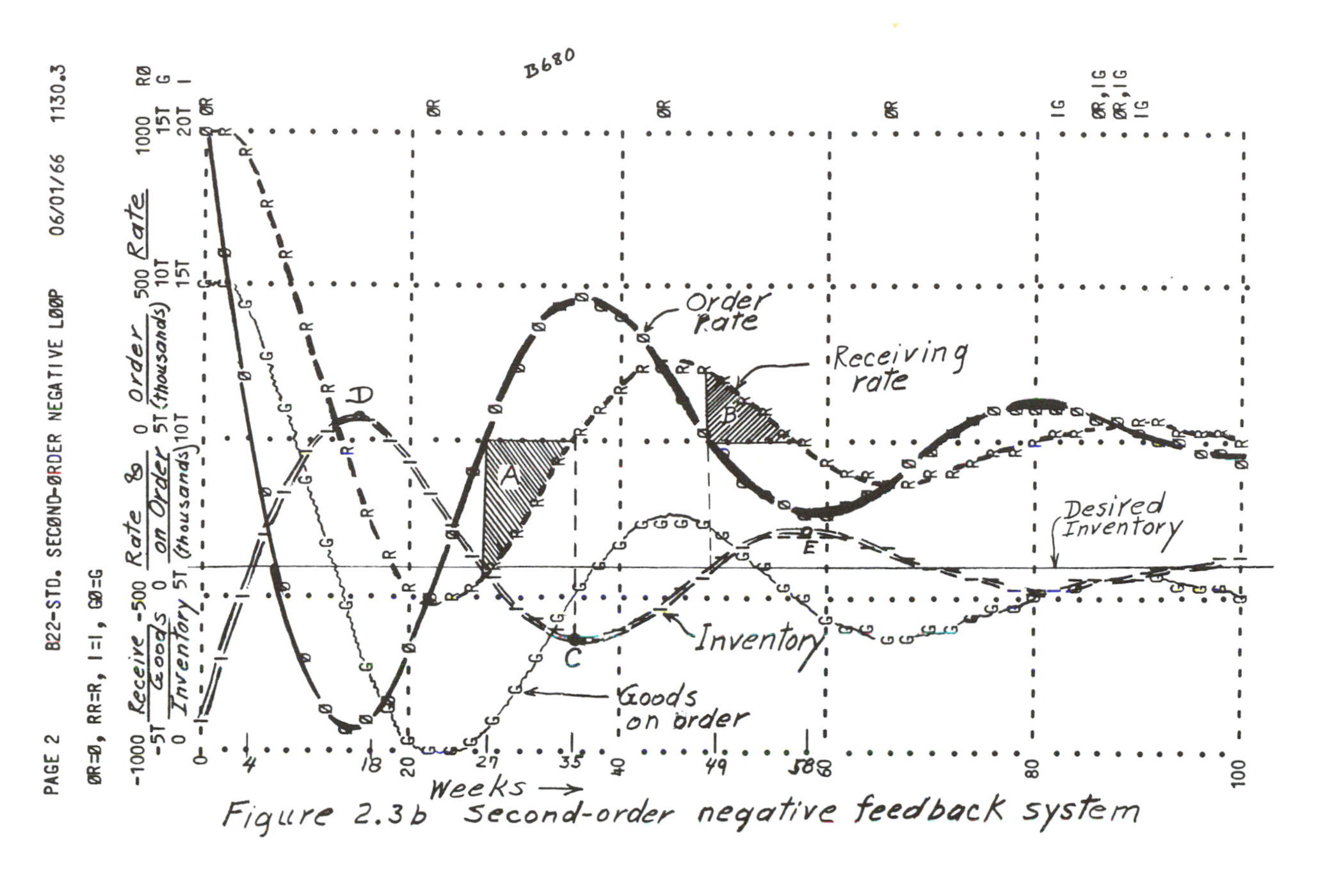

Figure 2.3b Second-order negative feedback system

velocity, and potential energy represented by the pendulum height above its lowest point. A pendulum is also a second-order system (two level variables).

Why does the inventory system oscillate as in Figure 2.3b? The oscillation appeared after the addition of the delay created by the "supply line" as represented by goods on order and receiving rate. Observe in the figure how the receiving rate is delayed relative to the order rate. Peaks in the receiving rate lag after peaks in the order rate by about 10 weeks corresponding to the value of the delay in ordering DO in Equations 2.3-1 and 2.3-3.

Introducing a delay between order rate and the receiving rate causes inventory to respond less quickly than before. In the figure at 27 weeks, inventory has fallen to desired inventory and, corresponding to Equation 2.3-2, order rate is zero (zero axis in center of figure). However, receiving rate is still negative and has not yet reached zero (zero axis in center) and does not do so until the 35th week. The shaded area A between the receiving rate curve and the zero axis represents the goods removed from inventory by the receiving rate after inventory has reached the desired value of 6000 units. Area A accounts for the further fall of inventory from desired inventory to point C at 35 weeks.

While inventory falls between weeks 27 and 35, order rate rises, and, after a delay, is reflected in the positive receiving rate between weeks 35 and 58. Area B is the reverse of Area A whereby the receiving rate continues to increase the inventory even after the inventory has again reached the desired inventory at week 49. The system is overcorrecting for inventory error as it seeks the desired inventory goal.

Observe the time relationship between receiving rate and inventory. The height of the receiving rate curve determines the slope of the inventory curve, that is, the steepest part of the inventory curve occurs at the same time as the peak of the receiving rate. Conversely, when the receiving rate is zero, the inventory curve is not changing and is horizontal at a maximum or minimum as at points C, D, or E.

The reader should examine the equations, the values in Table 2.3, and the curves in Figure 2.3b until it is clear how the fluctuation in the values can occur.

(See Workbook Section W2.3)

2.4 Positive-Feedback Loop

A positive-feedback loop does not seek an externally determined goal as does the negative-feedback loop. Instead, the positive loop diverges or moves away from the "goal." The positive loop does not have the reversal of sign in traversing the loop that is found in the negative loop. Action within the positive loop increases the discrepancy between the system level and a "goal" or reference point. The diagram of the simplest positive-feedback loop as in Figure 2.4a looks, at first glance, very similar to the negative feedback loop of Figure 2.2a. However, the difference appears in the nature of the decision process.

Figure 2.4a represents the way a sales force might grow. Assume that new salesmen are located and trained by the existing salesmen. The larger the sales force, the more men there are who can train new salesmen.

The salesmen hiring rate SHR depends directly on the number of salesmen.

sales doubling time SDT

SHR Salesmen hiring rate

(+)

S Salesmen

Figure 2.4a Positive feedback.

In a manner similar to Figure 2.2b, Figure 2.4b shows possible ways that the hiring rate might depend on the current number of salesmen. Line A suggests a high hiring rate for each present

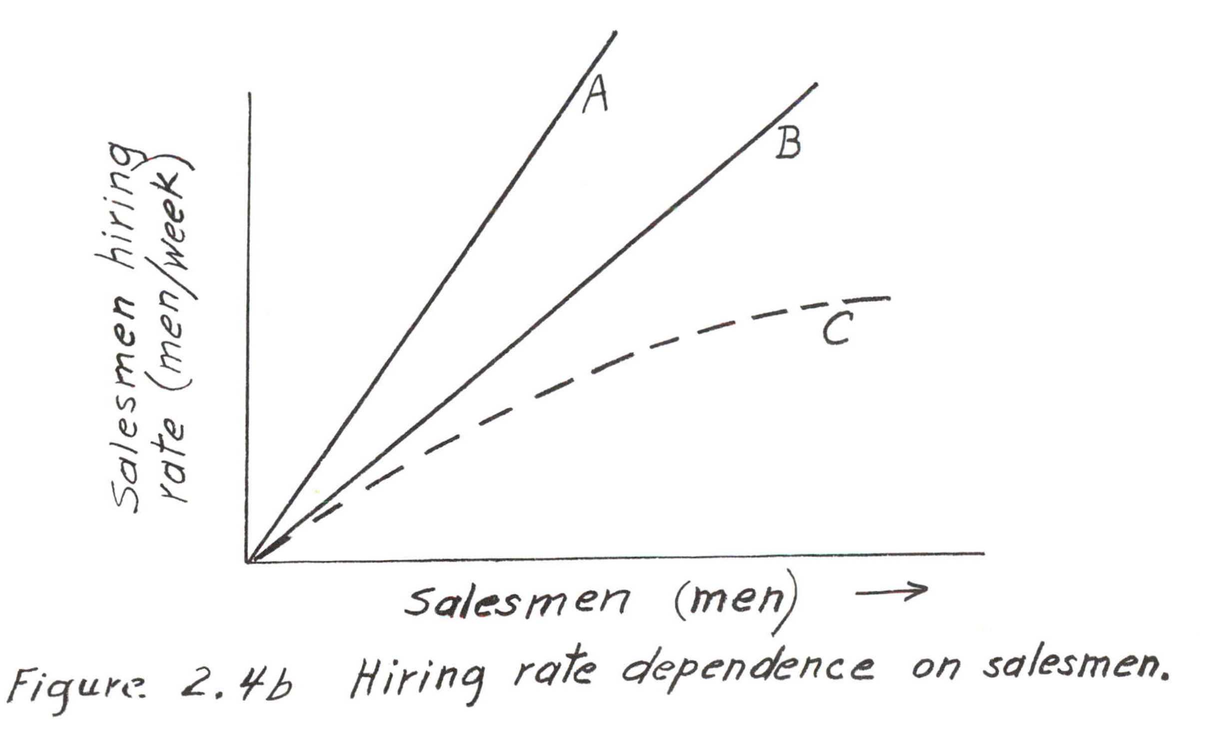

Figure 2.4b Hiring rate dependence on salesmen.

salesman. Line B represents a lower rate of hiring for each present salesman. Curve C depicts a nonlinear relationship in which hiring rate rises with the number of salesmen but then approaches a maximum, indicating perhaps that a larger group loses some of its average per capita ability to expand. Notice that the curves of Figure 2.4b all slope upward to the right instead of downward as in Figure 2.2b. The slope of the relationship is such that more salesmen cause a larger hiring rate which leads to still more salesmen. This simple system, without any limiting processes, would expand the number of salesmen at an ever increasing rate.

Linear relationships like Lines A or B can be represented by a hiring rate equation of the form:

$$SHR = \frac{1}{SDT}(S) \qquad \text{Eq. 2.4-1}$$

SHR--Salesmen hiring rate (men/week)
SDT--Sales doubling time (weeks)
S--Salesmen (men)

The equation states that in each week the new salesmen added are 1/SDT of the present sales force. If that rate were to persist (it doesn't because the number of salesmen is increasing) it would take SDT weeks to add as many salesmen as are now present, that is, the present hiring rate would double the present sales force in SDT weeks. The term 1/SDT has the dimensions of men per week hired per experienced man:

$$\frac{\text{men/week}}{\text{man}} = \frac{1}{\text{week}}$$

The sales doubling time SDT is the number of weeks worked by a salesman while locating and training another.

Equation 2.4-1 should be compared with Equation 2.2-1. In the negative feedback loop, the variable I entered the equation with a negative sign. Here in the positive feedback loop, the variable S enters with a positive sign. The negative loop in Equation 2.2-1 contains the system goal DI giving the desired inventory. The positive loop has a corresponding reference point, here zero, from which the number of salesmen is departing.

To compute the behavior through time of this positive feedback loop one must, as before, have values for the parameters (here SDT) and initial values for the level variables (here the initial number of salesmen). Assume that 50 weeks are required for one salesman to find and train a second, and that we start with a nucleus of six salesmen, so,

SDT = 50 weeks
S = 6 salesmen, initial value

Equation 2.4-1 then becomes

$$SHR = \frac{1}{50} (S) \qquad \text{Eq. 2.4-2}$$

SHR--Salesmen hiring rate (men/week)
S--Salesmen (men)

Table 2.4 shows the progression of the system as the salesmen train salesmen who train more salesmen. Examine the development of the numbers. The initial value of six salesmen, gives, according

(1)	(2) Change in salesmen (men)	(3) Salesmen (men)	(4) Salesmen hiring rate (men/week)
TIME	CS	S	SHR
.00		6.0	.12
20.00	2.4	8.4	.17
40.00	3.4	11.8	.24
60.00	4.7	16.5	.33
80.00	6.6	23.0	.46
100.00	9.2	32.3	.65
120.00	12.9	45.2	.90
140.00	18.1	63.2	1.26
160.00	25.3	88.5	1.77
180.00	35.4	124.0	2.48
200.00	49.6	173.6	3.47
220.00	69.4	243.0	4.86
240.00	97.2	340.2	6.80
260.00	136.1	476.2	9.52
280.00	190.5	666.7	13.33
300.00	266.7	933.4	18.67

mod B23
B696

Table 2.4 Positive feedback.

to Equation 2.4-2, a hiring rate in the first row of the fourth column of 0.12 man per week or one man in about eight weeks. In the second line, the second column, change in salesmen is the hiring rate from the previous line multiplied by the 20 weeks between successive computations, giving 2.4 men. The 2.4 new men are added to the initial six to give 8.4 salesmen in the third column at the end of 20 weeks. In a similar way, the successive lines of the table can be computed. Trace the computation until the procedure is clear.

The full impact of this "exponential" growth is better seen in Figure 2.4c than in a table of numbers. The behavior is of the same class as Curve C in Figure 2.1. It is the shape of curve that one

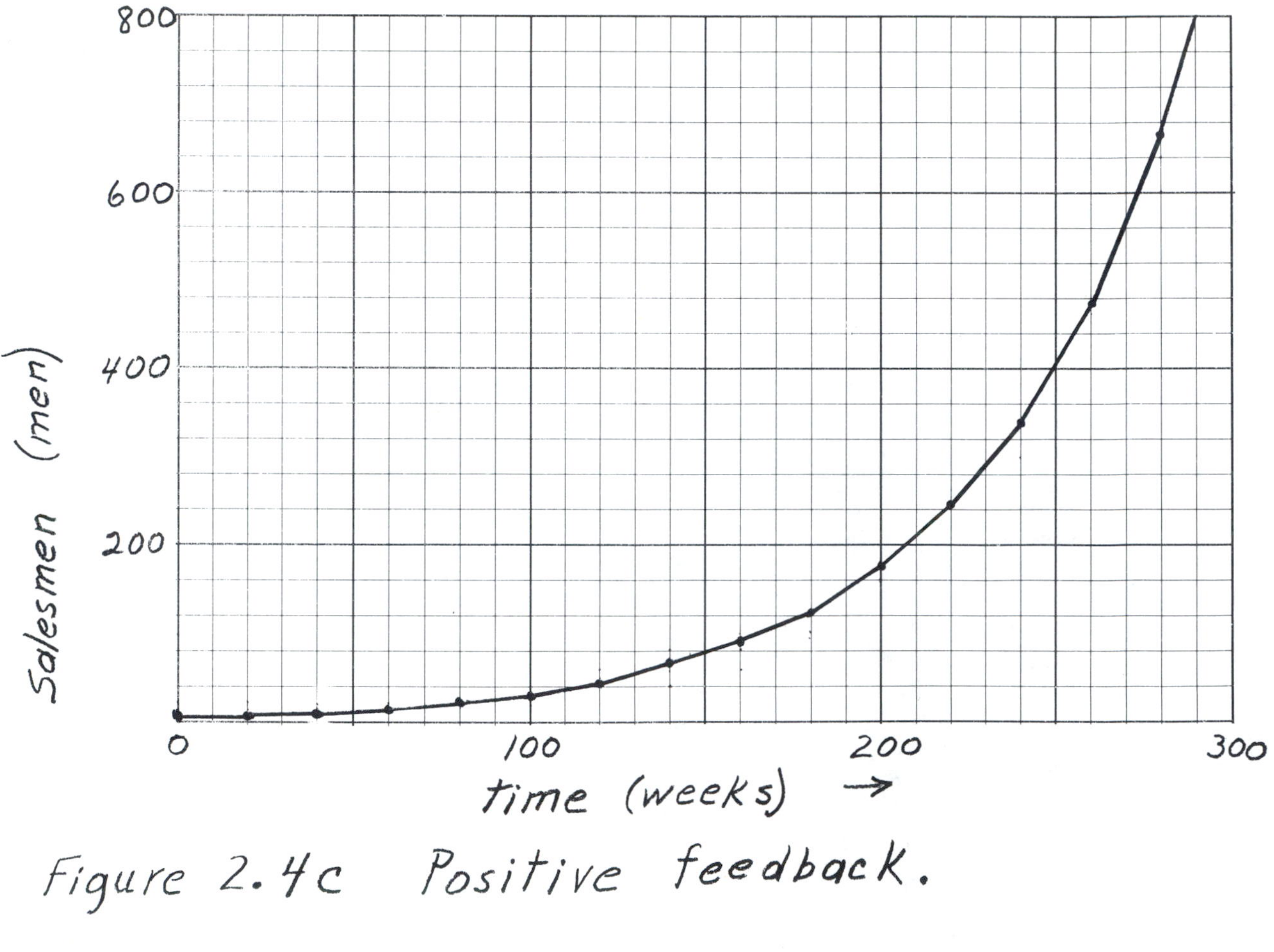

Figure 2.4c Positive feedback.

sees for the world population explosion, or the growth in technical knowledge. In such processes, the level to which the state of the system has already risen determines the rate of still further increase. The bigger it is the faster it grows--until something happens to alter the parameter values in the positive feedback loop.

(See Workbook Section W2.4)

2.5 Coupled Nonlinear Feedback Loops

(Note to the reader: This section may seem long and forbidding at such an early point in the book. But be of good cheer. It gives a preliminary view of some of the more complex interactions within systems as a background for simpler introductory material in later chapters.)

Growth processes show positive feedback. But exponential growth would reach overwhelming proportions if unchecked. In the positive feedback of a chemical explosion, the fuel is either consumed or the explosion destroys the system so that the process terminates. In biological growth, the consequences of growth alter the rate of future growth. Consider the multiplication of an animal population that can double its numbers each six months. If a pair of animals occupy one square foot, at the end of six months the population occupies two square feet and after a year, four square feet. In seven years they still live within an acre. But in about 80 years they cover one-sixteenth of the earth's surface, in one more year they cover a quarter of the surface, and in one additional year they cover the entire surface of the earth--assuming the original multiplication rate had continued. But growth interacts with parts of the surrounding system to modify the growth process. Growth toward an upper limit is illustrated by curve D in Figure 2.1.

Such growth toward a limit often occurs in the introduction of a new product to a market. At first, the product is successful, sales produce revenue that supports more sales effort to produce still more revenue. But at some point, sales become increasingly difficult--the market approaches saturation, or the easy initial sales cause the supplier to become careless and neglect quality, or the manufacturing capacity for producing the product becomes overloaded.

Sales growth which is eventually suppressed by overloading the manufacturing facilities often occurs through a system of relationships shown in Figure 2.5a. The left-hand loop controls the number of salesmen and is an elaboration on the positive loop in Figure 2.4a. The right-hand loop is a second-order, negative-feedback loop, as was first seen in Figure 2.3a. Here in Figure 2.5a the two levels in the major negative loop are the backlog and the delivery delay recognized. Within both the left and right loops are small, subordinate, first-order, negative-feedback loops having the simple structure of

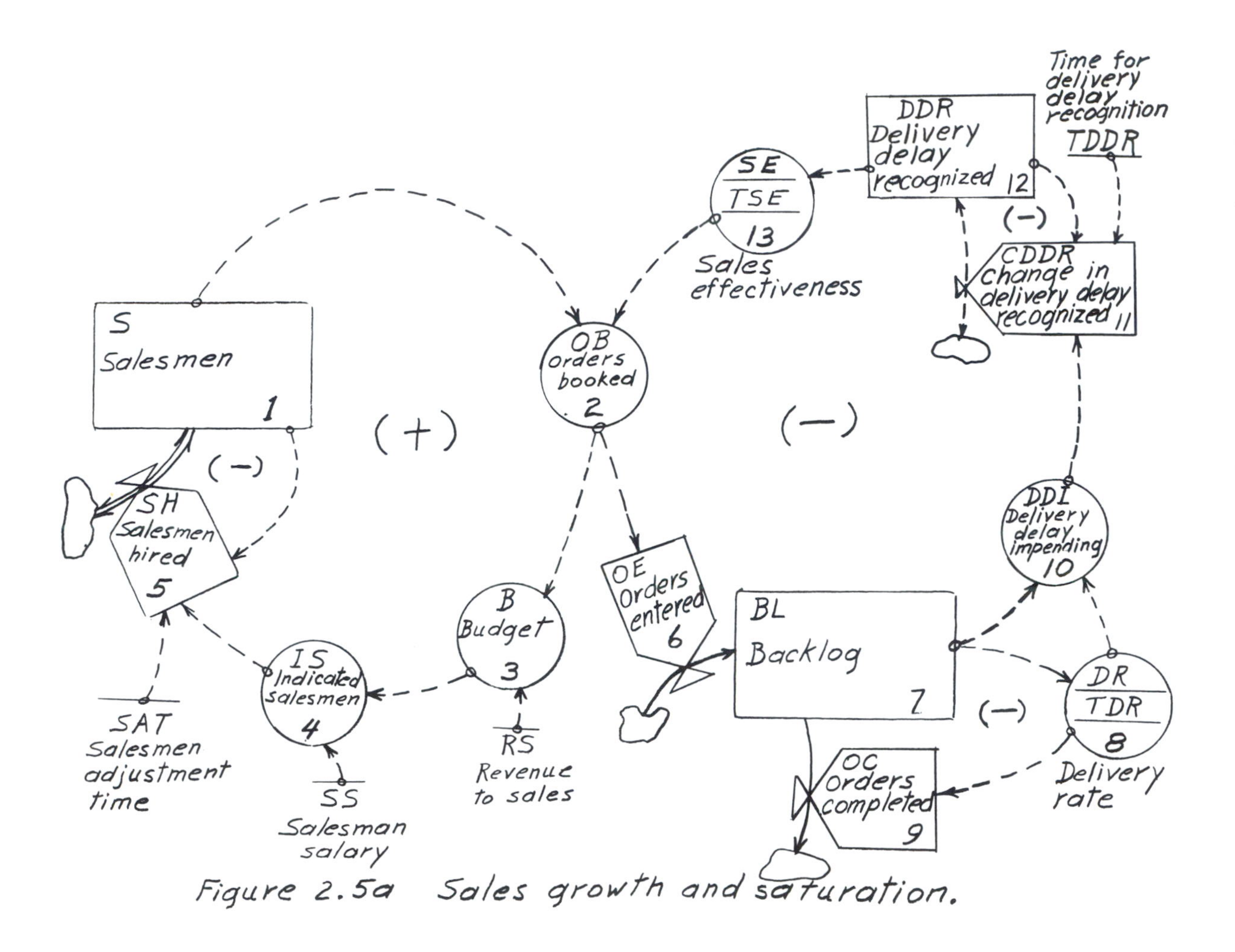

Figure 2.5a Sales growth and saturation.

Figure 2.2a. The system consists of five feedback loops--a major positive loop, a major negative loop, and three minor negative loops. In Figure 2.5a, the numbers within the symbols refer to the equation numbers in the following pages.

The Positive Loop

In the positive loop, salesmen book orders, a fraction RS of the revenue from the orders is available for the budget to pay the expenses of salesmen. The "indicated" salesmen IS are the number who could be supported by the budget. Salesmen are hired (or released) to adjust the actual salesmen S toward the number of indicated salesmen. In operation, if the salesmen sell more than enough to pay their own expenses, then an expansion of the sales force can occur. Salesmen produce revenue to hire more salesmen. The equations for the steps around the salesman loop can now be written to state, in symbolic form, what has already been described.

The number of salesmen at present are equal to the number when previously calculated plus the change since the previous calculation. The change in the number of salesmen can be computed as the product of the rate at which they have been added multiplied by the length of time that the rate has been flowing since the previous computation. In equation form, the salesmen can be stated as

$$S_{present}=S_{previous}+(\text{solution interval})(SH) \qquad \text{Eq. 2.5-1}$$

S--Salesmen (men)
solution interval--time between successive computations of the equations (months)
SH--Salesmen hiring (men/month)

The orders booked OB depend on the number of salesmen and on the sales effectiveness SE. Sales effectiveness is defined as the units of product sold each month by each salesman. In this

$$OB = (S)(SE) \qquad \text{Eq. 2.5-2}$$

OB--Orders booked (units/month)
S--Salesmen (men)
SE--Sales effectiveness (units/man-month)

equation, sales effectiveness is a variable that changes in value depending on how long the customer must wait for product delivery. When the product is available for immediate delivery the salesman has an easier job and can sell more per month than if the customer must wait for slow delivery.

The budget B for salesmen's monthly expenses is computed from orders booked multiplied by the dollars per unit that are allocated to selling cost. An arbitrary choice of ten dollars per unit will here be used for RS.

$$B = (OB)(RS) \qquad \text{Eq. 2.5-3}$$
$$RS = 10$$

B--Budget (dollars/month)
OB--Orders booked (units/month)
RS--Revenue to sales (dollars/unit)

The indicated salesmen IS in the diagram is computed by dividing the monthly budget by the monthly expense of each salesman, here taken as 2000 dollars per man-month. Indicated salesmen are the number who can be justified by the present rate at which new orders are being booked.

$$IS = \frac{B}{SS} \qquad \text{Eq. 2.5-4}$$

$$SS = 2000$$

IS--Indicated salesmen (men)
B--Budget (dollars/month)
SS--Salesman salary (dollars/man-month)

Salesmen hired SH adjusts the number of salesmen S toward the indicated number. Salesmen and salesmen hired form a small negative loop similar to that for inventory in Figure 2.2a. In Figure 2.2a the order rate acted to adjust inventory toward the desired inventory. In Figure 2.5a the salesmen hired rate SH adjusts the salesmen toward the indicated salesmen IS who can be supported by the budget. The form of the equation for SH can be the same as that for OR in Equation 2.2-1. Here the adjustment time for changing the number of salesmen will be taken as 20 months, meaning that one-twentieth of the salesmen discrepancy will be corrected each month. In Equation 2.2-1, the desired inventory goal DI was a

$$SH = \frac{1}{SAT}(IS-S) \qquad \text{Eq. 2.5-5}$$

$$SAT = 20$$

SH	Salesmen hiring	(men/month)
SAT	Salesmen adjustment time	(months)
IS	Indicated salesmen	(men)
S	Salesmen	(men)

constant specified from outside the negative loop. Here in Eq. 2.5-5, the indicated salesmen goal IS comes from outside the negative loop but is a variable created by the positive loop. Notice that it is the difference, indicated salesmen IS minus the (actual) salesmen S, that gives the incentive to hire more salesmen. The salesmen hiring rate adjusts the sales force toward the authorized level.

The Negative Loop

In the major negative loop on the right of Figure 2.5a, orders entered OE are taken equal to orders booked OB. Orders entered are placed in a backlog of unfilled orders BL which is depleted by the orders completed OC. The ratio of backlog to delivery rate implies a delivery delay DDI here called delivery delay impending because it has not yet had time to be recognized and to become effective in influencing the market's willingness to buy. This distinction between delivery delay impending and delivery delay recognized illustrates the contrast between actual and apparent conditions as discussed in Section 1.4. A time delay TDDR intervenes before the delivery delay is recognized in the market at DDR. The sales effectiveness SE depends on delivery delay recognized DDR in such a manner that short delivery delay makes the product easier to sell and slower delivery delay makes the product increasingly hard to sell.

Delivery rate DR depends on backlog so as to represent the fact that manufacturing capacity is limited.

The negative loop in its entirety tends to adjust the order booking rate OB to equal the maximum delivery rate (production capacity), whenever the sales force is large enough to support full production. If order booking rate OB is larger than the maximum delivery rate DR, the order backlog BL will increase. Increasing backlog will then increase the delivery delay impending DDI. After a time interval, the delivery delay recognized DDR will increase, causing the sales effectiveness SE to decline and thus reduce orders booked toward the delivery rate. If the orders booked are below delivery rate, the reverse sequence will tend to increase orders booked. The preceding verbal description can now be translated into explicit symbols and equations to give a clearer and more compact statement of the system relationships.

The rate of orders entered OE in the backlog will be taken equal to orders booked, recognizing no time delay between the two.

$$OE = OB \qquad \text{Eq. 2.5-6}$$

OE--Orders entered (units/month)
OB--Orders booked (units/month)

The backlog is a system level representing the net accumulation of orders entered minus orders completed. Like salesmen in Equation 2.5-1, backlog is computed as the previous value plus any change. The solution interval (difference in time) will be abbreviated as DT.

$$BL_{present} = BL_{previous} + (DT)(OE-OC) \qquad \text{Eq. 2.5-7}$$

BL--Backlog (units)
DT--solution interval between successive computations of the equations (months)
OE--Orders entered (units/month)
OC--Orders completed (units/month)

The system of this example has been simplified to omit inventories, implying that the product is manufactured to order. Even if the manufacturing capacity is not overloaded, each order would encounter a delay before shipment while the order is being processed. The diagram suggests that delivery rate DR is to depend on the backlog BL. Is there a relationship between backlog and delivery rate that will produce a constant manufacturing delay when the demand is less than manufacturing capacity, and will cause delivery rate to equal manufacturing capacity when orders exceed maximum capacity? First consider behavior below the upper limit of manufacturing capacity. Suppose the normal production delay is two months. When the orders booked and the delivery rate have been constant and equal, the backlog should equal the delivery rate multiplied by the production delay. For example, if the backlog is 10,000 units, the delivery rate would need to be 5000 units per month to reach the latest order in the backlog in two months. A 20,000 unit backlog would require a 10,000 unit per month delivery rate to maintain the same two-month delivery delay.

$$\text{Backlog} = (\text{Delivery rate})(\text{Delivery delay})$$

or the delivery rate can be written as

$$(\text{Delivery rate}) = \frac{\text{Backlog}}{\text{Delivery delay}}$$

This means that, for a constant delivery delay, the delivery rate should be proportional to the backlog as shown in the straight rising section of Figure 2.5b. But as the delivery rate begins

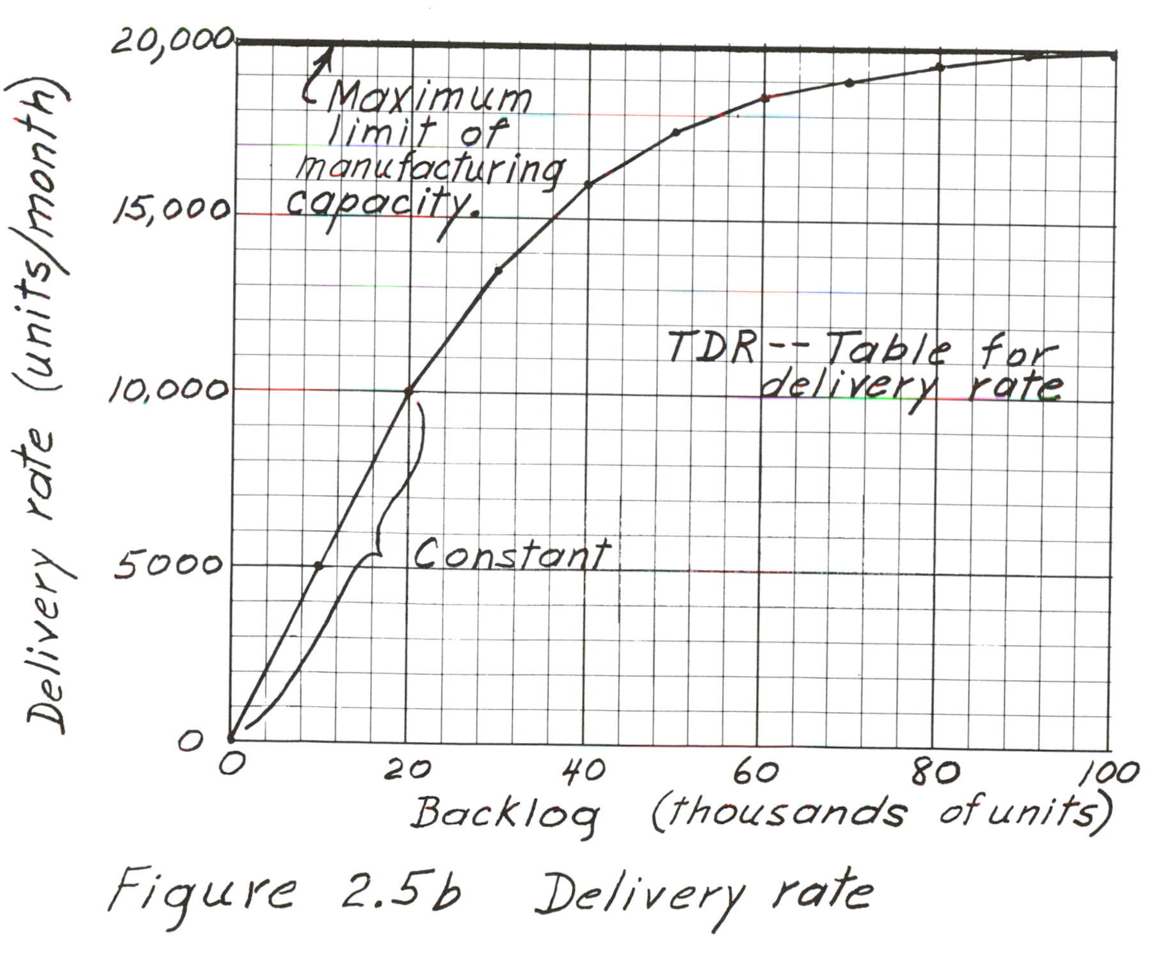

Figure 2.5b Delivery rate

to approach the maximum manufacturing capacity, the work load is difficult to increase evenly on each machine and some parts of the production process become overloaded before other parts. Delivery rate no longer rises in proportion to backlog and, as the maximum capacity is reached, no amount of further backlog increase can cause greater production. Greater production can come only from increasing the factory capacity and employees. As the production capacity is approached, a rate of orders booked in excess of delivery rate will only cause backlog to increase without any further increase in delivery rate. The delivery rate will now merely be indicated as obtained from a table of values, where the relationship is given in Figure 2.5b. In later chapters the computer symbolism for specifying such a table function will be explained.

DR = TABLE[Figure 2.5b for DR vs. BL]   Eq. 2.5-8

DR--Delivery rate (units/month)
BL--Backlog (units)

Delivery rate of goods shipped causes the corresponding depletion of the backlog, so

OC = DR   Eq. 2.5-9

OC--Orders completed (units/month)
DR--Delivery rate (units/month)

As discussed above, the relationships between the backlog, the minimum time to process an order, and the maximum possible production capacity determine the delivery rate. From the backlog and the delivery rate it is possible to deduce the actual delivery delay in filling an order. The delivery delay will be the size of the backlog divided by the delivery rate, giving the time necessary for the present delivery rate to work through the present backlog.

$$DDI = \frac{BL}{DR}$$ Eq. 2.5-10

DDI--Delivery delay impending (months)
BL--Backlog (units)
DR--Delivery rate (units/month)

But the "true" delivery delay implied by the backlog and delivery rate is not generally known to the customer. Even after the customer learns the delivery delay status of his supplier, it usually takes time to redirect procurement to another source. Therefore, there is a time lag or delay between the delivery delay status at the factory and when this affects the customer's willingness to order. The delivery delay recognized DDR is appropriately represented as a delayed version of delivery delay impending DDI (note that we are using "delay" in two different senses; there is the time lag or delay TDDR that occurs in recognizing at DDR the information about the delay in the delivery of goods given by DDI). This delay TDDR in the recognition of information can be created by a process that gradually adjusts the recognized information toward the source value. A fuller explanation will appear in a later chapter. The delay will be created in two steps, first the rate at which the recognized delay is changing, and then the recognized delay itself. In a simple but effective representation of the lag in the recognized situation, the recognized situation changes toward the true situation at a rate that depends on the discrepancy between the true and the recognized situations. Such a rate of change in the recognized delivery delay can be stated as

$$CDDR = \frac{1}{TDDR}(DDI-DDR) \qquad \text{Eq. 2.5-11}$$

$$TDDR = 6$$

CDDR--Change in delivery delay recognized (months/month)
TDDR--Time for delivery delay recognition (months)
DDI--Delivery delay impending (months)
DDR--Delivery delay recognized (months)

The preceding equation says that the rate of change in the recognized delay CDDR is proportional to the existing difference between the impending and recognized delays. The proportionality term TDDR describes how quickly DDR can adjust to the discrepancy between DDI and DDR. For the computations to follow, TDDR will be taken as 6 months, to represent the total of the time necessary for the salesmen to be informed of changes in delivery delay, the time for the customers to learn from the salesmen, and the ~~time~~ for the customers to plan changes in their supplier.

The delivery delay recognized is one of the system levels and results from accumulating the changes described by CDDR.

$$DDR_{present} = DDR_{previous} + (DT)(CDDR) \qquad \text{Eq. 2.5-12}$$

DDR--Delivery delay recognized (months)
DT--Solution interval between successive computations of the equations (months)
CDDR--Change in delivery delay recognized (months/month)

As we will see in the computed results and plotted curves that follow, the combination of Equations 2.5-11 and 2.5-12 cause the recognized delay DDR to lag behind the impending delay DDI.

There remains in the flow diagram of Figure 2.5a only the sales effectiveness SE. If the goods were available for immediate delivery, each salesman in a month could, on the average, sell an amount that is determined by such characteristics as price, quality, suitability to the customer's needs, manufacturer reputation, and selling skill. All these are constant in this example and are combined into the value of sales effectiveness when there is no delivery delay. In keeping with the other numerical values already selected, we will assume a sales effectiveness of 400 units per man-month at zero delivery delay. But as the customer recognizes that he must wait for delivery, more and more customers refuse to buy, thus reducing the average sales effectiveness. Such a relationship is shown in Figure 2.5c.

The figure suggests values suitable to a capital equipment item for which an industrial customer is willing to plan ahead. Even with a six-month delivery delay one-quarter as many units can be sold as could be with immediate delivery. Clearly the curve does not apply to normal consumer goods where alternate choices of competitive products are readily available.

Figure 2.5c Sales effectiveness.

In symbolic notation, the table in Figure 2.5c can be indicated in the same way as was done in Equation 2.5-8.

SE = TABLE[Figure 2.5c for SE vs. DDR] Eq. 2.5-13

SE--Sales effectiveness (units/man-month)
DDR--Delivery delay recognized (months)

Computation of System Operation

As with the simpler examples in the preceding sections, we can unfold the behavior of the system of Figure 2.5a using the equations that define system action. Initial values of the three level

(1)	(2)	(3)	(4)	(5)	(6)	(7)	(8)	(9)	(10)
	Sales-men (men)	Back-log (units)	Deliv-ery delay recog-nized (months)	Deliv-ery rate (units/ month)	Deliv-ery delay impend-ing (months)	Sales effect-iveness (units/ man-month)	Orders booked (units/ month)	Indi-cated sales-men (men)	Sales-men hired (men/ month)
TIME	S	BL	DDR	DR	DDI	SE	ØB	IS	SH
.00	10.0	8000.	2.00	4000.	2.00	350.	3500.	17.5	.38
2.00	10.8	7000.	2.00	3500.	2.00	350.	3763.	18.8	.40
4.00	11.6	7525.	2.00	3763.	2.00	350.	4045.	20.2	.43
6.00	12.4	8089.	2.00	4045.	2.00	350.	4348.	21.7	.47
8.00	13.4	8696.	2.00	4348.	2.00	350.	4674.	23.4	.50
10.00	14.4	9348.	2.00	4674.	2.00	350.	5025.	25.1	.54
12.00	15.4	10049.	2.00	5025.	2.00	350.	5402.	27.0	.58
14.00	16.6	10803.	2.00	5402.	2.00	350.	5807.	29.0	.62
16.00	17.8	11613.	2.00	5807.	2.00	350.	6242.	31.2	.67
18.00	19.2	12484.	2.00	6242.	2.00	350.	6710.	33.6	.72
20.00	20.6	13421.	2.00	6710.	2.00	350.	7214.	36.1	.77
22.00	22.2	14427.	2.00	7214.	2.00	350.	7755.	38.8	.83
24.00	23.8	15509.	2.00	7755.	2.00	350.	8336.	41.7	.89
26.00	25.6	16672.	2.00	8336.	2.00	350.	8961.	44.8	.96
28.00	27.5	17923.	2.00	8961.	2.00	350.	9634.	48.2	1.03
30.00	29.6	19267.	2.00	9634.	2.00	350.	10356.	51.8	1.11
32.00	31.8	20712.	2.00	10249.	2.02	350.	11133.	55.7	1.19
34.00	34.2	22479.	2.01	10868.	2.07	350.	11953.	59.8	1.28
36.00	36.8	24651.	2.03	11628.	2.12	348.	12802.	64.0	1.36
38.00	39.5	27000.	2.06	12450.	2.17	347.	13679.	68.4	1.45
40.00	42.4	29458.	2.10	13310.	2.21	344.	14587.	72.9	1.53
42.00	45.4	32012.	2.13	14003.	2.29	342.	15532.	77.7	1.61
44.00	48.6	35071.	2.18	14768.	2.37	339.	16487.	82.4	1.69
46.00	52.0	38510.	2.25	15627.	2.46	335.	17435.	87.2	1.76
48.00	55.5	42124.	2.32	16319.	2.58	331.	18373.	91.9	1.82
50.00	59.2	46232.	2.41	16935.	2.73	326.	19265.	96.3	1.86
52.00	62.9	50893.	2.51	17589.	2.89	319.	20069.	100.3	1.87
54.00	66.6	55851.	2.64	18085.	3.09	312.	20759.	103.8	1.86
56.00	70.4	61199.	2.79	18560.	3.30	303.	21288.	106.4	1.80
58.00	74.0	66655.	2.96	18833.	3.54	292.	21629.	108.1	1.71
60.00	77.4	72248.	3.15	19112.	3.78	278.	21495.	107.5	1.50
62.00	80.4	77014.	3.36	19351.	3.98	261.	20986.	104.9	1.23
64.00	82.8	80285.	3.57	19509.	4.12	245.	20261.	101.3	.92
66.00	84.7	81790.	3.75	19554.	4.18	230.	19476.	97.4	.63
68.00	86.0	81635.	3.89	19549.	4.18	218.	18777.	93.9	.40
70.00	86.8	80090.	3.99	19503.	4.11	211.	18299.	91.5	.24
72.00	87.2	77682.	4.03	19384.	4.01	208.	18172.	90.9	.18
74.00	87.6	75258.	4.02	19263.	3.91	209.	18283.	91.4	.19
76.00	88.0	73299.	3.98	19165.	3.82	211.	18594.	93.0	.25
78.00	88.5	72157.	3.93	19108.	3.78	216.	19073.	95.4	.34
80.00	89.2	72087.	3.88	19104.	3.77	220.	19588.	97.9	.44
82.00	90.0	73053.	3.84	19153.	3.81	223.	20034.	100.2	.51
84.00	91.1	74816.	3.83	19241.	3.89	223.	20331.	101.7	.53
86.00	92.1	76996.	3.85	19350.	3.98	222.	20434.	102.2	.50
88.00	93.1	79163.	3.89	19458.	4.07	218.	20341.	101.7	.43
90.00	94.0	80929.	3.95	19528.	4.14	214.	20093.	100.5	.32
92.00	94.6	82059.	4.02	19562.	4.19	209.	19778.	98.9	.21
94.00	95.1	82492.	4.08	19575.	4.21	205.	19528.	97.6	.13
96.00	95.3	82399.	4.12	19572.	4.21	203.	19318.	96.6	.06
98.00	95.4	81890.	4.15	19557.	4.19	201.	19175.	95.9	.02
100.00	95.5	81127.	4.16	19534.	4.15	200.	19116.	95.6	.00

Table 2.5 Sales growth and stagnation.

variables--salesmen, backlog, and delivery delay recognized--are needed as a point of departure. We will start with arbitrary choices of 10 salesmen, a backlog of 8000 units and a market expectation of a delivery delay DDR of 2 months which are chosen simply as a reasonable set of numbers for this example. The computation will be made at 2-month intervals through 100 months and is shown in Table 2.5. Columns (2), (3), and (4) are the level variables of the system. On the first line these are given the arbitrarily selected initial conditions. The value of DR in column (5) is found from the value of backlog BL using the relationship in Figure 2.5b. In column (6) the value of delivery delay impending is computed by Equation 2.5-10 using the values of backlog BL and delivery rate DR already available on the same line. In column (7) the sales effectiveness is found by using the delivery delay recognized DDR from column (4) as the entry value to the curve in Figure 2.5c. Orders booked in column (8) is computed by Equation 2.5-2 using the values of salesmen S from column (2) and sales effectiveness SE from column (7). For computing indicated salesmen IS in column (9) the two Equations 2.5-3 and 2.5-4 can be used or the intermediate step of computing budget B can be eliminated by combining the two steps to become

$$IS = \frac{(OB)(RS)}{SS} = \frac{(OB)10)}{2000} = \frac{OB}{200}$$

IS--Indicated salesmen (men)
OB--Orders booked (units/month)
RS--Revenue to sales (dollars/unit)
SS--Salesman salary (dollars/man-month)

Indicated salesmen IS depends only on orders booked OB from column (8) and on constants. Salesmen hired SH in column (10) is computed from indicated salesmen IS from column (9) and salesmen S from column (2) using Equation 2.5-5.

The first three entries on a new line are obtained by computing the level variables of the system. Each of these, as in the earlier examples, depends on its preceding value and on the changes that have occurred during the intervening time interval. The new number of salesmen S in column (2) is found as

$$S_{present} = S_{previous} + (\text{solution interval})(SH)$$
$$= 10.0 \text{ men} + (2 \text{ months})(.38 \text{ men/month})$$
$$= 10.76 \text{ men}$$

The value is rounded from 10.76 to 10.8 in the table.

In a similar way the backlog BL in column (3) is found from the previous value plus the units added by orders booked OB from the previous line of column (8) minus the units removed by delivery rate DR from the previous line of column (5).

$$BL_{present} = BL_{previous} + (\text{solution interval})(OB\text{-}DR)$$
$$= 8000 \text{ units} + (2 \text{ months})(3500 \text{ units/mo.} - 4000 \text{ units/mo.})$$
$$= 7000 \text{ units}$$

The delivery delay recognized DDR is computed by Equations 2.5-11 and 2.5-12 using delivery delay impending DDI from column (6) and the previous delivery delay recognized from column (4).

$$CDDR = \frac{1}{TDDR}(DDI\text{-}DDR)$$
$$= \frac{1}{6 \text{ months}}(2 \text{ months} - 2 \text{ months}) = 0$$
$$DDR_{present} = DDR_{previous} + (DT)(CDDR)$$
$$= 2 \text{ months} + (2 \text{ months})(0)$$
$$= 2 \text{ months}$$

There is no change in DDR until the 34th month because DDR and DDI are equal until DDI begins to rise at the 32nd month.

The reader should compute and verify the values in the table for enough lines to make the procedure obvious. It is important at this point to see clearly the relation between the results, the equations, the functional relationships of Figures 2.5b and 2.5c, and the flow diagram of Figure 2.5a.

Time Plot of System Operation

The behavior of the system is easier to comprehend in a time plot as in Figure 2.5d than in a table of numbers. Here we see families of

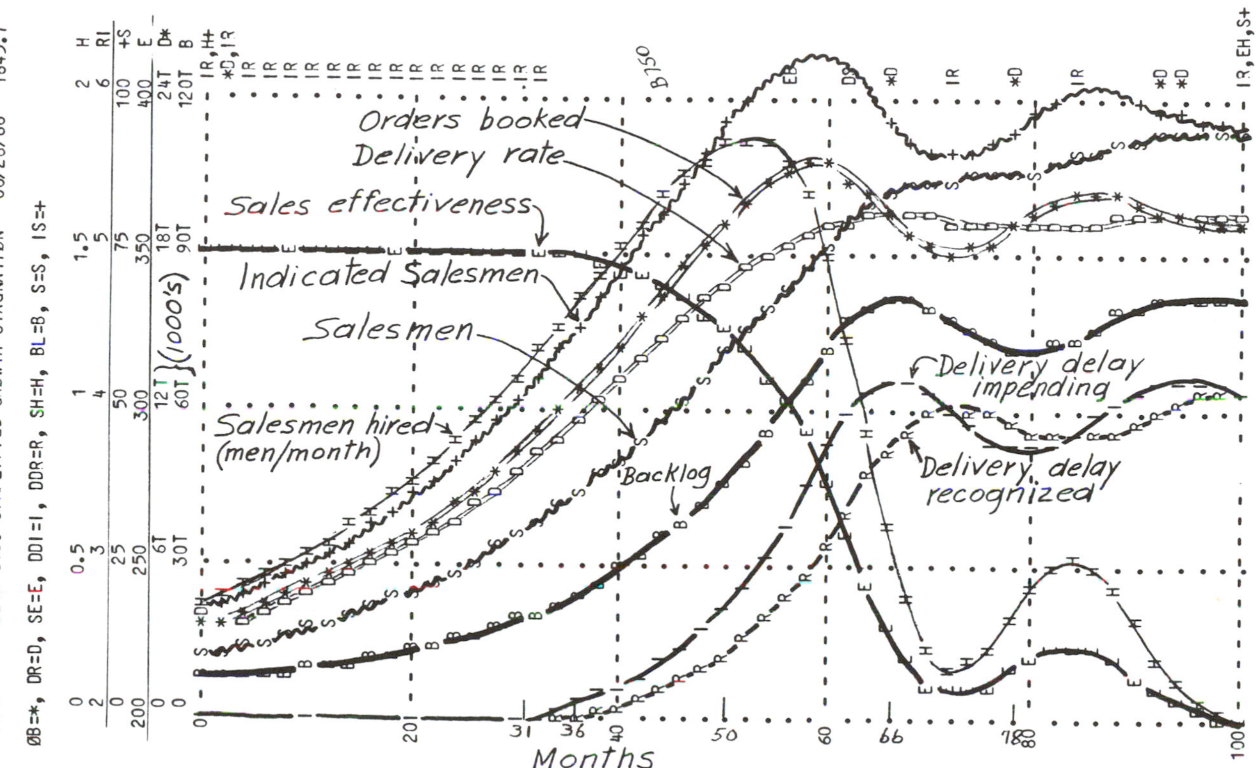

Figure 2.5d. Sales growth limited by factory capacity.

curves having the shape of Curve D in Figure 2.1. Two separate stages of system response are evident. The first stage between time 0 and time 50 months is positive feedback growth as was seen in Figure 2.4c. The second stage from time 50 months to the end of the plot shows a decreasing fluctuation as in Figure 2.3b. We will examine these two time periods separately to see why the behavior is so different.

The early growth phase is dominated by the positive loop that determines the number of salesmen. The two loops interact through the orders booked OB. Sales effectiveness depends on the delivery delay but the delivery delay is constant at 2 months (while the system operates within the straight line section of Figure 2.5b) until the backlog has risen to 20,000 units. Until backlog rises to 20,000 units (it starts at 8000) there is no change within the major negative loop, delivery delay is constant, sales effectiveness is constant and there are no changes in the negative loop to affect the positive loop.

During the early stage, the sales effectiveness is 350 units per man-month which is well above the 200 units per man-month necessary to sustain any existing number of salesmen (if OB=S(200), then IS equals S and there is no difference between IS and S to create a salesman hiring rate SH in Equation 2.5-5.) With a sales effectiveness of 350 units per man-month, the indicated salesmen IS will at all times be 1.75 times the actual salesmen S, leading in Equation 2.5-5 to a salesmen hiring rate

$$SH = \frac{1.75S - S}{20} = \frac{.75}{20}\,(S)$$

$$= .0375\,(S)$$

This relationship says that 3.75% of the existing number of salesmen are being added per month during the initial build-up phase. From the table and the figure we see that adding salesmen at this rate causes salesmen to double in number each 19 months. But as salesmen increase, the orders booked increase, and the orders must be delivered by the manufacturing capacity. In Figure 2.5b it is evident that delivery rate fails to keep pace with backlog above a backlog of 20,000 units which is reached by the 31st month. At 31 months the delivery delay impending, as seen in Figure 2.5d, begins to increase from the initial 2-month delay in filling orders. Shortly thereafter the delivery delay recognized begins to increase in response to the rise in delivery delay impending. Delivery delay impending rises to 4.18 months at the 66th month. The increasing delivery delay, acting through the relationship in Figure 2.5c, causes sales effectiveness to start falling at the 36th month. As sales effectiveness falls, the budget for salesmen no longer exceeds by so wide a margin the budget needed to support the existing sales force. By the time that sales effectiveness drops to 200 units per man-month at 100 months, the driving pressure to increase salesmen has disappeared.

So, in summary, growth is rapid at first while delivery is prompt and selling is easy, but declining product attractiveness in the form of slow delivery begins to make the product harder to sell and growth stops when sales are just able to support the cost of selling. (The product may or may not still be profitable depending on whether or not profits are expended in still more sales effort in a futile attempt to resume growth. Of course in this example, growth can only result from greater production capacity once the capacity limit has been reached.)

After the rising delivery delay slows the growth process, the system comes under the dominance of the negative loop that couples orders booked, backlog, delivery delay, and sales effectiveness. In this loop the delivery rate can not exceed the maximum production capacity. Furthermore, the orders booked can not in the long run exceed the maximum delivery rate, otherwise the backlog and delivery delay would continue to grow without limit which would not be tolerated by the customers. The effect of the negative loop is to adjust the orders booked so that on the average they equal the delivery rate. But in the short run, orders booked do not equal delivery delay as seen after month 60 in Figure 2.5d where orders booked fluctuate above and below delivery rate.

The reason for this fluctuation in orders booked can be found in the two time delays produced by the order backlog and the delivery delay recognized. In Figure 2.5d backlog fluctuations occur later than those in orders booked. Likewise the fluctuations in delivery delay recognized lag behind those in backlog.

If the delivery delay impending were immediately influential in determining sales effectiveness, then orders booked would level off without overshoot as did the inventory in Figure 2.2c. But when the corrective action on sales effectiveness is delayed because of the time for the market to recognize changes in delivery delay, the orders booked rise too far and cause a temporary increase in backlog and delivery delay impending that is longer than can be sustained in system equilibrium. When the market finally recognizes the unsatisfactory slowness of delivery, orders drop off to below the delivery rate until delivery rate has time to reduce the backlog. Two delays in the loop (delay in recognizing delivery delay, and the delay created by the backlog) make fluctuating behavior possible just as the delay in goods on order plus the delay in inventory produced the fluctuation in Figure 2.3b.

Several important time relationships can be observed in the curves of Figure 2.5d. Notice that changes in a level variable, for example the backlog curve, are caused by the difference between the inflow and outflow rates. For the first 66 months, except for the first two months, orders booked are greater than delivery rate and the excess of orders booked causes the backlog to rise to a peak that coincides with the time at which orders booked and delivery rate cross at month 66. Between months 66 and 78, orders booked are less than delivery rate so the backlog falls to a minimum that coincides with the second crossing of orders booked and delivery rate at month 78.

Observe also the relationship between indicated salesmen and salesmen and how the difference between the two is related to salesmen hired. Salesmen hired is moving the salesmen curve toward the indicated salesmen curve at a rate that depends on how large is the discrepancy between salesmen and salesmen indicated.

The effect of Equations 2.5-11 and 2.5-12 in generating delivery delay recognized DDR is seen by comparing that curve with delivery delay impending in Figure 2.5d. Delivery delay recognized is always moving toward delivery delay impending (just as salesmen move toward indicated salesmen) and as a result exhibits a shape similar to but lagging in time after the curve for delivery delay impending.

In this chapter we have gone from the simplest possible single feedback loops to a system of interconnected loops. The last example which contains five feedback loops suggests how the feedback loop is the basic unit of which systems are composed. We have seen how the structure of the individual loop can give it a goal-seeking or a growth characteristic. Within the goal-seeking loops (negative feedback) the simplest loop shows a smooth gradual approach to the goal. Loops containing more than one level variable can show oscillation. Later chapters will reexamine these ideas in greater depth.

(See Workbook Section W2.5)

CHAPTER 3

MODELS AND SIMULATION

3.1 Models

A model is a substitute for an object or system. A model can be of many forms and can serve many purposes. We are all familiar with physical models that represent objects. Children's models of automobiles and soldiers fulfill a visual purpose in supporting imagination and play. An architectural model assists in visualizing space and arrangement. But more abstract models are even more common. Any set of rules and relationships that describe something is a model of that thing. In this sense, all of our thinking depends on models.

Our mental processes use concepts which we manipulate into new arrangements. These concepts are not, in fact, the real system that they represent. The mental concepts are abstractions based on our experience. This experience has been filtered and modified by our individual perception and organization processes to produce our mental models that represent the world around us. The equation sets that described behavior in Chapter 2 were more specific than our mental models, though not necessarily more accurate. All models--mental, mathematical, or descriptive--can represent reality with varying degrees of fidelity.

* *
* *

Principle 3.1-1. Abstract models.

Mathematical simulation models belong to the broad class of abstract models. These abstract models include mental images, literary descriptions, behavior rules for games, and legal codes.

* *
* *

The human mind is well adapted to building and using models that relate objects in space. Also, the mind is excellent at manipulating models that associate words and ideas. But the unaided human mind, when confronted with modern social and technological systems, is not adequate for constructing and interpreting dynamic models that represent changes through time in complex systems. This book develops a foundation for constructing computer models to aid our mental process in dealing with time-varying systems.

There are several major defects in mental models of dynamic systems that can be alleviated (not eliminated) by converting from mental models to models represented by explicit statements in the form of flow diagrams and equations:

1. Our mental models are ill-defined. We have many models; they serve various purposes; and the purposes are often unclear. As a result, we keep changing the content of a mental model without realizing we have done so. We are continuously changing the assumptions, the interpretation of real-life observations into model structure, and the consequences that we think the models imply. But the assumptions, structure, and implied consequences are not changed in synchronism. There can be a high degree of internal contradiction.

2. Assumptions are not clearly identified in the mental model. It is usually not clear what information and experience produced the mental model. It is not possible to review how the model was generated.

3. The mental model is not easy to communicate to others. The ill-defined and nebulous nature of the intuitive mental process is hard to put into words. Even when reduced to words these often do not mean the same to both writer and reader. Furthermore, the imprecise nature of language can be used to hide a clouded mental image from both the speaker and the listener.

4. Mental models of dynamic systems can not be manipulated effectively. We often draw the wrong conclusions about system behavior, even if we start with a correct model of the separate system relationships. Perhaps this incorrect dynamic interpretation occurs because we "solve" for system behavior, not by tracing actions and consequences as was done in the computations in Chapter 2, but by drawing conclusions by analogy to past experience. As we will see later, this solution by analogy is particularly unreliable in estimating the behavior of feedback systems. Our experience comes from observing the simplest, usually first-order, systems. When the same expectations are applied to more extensive systems the wrong results are often obtained. For example, Figure 2.2c showing the behavior of a linear, single-loop, first-order system does little to let us anticipate Figure 2.5d showing the behavior of a nonlinear, five-loop, third-order system. Because we can not mentally manage all the facets of a complex system at one time, we tend to break the system into pieces and draw conclusions separately from the subsystems. Such fragmentation fails to show how the subsystems interact.

(Go to the corresponding Workbook section after completing each text section.)

3.2 The Basis of Model Usefulness

The validity and usefulness of dynamic models should be judged, not against an imaginary perfection, but in comparison with the mental and descriptive models which we would otherwise use. We should judge the formal models by their clarity of structure and compare this clarity to the confusion and incompleteness so often found in a verbal description. We should judge the models by whether or not the underlying assumptions are more clearly exposed than in the veiled background of our thought processes. We should judge the models by the certainty with which they show the correct time-varying consequences of the statements made in the models compared to the unreliable conclusions we often reach in extending our mental image of system structure to its behavioral implications. We should judge the models by the ease of communicating their structure compared to the difficulty in conveying a verbal description.

By constructing a formal model, our mental image of the system is clearly exposed. General statements of size, magnitude, and influence are given numerical values. As soon as the model is so precisely stated, one is usually asked how he knows that the model is "right." A controversy often develops over whether or not reality is exactly as presented in the model. But such questions miss the first purpose of a model which is to be clear and to provide concrete statements that can be easily communicated.

There is nothing in either the physical or social sciences about which we have perfect information. We can never prove that any model is an exact representation of "reality." Conversely, among those things of which we are aware, there is nothing of which we know absolutely nothing. So we always deal with information which is of intermediate quality--it is better than nothing and short of perfection. Models are then to be judged, not on an absolute scale that condemns them for failure to be perfect, but on a relative scale that approves them if they succeed in clarifying our knowledge and our insights into systems.

* *
* *

Principle 3.2-1. <u>Model validity</u>.

Model validity is a relative matter. The usefulness of a mathematical simulation model should be judged in comparison with the mental image or other abstract model which would be used instead.

* *
* *

As we move toward models that represent people, their decisions, and their reactions to the pressures of their environment, it is well to keep in mind these relative rather than absolute measures of model utility. The representation need not be defended as perfect, but only that it clarifies thought, captures and records what we do know, and allows us to see the consequences of our assumptions, whether those assumptions be perceived ultimately as right or wrong. A model is successful if it opens the road to _improving_ the accuracy with which we can represent reality.

When a model is reduced to diagrams and equations, when its underlying assumptions can be examined, when it can be communicated to others, and when we can compute its time patterns to determine the behavior implied by the model, then we can reasonably hope to understand reality better.

If models become easier to communicate and become more reliable than those that now dominate thought and literature, we can also expect them to become more influential in solving the great problems of our social systems. It is toward this goal of better understanding, easier communication, and improved management of social systems that we proceed.

3.3 Simulation Versus Analytical Solutions

In Chapter 2 several systems were described in terms of equations that indicated how to start from any existing condition of the system and to compute the condition that would follow at a brief time later. In other words, the equations described how the system changes, and these changes were accumulated step-by-step to unfold the behavior pattern of the system. But the equations did not tell how to go directly to some distant future condition without first computing through all of the intermediate stages.

This process of step-by-step solution is called _simulation_. The equations, that is, the instructions, for how to compute the next time step are collectively called a "simulation model." The simulation model is used in place of the real system. The value of the model arises because it can rapidly and inexpensively give useful information about the dynamic, that is, time-varying, behavior of the real system that the model represents.

It is only recently that simulation models have received wide attention. In the past, most effort has been focused on the quite different approach of solving the equations that describe the behavior of the system to obtain an analytical solution. Such a solution to the equations would express the system condition in terms of any future time, not just in terms of the short time intervals between successive computations. For systems where an analytical solution is possible, one would then be able to substitute any particular value of future time and evaluate the future system condition without first proceeding through the intervening conditions. Furthermore, the form of the solution would tell much about the general nature of the system response even without any numerical computation.

For an example, we can refer to the simple negative-feedback loop of Section 2.2. It was described by the order rate Equation 2.2-1 and by the process of accumulating the order flow to compute the next value for inventory. The step-by-step computation appeared in Table 2.2. Because this system is sufficiently simple, it is possible to obtain an analytical solution which is*

$$I = (DI) - \left[(DI) - I_o\right] e^{-t/(AT)} \qquad \text{Eq. 3.3-1}$$

$DI = 6000$

$I_o = 1000$

$AT = 5$

I--Inventory (units)
DI--Desired inventory (units)
I_o--Initial value of inventory at start (units)
e--base of natural logarithms = 2.718+
t--time measured from the beginning of the period of interest (weeks)
AT--Adjustment time (weeks)

---

*This is the solution to the corresponding continuous equations, assuming that the time interval between successive simulation computations has become vanishingly small.

Substituting the values of DI, I_o and AT gives

$$I = 6000 - 5000e^{-t/5} \qquad \text{Eq. 3.3-2}$$

This equation is quite different from any in earlier sections. It contains time, t, explicitly. In other words, we can substitute any number of weeks from the start of the system and compute directly the inventory without going through the intermediate step-by-step computation of a large number of intervening values. For example, suppose we wish to determine directly the value of inventory at 20 weeks. It is not necessary to compute the intervening steps as in Table 2.2. With time, t, equal to 20 in Equation 3.3-2,

$$I = 6000 - 5000e^{-20/5}$$
$$= 6000 - 5000e^{-4}$$

The negative exponent of e indicates, of course, the reciprocal of e to the same positive exponent, that is, $e^{-4} = 1/e^4$. Powers of e can be found in printed tables of mathematical values from which

$$e^{-4} = 0.01832$$

so

$$I = 6000 - 5000(0.01832)$$
$$= 6000 - 92.$$
$$= 5908 \text{ at time } t = 20.$$

The discrepancy of about one per cent between 5908 and the value of 5970 from Table 2.2 arises because the analytical solution in Equation 3.3-2 corresponds to a vanishingly small solution interval, whereas Table 2.2 was computed using the rather large time interval of 2 weeks between computations to reduce the amount of computing effort.

Not only does the analytical solution, as in Equation 3.3-1, allow us to compute directly the condition of the system for any specified time but it also divulges much information about the entire time pattern of system action. In Equation 3.3-1 we see immediately that inventory I has the correct initial value when t = 0. For t = 0 the value of $e^{-t/(AT)}$ is 1 (any number to the zero power is one). So the equation becomes $I = I_o$. As t becomes larger and larger, the exponential term, $(e^{-t/(AT)})$, approaches zero so that the long-term equilibrium of the system is I = (DI).

Figure 3.3a is a repeat of Figure 2.2c to which have been added the concepts of Equation 3.3-1. The equation says that inventory

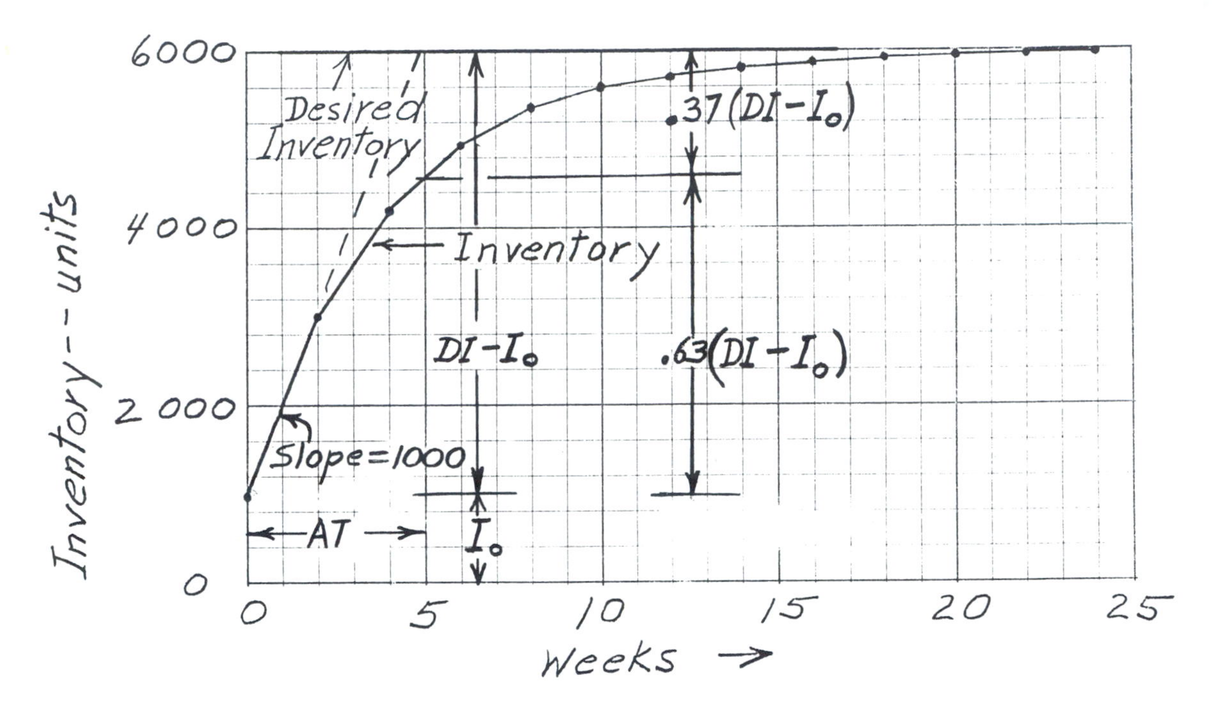

Figure 3.3a First-order system response.

equals the desired inventory DI minus a quantity that starts with a value of $\left[(DI)-I_o\right]$ and the latter quantity declines as time increases in accordance with the shape of the exponential term. The exponential term begins with the slope 5000/AT = 1000 when t = 0. When time equals AT, which equals 5, the exponential is e^{-1} which has the value of approximately 0.37. This means that only 0.37 of the subtracting term remains or that 1-.37 = .63 of the distance to the final value has already been traversed.

In a similar manner and with enough effort, the analytical solution can be obtained to the second-order system involving inventory and ordering delay that appears in Section 2.3. The solution has the form

$$I = C_1 + C_2 e^{-t/C_3} \sin\left(\frac{2\pi}{C_4} t + C_5\right) \qquad \text{Eq. 3.3-3}$$

I--Inventory (units)
e--base of natural logarithms = 2.718+
t--time (weeks)

C_1, C_2, C_3, C_4, and C_5 are constants composed of complicated expresions involving the known constants consisting of desired inventory DI, the adjustment time AT, the delay in ordering DO, the initial value of inventory I_o, and the initial value of goods on order G_o. The solution consists of the constant C_1, and a sinusoidal term of amplitude C_2 which declines in amplitude as controlled by the exponential term e^{-t/C_3}. C_3 is the time constant of decay of the exponential. C_4 is the period (time between peaks) of the fluctuation. C_5 is the phase shift that tells how far the fluctuating sine term is displaced in time (determining how far the first positive-going zero value of the sine is displaced from time = 0). The expressions for C_1 through C_5 are too complicated to be worth including here. But with them, with the numerical values of DI, AT, DO, I_o, and G_o, and with mathematical tables of the values of the exponential and sine functions, it would be possible to compute inventory directly at any specified time for this second-order system in a manner similar to that above for the simpler first-order system in Equation 3.3-1.

Because the analytical solution of system behavior is so informative and because it allows direct computation of system condition at any specified time, one might presume that an analytical solution should always be obtained in every system study. However, such can not be done. Only a few people can keep at their fingertips even the mathematical skill necessary to get the two preceding solutions. A slightly more difficult problem will tax the most skilled mathematician. By the time one approaches a system of even the limited scope of the one in Section 2.5, an analytical solution has become impossible.

When we must deal with systems whose analytical solutions are beyond the reach of today's mathematics, we turn to the process of simulation. Simulation does not give the general solution. It does not tell all the possible behavior patterns. Instead, simulation gives one time history of system operation corresponding to the coefficients and initial conditions whose numerical values were selected. For more information based on different conditions, another full step-by-step computation of a system time-response must be made. Because of the extensive computation that simulation studies require, they have had only limited use until the electronic digital computer became available.

* *
* *

Principle 3.3-1. Simulation solutions.

Most dynamic behavior in social systems can only be represented by models that are nonlinear and so complex that analytical mathematical solutions are impossible. For such systems, only the simulation process using step-by-step numerical solution is available.

* *
* *

Although the use of simulation has been limited by the time and cost involved, it has had a long and illustrious history for problems where the economic justification has been great enough. For example, during the development of world commerce beginning several hundred years ago, geographical explorers needed to know the relative positions of sun, earth, and moon. But even this relatively simple, three-body, celestial-mechanics problem is beyond the reach of analytical solution by presently available mathematics. However, the gravitational and inertial laws that determine the forthcoming locations of the bodies, given their present positions, have long been known. A step-by-step simulation solution was possible as a means of tracing the future condition of the solar system. By the 1600's some men were spending a lifetime computing navigation tables by the simulation procedure.

But throughout the development of science up until about 1955, the cost of computation was so great that most effort was applied to finding analytical solutions to simple systems and the more complex systems were ignored. Recently the economics of computation has been changing drastically. The cost of computation has been falling by about a factor of ten every five years since 1940. The emphasis on method has been reversed. Before 1940, the cost of simulation confined attention to the analytical solutions but these solutions could be obtained only to naively simple systems. Now the cost of computation has fallen to the point where repeated simulations of complex systems can be obtained inexpensively and quickly. Indeed, the cost barrier was not alone in the past in discouraging the study of larger systems. Even where the cost might have been justified, the time required to carry out computation was so long that people were unwilling to wait for results. All is changed now that a lengthy simulation of a complex system can be obtained for a few cents in a few seconds.

(See Workbook Section W3.3)

CHAPTER 4

STRUCTURE OF SYSTEMS

Section 1.2 introduced the importance of structure. The structure of a subject guides us in organizing information. If one knows a structure or pattern on which he can depend, it helps him to interpret his observations. An observation may at first seem meaningless, but knowing that it must fit into one of a limited number of categories helps in the identification. Structure exists in many layers or hierarchies. Within any structure there can be substructures. This chapter discusses the concepts of structure to be used in this book for organizing systems--the closed boundary; feedback loops; levels and rates; and, within a rate, the goal, apparent condition, discrepancy, and action. These concepts of structure organize into the following hierarchy of major and subordinate components:

I. The closed system generating behavior that is created within a boundary and not dependent on outside inputs.
  A. The feedback loop as the basic element from which systems are assembled.
    1. Levels as one fundamental variable type within a feedback loop.
    2. Rates (or policies) as the other fundamental variable type within a feedback loop.
      a. The goal as one component of a rate.
      b. The apparent condition against which the goal is compared.
      c. The discrepancy between goal and apparent condition.
      d. The action resulting from the discrepancy.

4.1 Closed Boundary

We are interested in systems as the cause of dynamic behavior. The focus is on interactions within the system that produce growth, fluctuation, and change. Any specified behavior must be produced by a combination of interacting components. Those components lie within a boundary that defines and encloses the system.

Figure 4.1a emphasizes the closed boundary concept. Formulating a model of a system should start from the question "Where is the boundary, that encompasses the smallest number of components, within which the dynamic behavior under study is generated?"

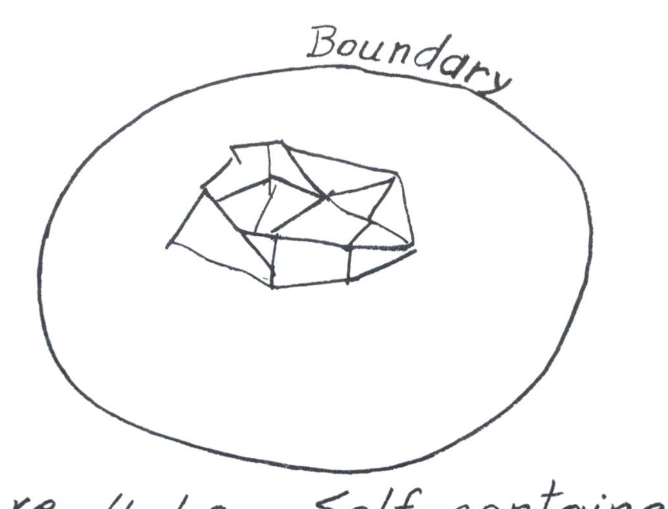

Figure 4.1a Self-contained system.

The essential idea in Figure 4.1a is the boundary across which nothing flows (except perhaps a disturbance for exciting the system so we can observe its reaction). Thinking in terms of an isolated system, forces one to construct, within the boundary of his model, the relationships which <u>create</u> the kinds of behavior that are of interest.

* *
* *

Principle 4.1-1. <u>Closed boundary</u>.

In concept a feedback system is a closed system. Its dynamic behavior arises within its internal structure. Any interaction which is essential to the behavior mode being investigated must be included inside the system boundary.

* *
* *

The application of Principle 4.1-1 appears in the systems in Chapter 2. There, the self-contained systems produced the reactions in Figures 2.2c, 2.3b, 2.4c and 2.5d.

4.2 Feedback Loop--Structural Element of Systems

Within the system boundary, the basic building block is the feedback loop. The feedback loop is a path coupling decision, action, level (or condition) of the system, and information, with the path returning to the decision point. Figure 4.2a, a repeat of Figure 1.4a, shows a loop structure.

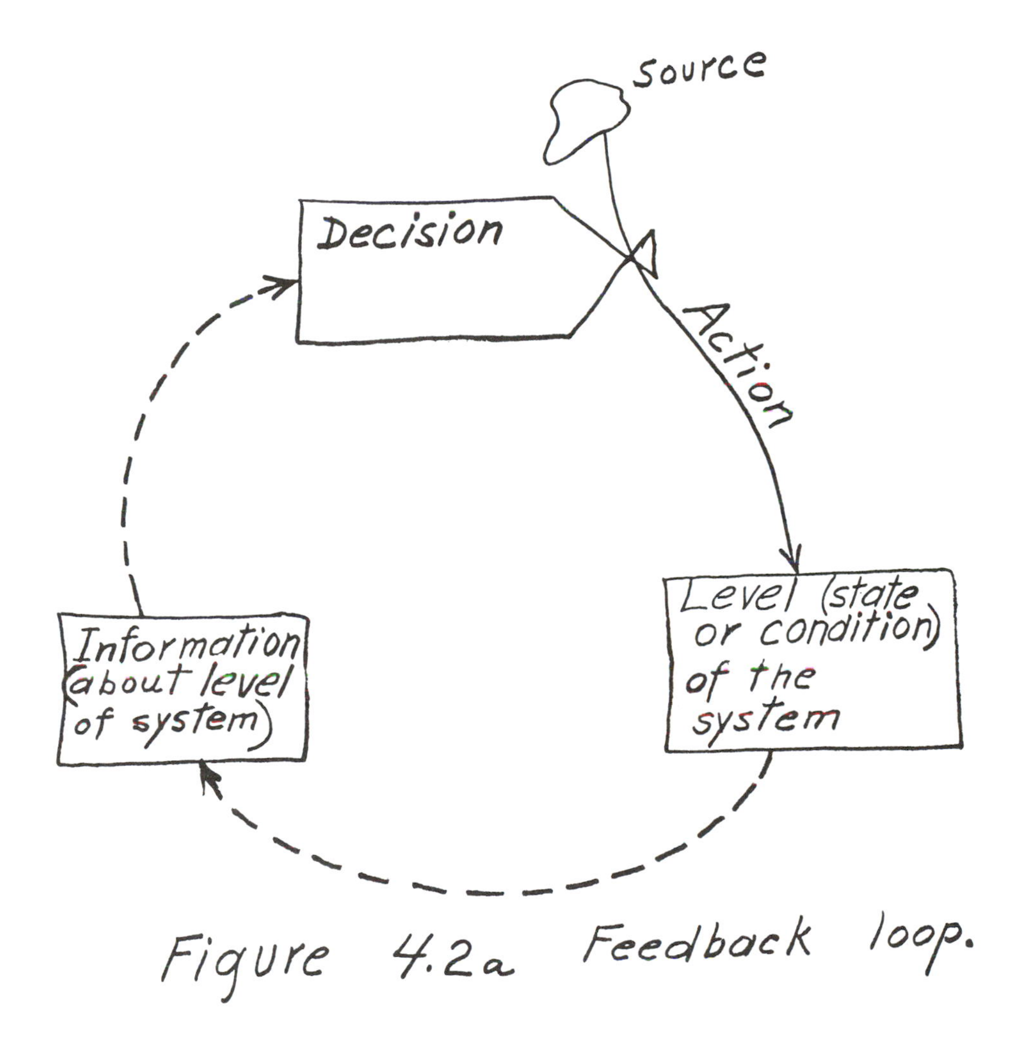

Figure 4.2a Feedback loop.

The "decision" process in Figure 4.2a implies a far broader concept than merely human decision-making in accordance with the common usage of the word "decision." As used here, a decision process is one that controls any system action. It can be a clear explicit human decision. It can be a subconscious decision. It can be the governing processes in biological development. It may be the valve and actuator in a chemical plant. It can be the natural consequences of the physical structure of the system. Whatever the nature of the decision process, it is always imbedded in a feedback loop. The decision is based on the available information; the decision controls an action that influences a system level; and new information arises to modify the decision stream.

* *
* *

Principle 4.2-1. <u>Decisions always within feedback loops</u>.

Every decision is made within a feedback loop. The decision controls action which alters the system levels which influence the decision. A decision process can be part of more than one feedback loop.

* *
* *

The preceding principle stating the circular or re-entrant nature of the decision-making environment focuses attention on the feedback loop. Decision processes are usually easy to discern. The associated feedback loops are not always evident. However, Principle 4.2-1 suggests that we must search for the system loop structure that returns each decision process to its own information inputs.

A system may be a single feedback loop or interlocking feedback loops. Each loop contains one or more decision points that control action and one or more system levels that result from the action. A system can be so simple that it has only a single system level as in Figures 2.2a and 2.4a or more complex as in the five interconnected loops containing three system levels as in Figure 2.5a. Beyond the three-level system of Figure 2.5a, simulation analysis is now being applied to systems of a hundred levels and more. Actual systems can, of course, far exceed even that complexity.

* *
* *

Principle 4.2-2. <u>Feedback loop--the structural element of systems</u>.

The feedback loop is the basic structural element in systems. Dynamic behavior is generated by feedback. The more complex systems are assemblies of interacting feedback loops.

* *
* *

4.3 Levels and Rates--the Substructure Within Feedback Loops

Interconnecting feedback loops form any system. But at a lower hierarchy, each feedback loop contains a substructure. There are two fundamental types of variable elements within each loop--the levels, and the rates. Both are necessary. The two are sufficient.

The level (or state) variables describe the condition of the system at any particular time. The level variables accumulate the results of action within the system. They are represented by the "level equations" to be discussed in the next chapter. In Table 2.5 the level variables are in columns 2, 3, and 4 and are set apart between the vertical double lines. Computing a new value of a level variable involves the previous value of that level variable itself, the rates (actions) that cause the level to change, and the length of the time interval since the last computation of the level. The computation of the new value of a level does not involve the values of any other level variables. The level variables accumulate the flows described by the rate variables. The level equations perform the process of integration (the mathematical process defined in calculus).

On the other hand, the rate (action) variables are quite different. The rate variables tell how fast the levels are changing. The rate variables determine, not the present values of the level variables, but the slope (change per time unit) of the level variables. The rate variables are defined by the "rate equations" to be discussed in the next chapter. The rate equations are the policy statements that describe action in a system, that is, the rate equations state the action output of a decision point in terms of the information inputs to that decision. Computing the value of a rate variable is done using only the values of level variables and constants. The rate variable does not depend on its own past value, nor on the time interval between computations, nor on other rate variables.

* *
* *

Principle 4.3-1. <u>Levels and rates as loop substructure</u>.

A feedback loop consists of two distinctly different types of variables--the levels (states) and the rates (actions). Except for constants, these two are sufficient to represent a feedback loop. Both are necessary.

* *
* *

The level variables are the accumulations in a system. The levels accumulate (or integrate) the net difference between inflow and outflow rates. For example, the water in a pail is a level variable. It is changed by the rate at which water is added and drained. But the water at this instant in a particular pail is not dependent on the water at this same instant in some other pail. The two pails are coupled only if there is a flow rate between them.

* *
* *

Principle 4.3-2. Levels are integrations.

The levels integrate (or accumulate) the results of action in a system. The level variables can not change instantaneously. The levels create system continuity between points in time.

* *
* *

The level variable is of such a nature that it depends only on the accumulation of past rates of flow. The present rates of flow do not determine the present level, but only the rapidity with which the level is changing. A level variable, being dependent only on the past accumulation of its associated rates of flow, is not directly dependent on any other level variable. Any interdependence between two levels must arise because they have in the past been coupled by a connecting flow rate.

* *
* *

Principle 4.3-3. Levels are changed only by the rates.

A level variable is computed by the change, due to rate variables, that alters the previous value of the level. The earlier value of the level is carried forward from the previous period. It is altered by rates that flow over the intervening time interval. The present value of a level variable can be computed without the present or previous values of any other level variables.

* *
* *

Just as the level variables are independent of one another, the structure of a system and the behavior of natural processes are such that the rate variables likewise can not interact directly. As used here, the flow rates are defined, not as averages over some time period, but as the instantaneous flow rates in the action channels of the system. Rates can not be directly interacting because they can act only through their influence on system levels.

No rate of flow can be measured instantaneously. All instruments that purport to measure rates actually require time for their functioning. They measure, not the instantaneous rate, but rather the average rate over some time interval. Likewise, we see that time is required to observe action in social systems and that measured rates are actually observed averages. As we will see later, an _averaged_ rate is a system level variable, not a rate variable. The true rate is the instantaneous action stream that is being averaged.

The units of measure of a variable do not indicate whether the variable is a level or a rate. As an illustration, a rate and the level that is an average of that rate will have the same units of measure. For example, the instantaneous variable, orders per week, that actually changes the inventory in a warehouse is measured in the same units as the average variable, orders per week, that influences the reorder decision.

* *
* *

Principle 4.3-4. _Levels and rates not distinguished by units of measure._

The units of measure of a variable do not distinguish between a level and a rate. The identification must recognize the difference between a variable created by integration and one that is a policy statement in the system.

* *
* *

Principle 4.3-4 is a warning. Of more help is a rule-of-thumb to distinguish levels and rates. The rates are action variables, they cease when action stops. The levels are the accumulations of the effect of past action and continue to exist and can be observed even if there is no present activity. So, as a test of levels and rates, imagine that all activity in a system is brought to rest. Only the level variables would remain and be observable. In a stationary system, all action would be frozen but all levels would continue to exist. A tree would stop growing but the level of its accumulated height would be visible. In a factory, activity would have stopped but the levels representing number of employees, work in progress, capital equipment, and bank balance would be measurable. The more intangible levels would likewise remain--employee morale, company reputation, and quality of the product. Current instantaneous sales rate would have disappeared, but the knowledge of average sales rate for the past year would remain as a system level.

Because some interval of time is necessary to measure and transmit information about any rate, we can argue that no rate at one instant can depend on other rates at the same instant. Practical, rather than theoretical, considerations lead to the same principle as a working guide to model formulation. Usually, regardless of the nature of the system, the time required to observe rates is significant compared to the delays inherent in other parts of the system. Rates do not act directly on other rates but only by first being averaged (and these averages contain accumulations or integrations and involve level variables).

* *
* *

Principle 4.3-5. <u>Rates not instantaneously measurable</u>.

No rate of flow can be measured except as an average over a period of time. No rate can, in principle, control another rate without an intervening level variable.

* *
* *

Although Principle 4.3-5 is true as a basic concept, the averaging time necessary for measuring a rate is sometimes very short compared to other time delays within a system. The relative rapidity of measurement occasionally permits a simplification by coupling one rate to another without seriously affecting dynamic behavior of a model. Valid opportunities for this short cut are infrequent. The beginner should strictly avoid any rate-to-rate coupling in a model.

* *
* *

Principle 4.3-6. Rates depend only on levels and constants.

The value of a rate variable depends only on constants and on present values of level variables. No rate variable depends directly on any other rate variable. The rate equations (policy statements) of a system are of simple algebraic form; they do not involve time or the solution interval; they are not dependent on their own past values.

* *
* *

The preceding principles say that level variables feed information only to rate variables and that rate variables cause changes only in level variables. As a corollary, it follows that level and rate variables must alternate along any path through a system structure.

* *
* *

Principle 4.3-7. Level variables and rate variables must alternate.

Any path through the structure of a system encounters alternating level and rate variables.

* *
* *

Principles 4.3-2 through 4.3-7 imply one of the functions of a system level (or state) variable. The system level decouples rates of flow into and out of the level. Since one flow rate can not directly control another, the intervening level allows the two flow rates to differ. For example, a bank balance is needed to absorb the short-term differences between cash flowing in and cash flowing out. Likewise inventories exist because the supply rate and the demand rate can differ and the inventory absorbs the difference in rates.

The distinction between level and rate variables is recognized in several intellectual disciplines. In financial accounting for example, a clear separation is made between the balance sheet and the profit-and-loss statement. The balance-sheet variables are levels, giving the financial condition of the business system at one point in time. The balance-sheet levels show the effect of accumulating the rates of flow over all past time. The profit-and-loss statement, by contrast, gives the rates of flow that have existed since the previous balance sheet. The profit-and-loss rates cause the changes from the previous balance sheet to the present.

A rate of flow determines how fast a system level is changing but the present rate does not determine the present level. The present level is the result of having accumulated the in and out flow rates over all past time. The level variables of the system carry the continuity of the system from the past to the present. The levels contain all the remaining and presently available history of the system. If the levels are known, the rates can be determined. It follows that the levels fully determine the condition of the system. The model of a system must contain one level for each quantity needed to describe the condition of the actual system.

* *
* *

Principle 4.3-8. Levels completely describe the system condition.

Only the values of the level variables are needed to fully describe the condition of a system. Rate variables are not needed because they can be computed from the levels.

* *
* *

The start of a simulation run must begin from a specified condition of the system. Principle 4.3-8 indicates that only the levels (states) of the system need be specified as initial conditions. For example, in Table 2.5 it was necessary to have values for only salesmen, backlog, and delivery-delay-recognized in order to launch the computation.

The preceding principles are all illustrated in the system examples of Chapter 2. Figures 2.2a, 2.3a, 2.4a, and 2.5a show the levels and rates as the feedback loop substructure. The levels appear in the flow diagrams as the rectangles, the rates as the valve symbols controlling a flow stream. In the diagrams of Chapter 2, levels are influenced only by rates and rates receive information only from levels. (In Figure 2.5a the circle symbols should be ignored. As will be seen in the next chapter they are auxiliary equations which are actually part of the rate equations.) In Figure 2.5a any path moving forward along the direction of the flow lines encounters alternating levels and rates.

4.4 Goal, Observation, Discrepancy, Action--Sub-substructure within a Rate.

In the preceding sections, the structure of a system has been developed in three layers--the closed boundary, the feedback loop structure, and the level and rate substructure. One can now look for a sub-substructure within the levels and rates.

It does not seem useful to create a subdivision within the concept of a level. The structure within a level computation is straightforward. It is simply arithmetic to add the change in the level to its previous value. Skill and judgment are required to decide what levels should be incorporated into a system model, but not to handle the equation and computations for a level once that level has been identified along with its pertinent flows inward and outward.

But a rate equation does contain an important and helpful internal structure. We will refer to the components of the rate equation as the sub-substructure of a system, because the rate equation is itself one of the substructural elements. But, before discussing the sub-substructure within a rate equation, the meaning of a "rate equation" should be clarified.

A rate equation is a policy statement. That is, the rate equation tells how a "decision stream" (or "action stream") is generated. "Rate equation" and "policy," as used here, have the same meaning. A policy describes how the available information is used to generate decisions. "Decision stream" and "action stream" are equivalent because, as used here, the decision and the action are one and the same. Any delays and discrepancies between the deciding and the doing that we might expect from the common usage of the words would involve level equations in a model. So the policy, or rate equation, tells how to compute the rate (the flow into some level) based on the values of levels and constants.

Rate equations have already been encountered in Chapter 2. For example, Equation 2.2-1 describes how the order rate is to be computed, given the varying inventory and the constants for desired inventory and adjustment rate. Equation 2.3-1 for receiving rate and Equation 2.3-2 for order rate define the two rate equations (policies) necessary for the operation of the system in Section 2.3. In Section 2.4, Equation 2.4-1 is the rate equation that tells how the number of salesmen determine the hiring rate for new salesmen.

In Section 2.5 there are four rate equations. Equation 2.5-5, along with its subdivisions in Equations 2.5-4, 2.5-3, 2.5-2, and the relationship in Equation 2.5-13 as illustrated in Figure 2.5c, depends on the levels for salesmen and delivery delay recognized and determines the rate of flow of salesmen hired. The rate of orders entered flowing into backlog is determined by the rate Equation 2.5-6 which depends on the level of salesmen and on the level of delivery delay recognized, the latter acting through the relationships in Equations 2.5-2 and 2.5-13. The orders completed flowing out of backlog is determined by the level of backlog as shown in Figure 2.5b. The rate of change in delivery delay recognized is defined by Equation 2.5-11 which depends on the levels for delivery delay recognized and backlog.

The rate equation is instantaneous in its behavior (neglecting the delays introduced by the interval between computations which must be short enough to be of no consequence). It is a pure algebraic expression that states the present flow rate in terms of the present information. Any delays that we might commonly anticipate in a decision process actually imply the presence of intermediate levels with a cascade of alternating levels and rates. The rate equation is algebraic, it is free of lags and time-dependent distortion. All time-dependent changes in the character of a flow are created by the level equations.

We turn now to the components of a rate equation--the sub-substructure of a system. Four concepts are to be found within the rate equation (that is, a policy statement):

1. A goal
2. An observed condition of the system
3. A way to express the discrepancy between goal and observed condition
4. A statement of how action is to be based on the discrepancy.

These four are related as in Figure 4.4a and are clearly visible in the rate Equation 2.2-1, repeated here:

$$OR = \frac{1}{AT} (DI-I) \qquad \text{Eq. 4.4-1}$$

OR--Order rate (units/week)
AT--Adjustment time (weeks)
DI--Desired inventory (units
I--Inventory (units)

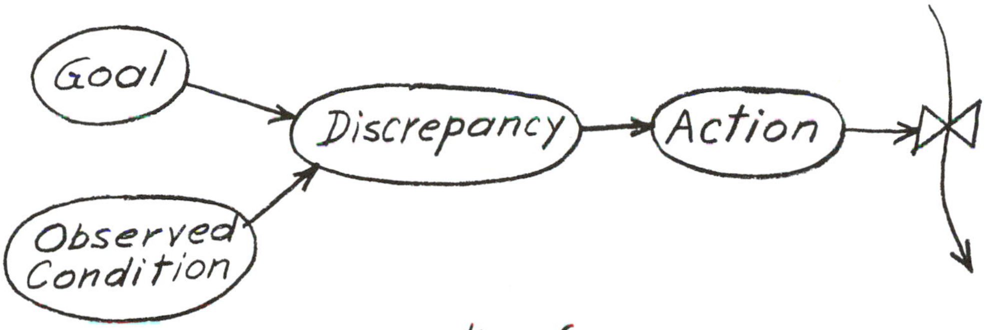

Figure 4.4a Components of a rate equation (or policy).

In this equation, the goal is the desired inventory DI. The order rate acts to move inventory toward the goal. The observed condition of the system is the inventory I. In the simple system of Section 2.2, there was no distinction between actual and apparent inventory. The discrepancy between the desired and apparent conditions of the system are here taken as the simple difference (DI-I). The action in the above rate equation is stated to be 1/AT of the discrepancy.

In many rate equations the four components of sub-substructure may be obscured. In Equation 2.5-6 for orders entered,

OE = OB = (S)(SE)

OE--Orders entered	(units/month)
OB--Orders booked	(units/month)
S--Salesmen	(men)
SE--Sales effectiveness	(units/man-month)

the goal, the observed condition, the discrepancy, and part of the action statement are all combined into the curve of Figure 2.5c. There is a concept of "equilibrium delay" which is the goal and is the delay which would produce an order rate equal to the production capacity. The observed condition is the delivery delay recognized. The discrepancy between the two is modified by two factors--one represented by the slope of the curve in Figure 2.5c and the other by the number of salesmen as entering through Equation 2.5-2--to define the action which is the rate of orders entered in the backlog.

In a positive feedback loop, the "goal" takes on a somewhat different meaning than in the negative feedback loop. In the negative loop the goal, as implied by common usage of the word, is the system condition that is being sought. In the negative loop, the rate equation creates a rate of flow that tries to move the condition of the system toward the goal. In the positive feedback loop, "goal" takes on the reverse meaning. In the positive loop the goal is that value from which the system is departing in its ever-increasing growth.

* *
* *

Principle 4.4-1. Goal, observation, discrepancy, and action--system sub-substructure.

A policy or rate equation recognizes a local goal toward which that decision point strives, compares the goal with the apparent system condition to detect a discrepancy, and uses the discrepancy to guide action.

* *
* *

The hierarchies in system structure can be summarized as:

- Closed boundary
  - Feedback loop structure
    - Level and rate substructure
      - Goal, observation, discrepancy, and action as the sub-substructure within rates.

CHAPTER 5

EQUATIONS AND COMPUTATION

In Chapter 2 were example computations for simulating the behavior of several systems using models that were described partly by equations and partly by verbal statements of the processes to be followed. But verbal statements are awkward, space consuming, and lack precision. Before going further with the modeling of dynamic systems we need a clear set of conventions to follow as a means of communicating the details of model structure. The conventions will be in the form of equations that can state precisely what each element in the model does. The equations grow out of the basic concepts of system structure that were discussed in Chapter 4 but many of the details that follow are arbitrary and should be accepted as a language to make communication easier.

5.1 Computing Sequence

When computing the successive time steps in the dynamic behavior of a system, we need a standardized sequence for the computation and a terminology to use in designating the procedure. The computation progresses in time-steps as in Figure 5.1a. The figure assumes that the computations at time 5 have been completed, ready to begin computing the condition of the system at the next solution period 5+DT. The symbol DT for "difference in time" is used for the length of the time interval between computations. The 5 and the 6 in the figure represent the units of time used in defining the system, for example, weeks or months, but the appropriate solution interval need not be the same as the unit of time measurement. The figure illustrates a situation where there are four computations of system condition in each unit of time.

As shown in Figure 5.1a, the arbitrary convention has been adopted of using "K" to designate the point in time to which the current computation applies. The time 5+DT is designated by the "K" as being the point in the time sequence now being evaluated. Correspondingly, "J" is used to designate the time at which the preceding

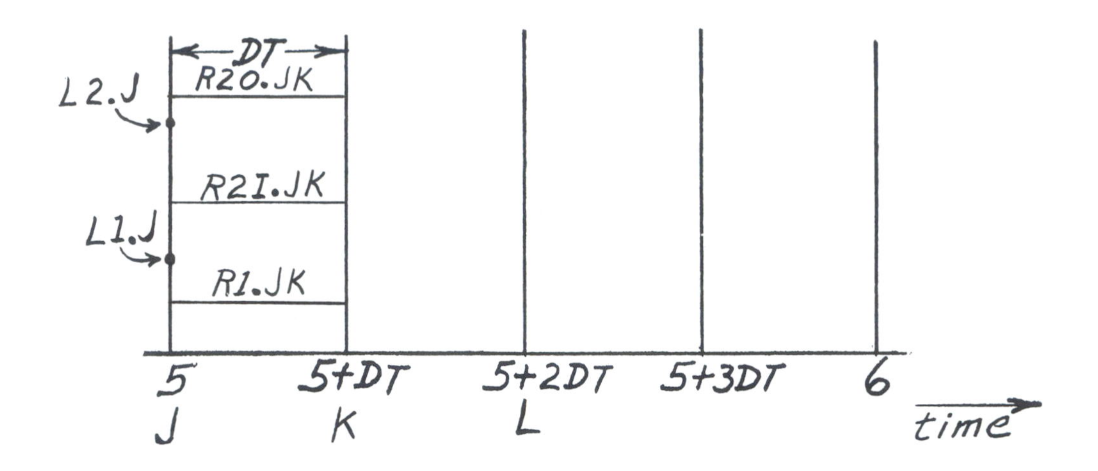

Figure 5.1a. Start of new computation sequence.

computation was made, and, "L" is used to designate the next point in time. The equations are so structured that no other points in time need enter into the computation process. The computation is confined completely to the time J, the interval JK from J to K, the time K, and the interval KL.

At the start of the computations at time K, there are available from the previous computations the levels at time J and the rates of flow which existed over the interval JK. In Figure 5.1a the levels L1.J and L2.J designate two values of levels (system states) at time J.* Also illustrated are three rates that existed over the time interval JK. The rate R1.JK flowed into level L1 and is the only rate that affected level L1. The rate R2I.JK flowed into level L2 while the rate R2O.JK flowed out of level L2.

The rates of flow are expressed in the time units of the system, such as dollars/week, (in Figure 5.1a, the time unit is from 5 to 6) not in terms of the solution interval DT. The selection of a

---

*Many books use subscripts such as "$L1_j$" to designate time position. But subscript notation is seldom available on the printers of computing machines so we will use postscripts as the time designator separated from the variable by a period.

solution interval DT is a technical matter to be discussed later and usually can be specified only after the model has already been constructed in terms of the time unit which is customarily used in the real system being represented.

In Figure 5.1a all the information is available that is needed to compute the new values of levels at time K. But the new values of rates for the KL interval can not yet be computed because they depend on the not-yet-available levels at time K. The constant rates of flow during the JK interval acted on the levels beginning at time J and caused the levels to change at a uniform slope over the interval. The new values of levels are found by adding and subtracting the changes represented by the rates. The changes are found by multiplying the rates by the solution time interval. For example, the change in inventory in a one-quarter-month solution interval caused by a production rate of 800 cartons per month would be 200 cartons added to inventory. All of the levels can be computed. The sequence of computation does not matter because each level depends only on its own old value and on rates in the JK interval. No level depends directly on any other level. Finishing the computation of levels creates the situation in Figure 5.1b

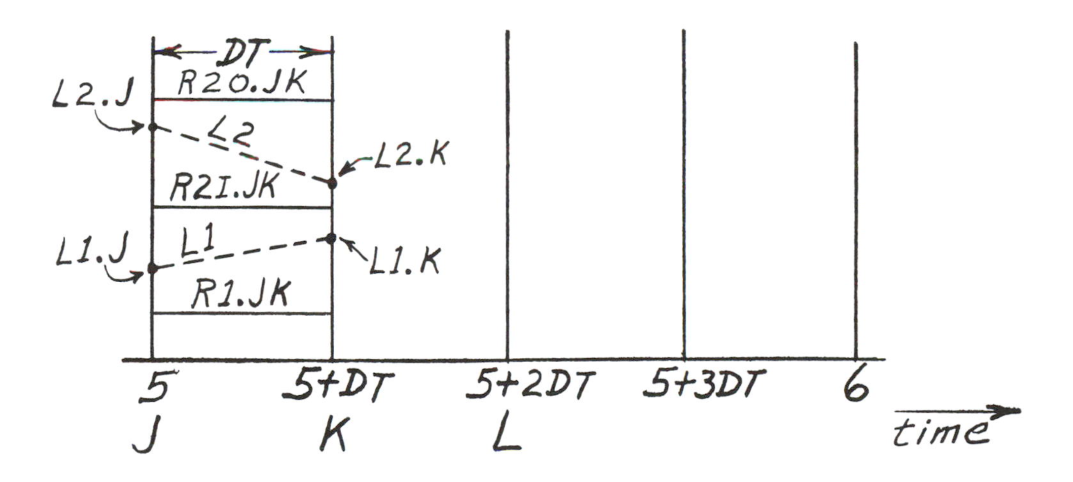

Figure 5.1b. After computation of levels.

where the new levels for time K are now available. Although no values are actually computed except at the discrete solution intervals separated by DT, the nature of the rate and level equations implies that the constant rates have caused continuously changing levels as shown by the dashed lines. Levels are continuous curves in the form of connected straight-line sections that can change slope at the solution times.

Only the present values of levels at time K are needed to compute the forthcoming rates that represent the action during the KL interval. The forthcoming action is based only on the currently available information at time K. Once all levels are computed, the rates can be computed. The order in which the rates are computed does not matter because they do not depend on one another. All of the information needed for all of the rates is available in the levels at time K.

Computation of the new rates brings the situation to that in Figure 5.1c. The rates of flow are thought of as constant over a

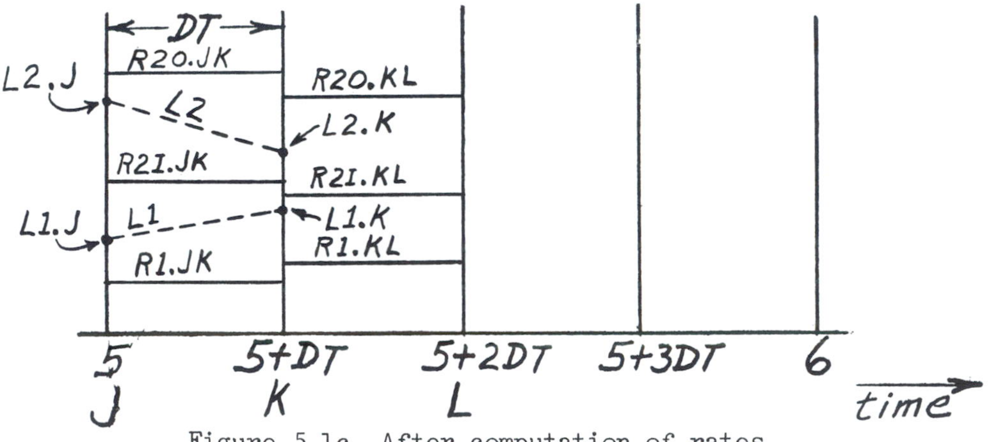

Figure 5.1c After computation of rates.

solution interval and change discontinuously to a new value at the solution time. The solution intervals are taken short enough that the step-wise discontinuities in rates are of no significance. Figure 5.1c shows the completion of the computations at time K.

The entire process is now repeated for the next point in time. To do this, the first step is to advance the time designators, J, K, and L, by one solution interval as shown in Figure 5.1d. Relative to the

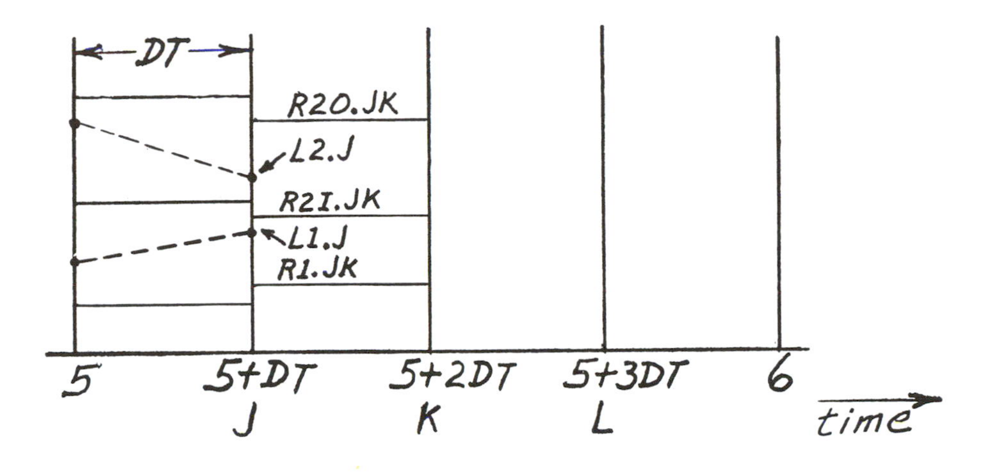

Figure 5.1d Time designators advanced to next solution interval.

new position of K, the conditions are the same as in Figure 5.1a. The K levels have become the J levels and the KL rates have become the JK rates. The levels at time J and the rates for the interval JK are available and the new values of levels can be computed.

A variation from the level-rate sequence of computation occurs at the beginning time, t = 0, for a simulation series. The initial values of all levels must be given. The rates before time = 0 are immaterial. With the levels already available, the computation begins by computing the rates for the interval from time = 0 to time = 0+DT. Thereafter, the full cycle of computing first the levels and then the rates is followed.

(See Workbook Section W5.1)

5.2 Symbols in Equations

The variables and constants in equations are represented by symbols (abbreviations). Practical considerations make it desirable to adopt a standardized style for the symbols.

The format for the standard symbol should permit enough variety so that the symbol bears an obvious relationship to the quantity it represents. On the other hand, the maximum length of the letter group that is used as a single symbol should be specified to ease the computer programming task. Computing machine printers do not offer superscript and subscript notation, so the symbol should be composed of characters at line level.

To satisfy the several demands, a standard symbol specification will be used in this text:

> A symbol to represent a constant or variable shall consist of a group of six characters or fewer, the first of which must be alphabetic.

In addition to the basic symbol, variables will be followed by a period and the time postscript. Levels will carry the single letter J or K indicating the point in time to which the value applies. Standard symbols for levels could include the following:

A.J	MEN.J
ABCDEF.K	INV8.K
B57L.J	B.J
CASH.K	EMPL5.K

Rates carry the time postscript JK or KL indicating the preceding or succeeding interval and might include:

FLOW3.JK	AS9P27.JK
RATE7.KL	EMPLH.KL
S.JK	MH.KL

Constants are identified by the absence of a time postscript, so would appear as:

AT	D
BCD	D27
EFFY8C	MPM

The DYNAMO Compiler for simulating model behavior has been designed to be consistent with the preceding conventions for symbols.

(See Workbook Section W5.2)

5.3 Level Equations

A level equation represents a reservoir to accumulate the rates of flow that increase and decrease the content of the reservoir. The new value of a level is calculated, by adding to or subtracting from the previous value, the change that has occurred during the intervening time interval. We will adopt the following format for a level equation:

L.K = L.J + (DT)(RA.JK - RS.JK) Eq. 5.3-1, L

L--Level (units)
L.K--New value of level being computed at time K (units)
L.J--Value of level from previous time J (units)
DT--The length of the solution interval between time J and time K (time measure)
RA--Rate being added to level L (units/time measure)
RA.JK--The value of the rate added during the JK time interval (units/time measure)
RS--Rate being subtracted from level L (units/time measure)
RS.JK--The value of the rate subtracted during the JK time interval (units/time measure)

Any number of rates, one or more, can be adding to or subtracting from a level. This is the only flexibility permissible in the standard level equation. The right hand side of the equation must contain the previous value of the level being computed. It must also contain the solution interval DT as a multiplier of the flow rates. The level equation is the only equation type which properly contains the solution interval DT.

The solution interval DT is a parameter of the computing process, not a parameter of the real system that the model represents. The flow rates of the system, measured in "units/time measure" (for example, pounds/hour, or men/month, or calories/second) are accumulated in steps or batches over the successive time intervals of DT in length. The solution interval DT, measured in units of time, converts the flow rates to a quantity of the item flowing. It is this product of flow rate multiplied by time that creates the correct units of measure for adding to the value of the level. The solution interval can be arbitrarily changed (if it does not become too long) without affecting the validity of the model. All other equations in the model are formulated in terms of the basic unit of time used in the real system. The solution interval DT should not appear in any equation other than a level equation.

* *
* *

Principle 5.3-1. Solution interval DT in all level equations and no others.

The solution interval multiplies the rates and is essential in the accumulating (or integrating) function of the level equation. All other equations should be formulated in terms of the unit of time measurement customary in the real system. After model formulation, the solution interval will be selected to insure stability within the computation process itself (to be distinguished from system stability).

* *
* *

The level equation performs the process of integration. In the notation of calculus and differential equations, Equation 5.3-1 would be written as follows:

$$L = L_o + \int_o^t (RA-RS)dt \qquad \text{Eq. 5.3-2}$$

L--the value of the level at any time t (units)
L_o--the initial value of the level at t = 0

$\int_o^t$--the operator indicating integration or accumulation from time = 0 until time = t of the difference in flow rates (RA-RS)
RA--the flow rate being added
RS--the flow rate being subtracted
dt--the differential operator representing the infinitesimally small difference in time that multiplies the flow rates. (corresponds to the coarser time steps represented by DT)

Equation 5.3-1 is also known as a first-order difference equation in the branch of mathematics dealing with step-by-step integration.

(See Workbook Section W5.3)

5.4 Rate Equations

Rate equations state how the flows within a system are controlled. The inputs to a rate equation are system levels and constants. The output of a rate equation controls a flow to, from, or between levels.

Following the time notation of Section 5.1, the rate equation is computed at time K, using the information from levels at time K, to find the forthcoming flow rates for the KL interval.

The form of a rate equation is:

$$R.KL = f(\text{levels and constants}) \qquad \text{Eq. 5.4-1}$$

where the right-hand side implies any function, or relationship, of levels and constants that describe the policy controlling the rate. Using the conventions that have now been established, some of the rate equations from Chapter 2 can be rewritten to follow the proper notation for rates and levels. Equation 2.2-1 becomes

$$OR.KL = \frac{1}{AT}(DI - I.K) \qquad \text{Eq. 5.4-2, R}$$

Equation 2.3-1 becomes

$$RR.KL = \frac{GO.K}{DO} \qquad \text{Eq. 5.4-3, R}$$

Equation 2.4-1 becomes

$$SHR.KL = \frac{1}{SDT}(S.K) \qquad \text{Eq. 5.4-4, R}$$

Unlike the level equations, the rate equations are unrestricted in form except for three prohibitions already implied in earlier sections:

1. A rate equation should not contain the solution interval DT. Except in the level equations, the solution interval has no significance in formulating model equations.

   The solution interval arises from the step-by-step computation process and is the only quantity appearing in the equations that does not have significance in the real system that the model represents.

2. There should be no rate variable on the right side of a rate equation, only levels and constants.

3. The left side of the equation contains the rate variable being defined by the equation. The value of the rate is for the KL interval that is immediately forthcoming after the time K at which the computation is being done.

The rate equations are policy statements that tell how "decisions" are made. The policy (rate equation) is the general statement of how the pertinent information is to be converted into a decision (or flow, or present action stream, all being synonymous terms). The rate equations tell how the system controls itself.

The words, "policy" and "decision," have broader meanings here than in common usage. They go beyond the usual human decisions and include the control processes that are implicit in system structure and in habit and tradition. A rate equation (or policy statement) might describe how the flow in a pipe depends on the valve position and the fluid pressure difference across the valve. A rate equation could also represent the subjective and intuitive responses of people to the social pressures within an organization. Or a rate equation might represent the explicit policies that control information flow in the real system where the processes are automatic under control of a digital computer program.

The rate equations are more subtle than the level equations. The rate equations state our perception of how the real-system decisions respond to the circumstances surrounding the decision point.

(See Workbook Section W5.4)

5.5 Auxiliary Equations

Very often, the clarity and meaning of a rate equation can be enhanced by dividing it into parts that are written as separate equations. These parts we will call auxiliary equations.

The presence of auxiliary equations in a model does not in any way contradict the concept that the structure of a system is composed only of levels and rates. The auxiliary equations are merely algebraic subdivisions of the rates.

Suppose that desired inventory in an inventory ordering equation is a variable that depends on the average sales rate. The rate equation for ordering and the accompanying auxiliary equation for desired inventory could then be:

$$OR.KL = \frac{1}{AT}(DI.K-I.K) \qquad \text{Eq. 5.5-1, R}$$

$$DI.K = (WID)(ASR.K) \qquad \text{Eq. 5.5-2, A}$$

OR--Order rate (units/week)
AT--Adjustment time (weeks)
DI--Desired inventory (units)
I--Inventory (units)
WID--Weeks of inventory desired (weeks)
ASR--Average sales rate (units/week)

In Equation 5.5-2, WID is a constant, the value of which states the desired inventory in terms of weeks of average sales. Average sales rate ASR is a level. Equation 5.5-2 can be substituted into Equation 5.5-1 to create the following rate equation that depends only on levels and constants:

$$OR.KL = \frac{1}{AT}\left[(WID)(ASR.K) - I.K\right] \qquad \text{Eq. 5.5-3, R}$$

In this form, the auxiliary equation has disappeared.

Auxiliary equations must be evaluated after the level equations on which they depend, and before the rate equations of which they are a part. When auxiliary equations exist, and they ordinarily will be numerous, the computation is in the sequence: levels, auxiliaries, rates.

Unlike the levels and rates, auxiliary equations can depend on other auxiliary equations in a chain, so, some groups of equations may have to be evaluated in a particular order. Consider the following equations from Section 2.5, now written in the proper time notation

$$SH.KL = \frac{1}{SAT}(IS.K-S.K) \qquad \text{Eq. 5.5-4, R}$$

$$IS.K = B.K/SS \qquad \text{Eq. 5.5-5, A}$$

$$B.K = (OB.K)(RS) \qquad \text{Eq. 5.5-6, A}$$

$$OB.K = (S.K)(SE.K) \qquad \text{Eq. 5.5-7, A}$$

$$SE.K = TABLE(TSE,DDR.K,0,6,.5) \qquad \text{Eq. 5.5-8, A}$$

In the above equation sequence, it is evident that the first is a rate equation because of the KL time notation. The others are auxiliary equations which carry the same time K designator as level equations but are not levels as is apparent because the equation form is not that of a level equation. Equation 5.5-8 is not an algebraic equation but rather is an operational instruction that says that sales effectiveness SE is a function of DDR. The format is that used by the DYNAMO compiler for a table function. TSE is the name of the table that contains values from the relationship in Figure 2.5c. The numbers in the last part of the statement say that the table spans the values of 0 to 6 at a spacing of 0.5 in the scale of measurement of DDR.

Starting from the bottom of the foregoing group of equations it is possible to substitute values in turn into each preceding equation. Starting with the values of levels for DDR and S, the auxiliary equations in the above group must be evaluated in the sequence SE, OB, B, and IS.

When there are interlinked chains of auxiliary equations, they must be evaluated in the sequence that permits successive substitution. Such a sequence will always exist in a properly formulated system. If there is a circular loop of auxiliary equations, it implies a set of simultaneous algebraic equations, a solution sequence does not exist, the auxiliaries are not simple links from the level equations to the rate equations. The circular loops of auxiliary equations will be found and rejected by the DYNAMO compiler as a system formulation error.

(See Workbook Section W5.5)

5.6 Constant and Initial Value Equations

Constants

A constant, represented by a symbolic name, is given a numerical value in a constant equation. The constant equations carry the type designator C after the equation number. A constant has no time postscript because it does not change through time.

XY.K = (AB)(Z.K) Eq. 5.6-1, A

AB = 15 Eq. 5.6-1.1, C

The value of the constant AB is given in the constant Equation 5.6-1.1. The equation number of a constant will be given as a decimal subdivision of the primary equation number in which the constant first appears.

Initially Computed Constants

It is often convenient to specify one constant in terms of another constant when the former depends on the latter and when the former should change in any simulation run where the latter is given a new value. Suppose that the constant CD is always to be 14 times the value of AB. AB is given above in Equation 5.6-1.1,C where its value might be altered from the original 15. CD could then be written as

CD = (14)(AB) Eq. 5.6-2.1, N

The type designator N (the same as for the initial value equations that give the initial values of levels) indicates that this equation need be evaluated only once at the beginning of the simulation computation, because the constants, by their very nature and definition, are values that are not to vary during any one simulation run. The equation number will be a decimal subdivision of the equation in which the initially computed constant first appears. For example, CD above is used in Equation 5.6-2 (not shown).

Initial Value Equations

All level equations must be given initial values at the start of the simulation computation. These level variables represent the complete condition of the system necessary for determining the forthcoming flow rates. All system history which influences present action is represented by <u>present</u> values of appropriate level variables. It is only the form of the system history existing at the present that can be effective. Recollections of the past are dimmed and modified by time. It is the present version of history represented in the present values of system levels that governs present action. Initial values for rate variables need not and should not be given because they are fully determined by the initial values of the level variables.

From the initial values of level variables, the rates of flow immediately following time zero can be computed and, with the initial values and the rates, the new values of levels at the end of the first time step can be computed. The initial value equation will carry the type designator N after the equation number. No time postscripts are used. The right side of the initial value equation is written in terms of numerical values, symbolically indicated constants, and the initial values of other levels. Two or more initial values can not mutually depend on one another because the result would be indeterminate. There must be a beginning point in any chain of initial values that is stated in terms of constants. An initial value equation is customarily written immediately following the corresponding level equation:

PT.K = PT.J+(DT)(M.JK-N.JK)   Eq. 5.6-3, L

PT = 8   Eq. 5.6-3.1, N

The number of the initial value equation will be given as a decimal subdivision of the number of its level equation--as in Equations 5.6-3,L and 5.6-3.1,N above.

The initial value equation above could also have been written in terms of constants as

PT = (3)(CD)   Eq. 5.6-3.2, N

or   PT = AB   Eq. 5.6-3.3, N

It is unambiguous and permissible to state an initial value of one level equation in terms of the initial value of some other level as long as the latter is independent of the first. For example, the following initial value depends on the initial value in Equation 5.6-3.1,N:

```
RS.K = RS.J+(DT)(ML.JK-NL.JK)        Eq. 5.6-4, L
  RS = PT                            Eq. 5.6-4.1, N
```

(See Workbook Section W5.6)

CHAPTER 6

MODELS--MISCELLANEOUS

Chapters 3 through 8 give background information which will be useful later when Chapter 9 returns to the dynamic behavior of systems. This chapter discusses dimensions, solution interval, the importance of not letting the "fine detail" of the computation obscure the broader sweep of the dynamic changes that are occurring in a model, and the relationship of differential equations to the integral equations used in this text.

6.1 Dimensions

In an equation, all terms must be measured in the same dimensions. "One can't add apples and oranges" is a common expression saying the same. "Dimensional analysis" is a subject that teaches the necessity of combining only like terms and shows how to determine the dimensional measure in the more subtle situations.

The identity of dimensions can be observed in each of the equations in earlier sections. As a further example, consider this equation for the production capacity of a factory.

PC.K = PC.J+(DT)(PCR.JK)

PC--Production capacity (pounds/month)
DT--Solution interval (months)
PCR--Production capacity received (pounds/month/month)

This is a level equation describing one of the system states. It says that the production capacity now (at time K) is equal to the previous capacity (at time J) plus the capacity which has been added during the intervening time interval DT. The capacity added is given by the rate at which capacity is being added PCR multiplied by the length of time DT since the last computation. The addition rate PCR can be negative, implying a reduction in capacity. In terms of dimensions of measure, the equation has the following form:

$$\frac{\text{pounds}}{\text{month}} = \frac{\text{pounds}}{\text{month}} + \left(\text{months}\right) \frac{\text{pounds/month}}{\text{month}}$$

Units of measure in a dimensional equation are handled and simplified in the same way as algebraic symbols. In the last term above, the first "months" measure can be used to cancel the "months" in the denominator leaving a "pounds/month" term having the same dimensions as in the term on the left side of the equation and in the first term to the right of the equals sign.

Likewise, all other types of equations must be in dimensional balance. The following auxiliary equation for delivery delay in terms of backlog and production rate again illustrates:

DD.K = BL.K/DRA.K

DD--Delivery delay (weeks)
BL--Backlog (automobiles)
DRA--Delivery rate average (automobiles/week)

where the dimensional equation is

$$\text{weeks} = \frac{\text{automobiles}}{\text{automobiles/week}} = \text{automobiles}\left(\frac{\text{weeks}}{\text{automobile}}\right) = \text{weeks}$$

* *
* *

Principle 6.1-1. Dimensional equality.

In any equation, every term must be measured in the same dimensions. Terms having different units of measure within an equation indicate a faulty equation formulation.

* *
* *

In simple equations and descriptions, it is obvious that only terms of like dimensions of measure can be added and subtracted. But practice often becomes careless, terms are not defined precisely, and soon incompatible items are being combined. Dimensional analysis of equations is a powerful safeguard for the model builder. Units of measure for each variable and constant should be explicitly defined in writing. Each equation should be checked for internal consistency. Only the inexperienced forego this easy and effective self checking.

6.2 Solution Interval

Section 5.1 has described the computing sequence for a dynamic model. The equations are recomputed at uniformly spaced points in time that are DT time units apart. The solution interval DT appears only in the level equation. (Recall Principle 5.3-1)

The proper length of the solution interval is related to the shortest delays that are represented in the model. If the solution interval is too long, instability is generated which arises from the computing process, not from any inherent dynamic characteristic of the system itself. If the solution interval is too short, the equations will be evaluated unnecessarily often, costing extra computer time.

As a practical rule-of-thumb, the solution interval should be half or less of the shortest first-order delay in the system. The reason for this can be seen by examining the behavior of a first-order feedback loop. As an example, different lengths of solution interval will be tried in the simple feedback loop of Section 2.2.

Figure 6.2a shows the structure of the loop. Two equations describe this simple system--the rate equation stating the policy for order rate and the level equation to generate the inventory. Using the Chapter 5 conventions for time notation, and changing the numerical values so that the time relationships are easier to see, these equations are:

$$OR.KL = \frac{1}{AT}(DI - I.K) \qquad \text{Eq. 6.2-1, R}$$

$$AT = 1 \qquad \text{Eq. 6.2-1.1, C}$$

$$DI = 1 \qquad \text{Eq. 6.2-1.2, C}$$

$$I.K = I.J + (DT)(OR.JK) \qquad \text{Eq. 6.2-2, L}$$

$$I = 0 \qquad \text{Eq. 6.2-2.2, N}$$

OR--Order rate (units/week)
AT--Adjustment time (weeks)
DI--Desired inventory (units)
I--Inventory (units)
DT--Solution interval (weeks)

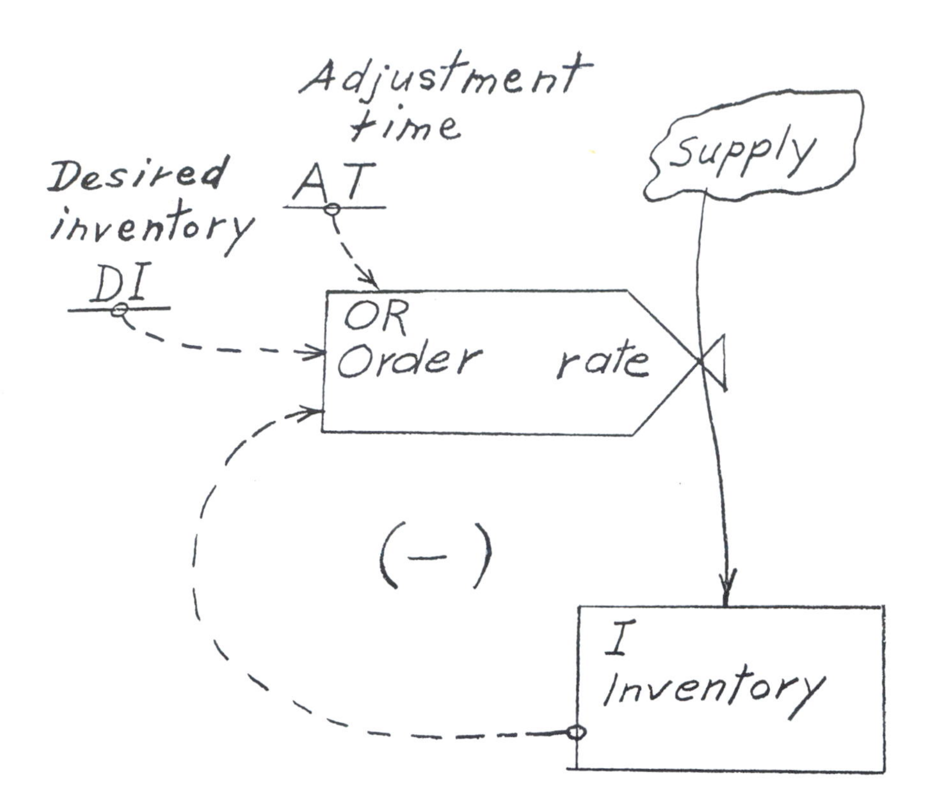

Figure 6.2a First-order negative feedback loop with delay = AT.

The preceding equations completely specify the system. They fully define the computation process except for the value of the solution interval DT. We will now examine how the length of the solution interval affects the computation. Equation 6.2-2.2 gives the initial value of inventory. The numerical values may not seem realistic for an inventory situation but are chosen so that ratios of solution interval to the feedback delay AT will be clear. The delay AT and the goal DI are both taken as unity for easy proportional comparisons.

Using the initial value (zero) of inventory, Equation 6.2-1 gives 1 unit per week as the first order rate. This first order rate persists until recomputed at the end of the first solution interval.

(Think of the "units" in these equations as representing thousands so that fractions of units and time increments of less than one week are both meaningful.) If the solution time interval before computation of a new inventory is very short, only a small inventory increase will have occurred in the interval. If the solution interval is less than one week, inventory will still be less than desired inventory, the discrepancy (DI - I.K) will still be positive but smaller than before, and the new order rate as based on the new discrepancy will be smaller than at first. As the inventory gradually builds toward the goal, the order rate gradually declines.

But if the solution interval is one week, the initial maximum order rate persists until the inventory reaches desired inventory, the discrepancy disappears, and the newly computed order rate will be zero ever after. Going further, if the solution interval is longer than a week, the inventory will rise above desired inventory before a new computation re-evaluates the order rate, the discrepancy is then negative, and the order rate will reverse to bring the inventory back down toward the goal.

Table 6.2a is a short section of computation for a solution interval of 0.2 week which is 0.2 of the system delay AT.

Time	Inventory	Order rate
0	0	1.000
.2	.200	.800
.4	.360	.640
.6	.488	.512

Table 6.2a. Computation of inventory for DT = 0.2 and AT = 1.

The reader should trace the computation until the procedure is clear in relationship to the system equations.

(1) Time	(2) DI	(3) I for DT=.2	(4) I for DT=.4	(5) I for DT=.8	(6) I for DT=1.0	(7) I for DT=1.6	(8) I for DT=2.4
.0	1.000	.000	.000	.000	.000	.000	.000
.2	1.000	.200					
.4	1.000	.360	.400				
.6	1.000	.488					
.8	1.000	.590	.640	.800			
1.0	1.000	.672			1.000		
1.2	1.000	.738	.784				
1.4	1.000	.790					
1.6	1.000	.832	.870	.960		1.600	
1.8	1.000	.866					
2.0	1.000	.893	.922		1.000		
2.2	1.000	.914					
2.4	1.000	.931	.953	.992			2.400
2.6	1.000	.945					
2.8	1.000	.956	.972				
3.0	1.000	.965			1.000		
3.2	1.000	.972	.983	.998		.640	
3.4	1.000	.977					
3.6	1.000	.982	.990				
3.8	1.000	.986					
4.0	1.000	.988	.994	1.000	1.000		
4.2	1.000	.991					
4.4	1.000	.993	.996				
4.6	1.000	.994					
4.8	1.000	.995	.998	1.000		1.216	-.960
5.0	1.000	.996			1.000		
5.2	1.000	.997	.999				
5.4	1.000	.998					
5.6	1.000	.998	.999	1.000			
5.8	1.000	.998					
6.0	1.000	.999	1.000		1.000		
6.2	1.000	.999					
6.4	1.000	.999	1.000	1.000		.870	
6.6	1.000	.999					
6.8	1.000	.999	1.000				
7.0	1.000	1.000			1.000		
7.2	1.000	1.000	1.000	1.000			3.744
7.4	1.000	1.000					
7.6	1.000	1.000	1.000				
7.8	1.000	1.000					
8.0	1.000	1.000	1.000	1.000	1.000	1.078	
8.2	1.000	1.000					
8.4	1.000	1.000	1.000				
8.6	1.000	1.000					
8.8	1.000	1.000	1.000	1.000			
9.0	1.000	1.000			1.000		
9.2	1.000	1.000	1.000				
9.4	1.000	1.000					
9.6	1.000	1.000	1.000	1.000		.953	-2.842
9.8	1.000	1.000					
10.0	1.000	1.000	1.000		1.000		

Table 6.2b Inventory when computed for solution intervals that are 0.2, 0.4, 0.8, 1.0, 1.6, and 2.4 of the loop delay.

Table 6.2b shows how the computation of inventory depends on the solution interval DT. The last six columns give inventory for six different solution intervals. The corresponding order rates do not appear. The first column is time, the second column is the desired inventory DI toward which inventory is adjusting. The third column repeats and extends the second column of Table 6.2a for a solution interval of 0.2; note that the inventory gradually approaches the final value. Columns (4) and (5) for solution intervals of 0.4 and 0.8 cause inventory to approach final value more quickly. In Column (6) the solution interval equals the loop delay AT, both being 1.0, and the inventory takes on its final value at the end of the first solution interval. Column (7) for DT = 1.6 shows inventory oscillating higher and lower than its final value but converging toward the desired inventory. In Column (8), when the solution interval of 2.4 is more than twice the system delay, the oscillation of inventory becomes larger and larger.

Figures 6.2b and 6.2c show the same results in graphical form. Figure 6.2b shows inventory for solution intervals 0.2, 0.4, 0.8, 1.0, and 1.6 times the loop delay. Figure 6.2c is plotted to a different vertical scale and repeats the solution intervals of 0.4 and 1.6 to compare with the growing fluctuation that occurs when the solution interval of 2.4 is more than twice the delay around the first-order loop.

As can be seen in the figures, all sequences start with the same slope. This slope represents the rate defined by the initial conditions at time = 0. The initial rate continues until the end of the first solution interval at which time a new value of inventory is computed and becomes the basis for a new order rate.

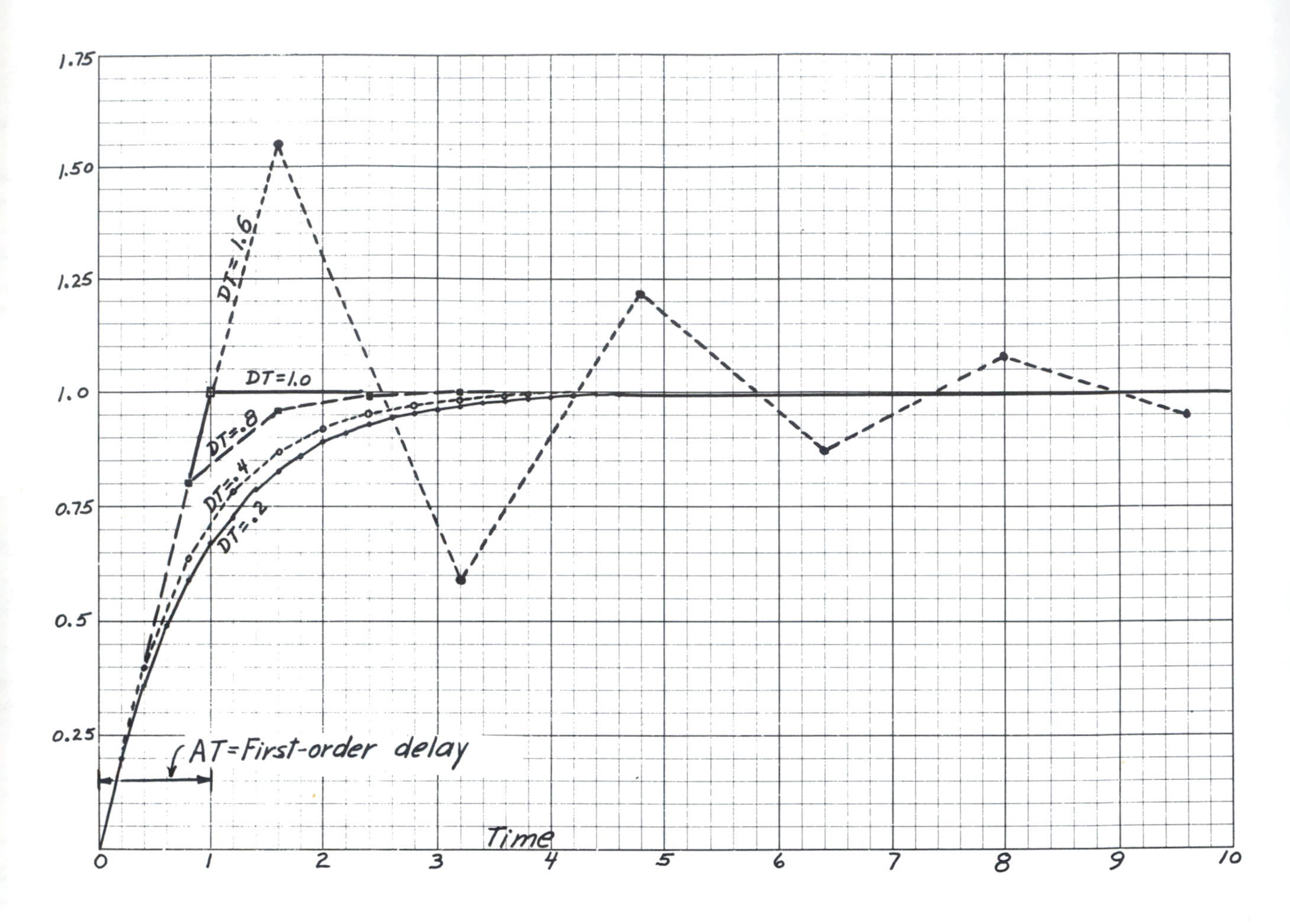

Figure 6.2b. Approach of the inventory to the desired inventory (=1.0), showing how the computation depends on the solution interval DT.

The curve for DT = 0.2 in Figure 6.2b is very close to the curve which would result from a vanishingly small solution interval and is correspondingly close to the behavior intended by the system equations. Doubling the interval to 0.4 causes the inventory to rise more quickly toward the final value but the departure from the lower curve would have negligible effect in most simulation models. The curve for DT = 1.0 is a special case in which the inventory has risen exactly to the final value at the end of the first solution interval, there is no difference between the inventory and the desired inventory to produce further change, and the value remains constant thereafter. The characteristic of gradual adjustment toward the final value has been lost. When the solution interval becomes longer than the loop delay, as for DT = 1.6, the computation procedure itself introduces a fluctuation into the results that does not arise from the intended behavior of the equations.

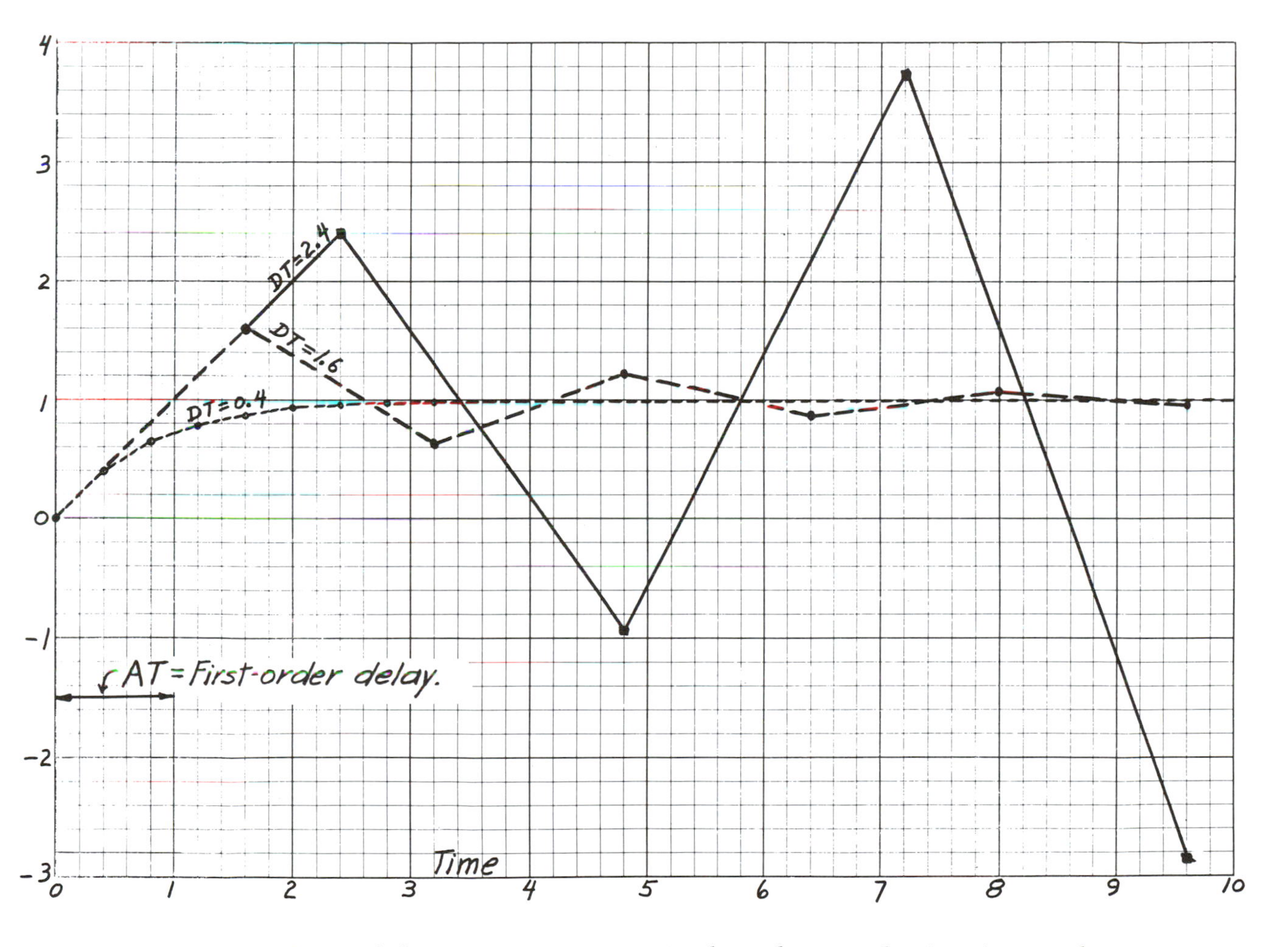

Figure 6.2c. Inventory, as it depends on solution interval, showing instability in the computation when the solution interval is longer than twice the loop delay AT.

* *
* *

Principle 6.2-1. Length of solution interval.

In any model, the solution interval DT should be less than half the shortest first-order delay in the system, but, to save computing time, the solution need not be less than one-fifth the shortest delay.

* *
* *

When the solution interval is much shorter than required by this principle, unnecessary computation will be expended in simulating system behavior with the model. As the solution interval is lengthened to equal the shortest delay, that part of the system fails to be represented properly. As the solution interval is made longer than the shortest delay, the computation for that delay becomes oscillatory. When the solution interval is more than twice as long as the delay, the oscillation in the computation process itself grows without limit leading to absurd results.

The solution interval is not part of the description of the real system. The solution interval is a technical detail arising in the simulation computation; it should not be long enough to affect the computed results. All of the dynamic behavior of the real system which the model is intended to capture should be represented in the equations themselves. None of the dynamic behavior should depend on the frequency of the successive computations.

6.3 Fine Structure of Computation to be Ignored.

Chapters 4, 5, and 6 have explained the details of the computation process. The reader may convert the detailed discussion into an undue concern with the computational interaction between rate and level equations. Although the computation process is such that the levels change as continuous variables that are formed of connected sections of straight lines, and the rates change discontinuously in stair-step fashion, the solution interval is always short enough that such detailed behavior is of no interest because it does not affect the system performance.

If the solution interval is chosen short enough in accordance with Section 6.2, and if the system structure of alternating levels and rates has been followed as in Section 4.3, then the model experimenter should concentrate on the broad sweep of system behavior. As in Figure 2.5d, continuous, smooth curves should be drawn through successive plotted points, whether the points represent levels or rates.

6.4 Differential Equations--a Digression*

Integration (or accumulation) shifts the time-character of action, produces delays between action streams, and creates the dynamic behavior in systems. Integration occurs naturally in both the physical and biological worlds. The integration processes of the real world are represented in our models by the level equations. The level equations are first-order difference equations, and, with a short enough solution interval, are entirely equivalent to the continuous process of integration.

But, those who have already studied some of the mathematics of dynamic systems have almost certainly done so in terms of differential equations rather than integral equations. Most of the mathematics of systems, developed originally in the physical sciences and engineering, has been cast in terms of differential equations. Even so, the differential equation view of systems seems to mislead many students and leaves them without a strong linkage between the real world and the world of mathematics. For those who have been introduced to systems by way of differential equations, this section may explain the integral equation emphasis of this book.

---

*Section 6.4 is addressed primarily to those who have already studied mathematics through differential equations.

Formulation of systems in terms of differential equations obscures for many students the direction of causality within systems, or, even worse, creates intuitive feelings of reversed cause and effect relationships. For example, consider the relationships between position, velocity, and acceleration. Thinking of velocity as the slope, or derivitive, of the position versus time curve can suggest that the changing position is responsible for the velocity rather than the other way. The direction of causality stands more clearly when the system description starts with the force that causes the acceleration, integrates acceleration to produce velocity, and integrates velocity to produce position.

Representing a system in terms of integral equations gives a more immediate and evident equivalence between the model and the real system. Such emphasis on integration is plausible when one notes that all the processes of nature are the processes of integration. Nowhere in natural processes does differentiation take place. True differentiation would depend on measuring a velocity instantaneously. But such is not possible. All natural and man-made devices that seem to measure velocity actually operate by a process of integration that in some sense measures the difference between past and present positions.

The "differential analyzer" illuminates the inescapable nature of the process of integration. The differential analyzer is a mechanical or electrical device for generating behavior in accordance with a set of differential equations. But the differential analyzer is built of integrators and, before it can be used, the differential equations must be converted to integral equations.

Except for the reorientation imposed on those whose experience has been entirely with differential equations, the quickest and simplest route to understanding dynamic systems seems to be through models depending on integration, avoiding completely the rather artificial concept of differentiation.

CHAPTER 7

FLOW DIAGRAMS

Feedback systems are elusive. Their structure and dynamic implications are hard to grasp and to keep in mind. One needs as many viewpoints as possible. From each viewpoint he may see something that was missed in a different exposure. A verbal description is one approach to a system; equations describing behavior of the separate parts is another. But to show the relationships between the parts and to accentuate the loop structure of a system, the flow diagram is best.

The flow diagram helps most when it provides new insights. It need not repeat the detail that lies within each equation, but it should give a broader perspective. The equations of a system focus on the composition of each level and rate. The flow diagram should show how levels and rates are interconnected to produce the feedback loops and how the feedback loops interlink to create the system.

The flow diagram should show the level, rate and auxiliary equations and how they are interconnected. The following symbols represent the elements of a system. Additional symbols in Chapter 8 stand for DYNAMO "functions" which are either special operations or frequently used assemblies of basic system equations.

Levels (Integrations) All level equations, and any special functions that also involve integration, will be represented by a rectangle. The simple level equation, as in Figure 7-1 is identified by the rectangle, the rates in and out that are being integrated, the symbol group representing the variable, the full name of the variable for easy communication, and the equation number as a cross reference to the formal definition of the model in the equation set.

$$I.K = I.J + (DT)(RR.JK - SR.JK) \qquad \text{Eq. 7-1, L}$$

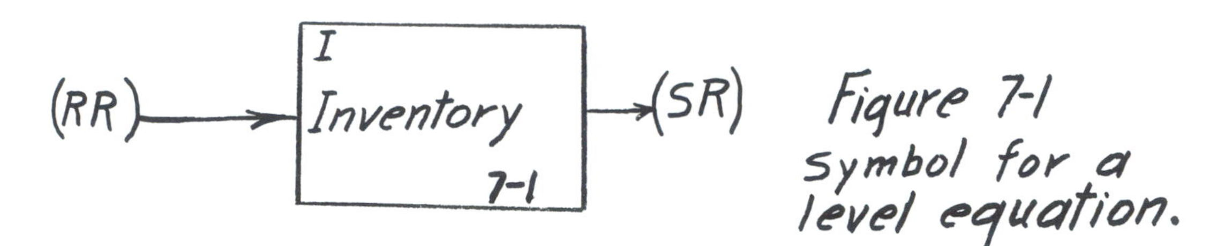

Figure 7-1
symbol for a
level equation.

Rates (Policies) The rate equations are the policy statements that define the flow streams in a system. The rate equation receives only information as its input and controls a rate of flow. As such, it serves like a valve in a hydraulic system and a symbolic valve is used to represent a rate. The rate is identified as in Figure 7-2 which shows an equation and its valve symbol in the flow diagram. The symbol should show the letter group representing the variable and its full name, the equation number, and the information inputs on which the rate depends.

$$SR.KL = \frac{BL.K}{DD}(IA.K) \qquad \text{Eq. 7-2, R}$$

DD
(BL)
(IA)
SR
Sales rate
7-2

Figure 7-2
symbol for a
rate equation.

Auxiliary Variables The auxiliary variables lie in the information channels between the level variables and the rates. They are parts of the rate equations, subdivided and separated because they express concepts that have independent meaning. Figure 7-3 shows an auxiliary equation and corresponding flow diagram symbol. The symbol shows an auxiliary equation by the circle, the abbreviation of the variable and its name, the equation number, and the input and output information streams.

$$B.K = (AS.K)(P) \qquad \text{Eq. 7-3, A}$$

P
B
Budget
7-3
(AS)

Figure 7-3
Symbol for an auxiliary equation.

Flow Lines A variety of flow lines clarify a diagram by distinguishing between classes of variables. Flows occur in subsystems representing the different "conserved" variables showing quantities that are moved from place to place in the system. Information connections occur in a "nonconserved" subsystem in which information can be used without depleting the source. The six flow lines in Figure 7-4 are useful to represent as many classifications of real-system variables. The information network has a unique status. But the other five are arbitrary subdivisions into which actual management-system variables usually fit conveniently and which may need redefining for other kinds of systems.

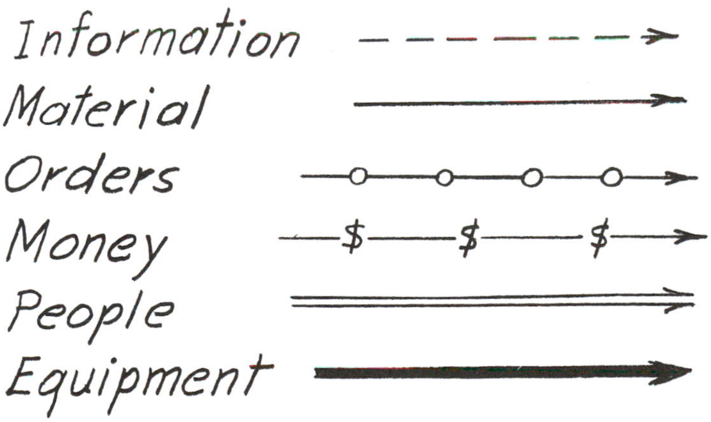

Information
Material
Orders
Money
People
Equipment

Figure 7-4
Flow lines.

Information Take-off Lines that indicate information transfer from a level should be distinguished from lines that represent flow of the content of the level. A flow connection moves a quantity from one location to another and is controlled by a rate equation. But information about a variable can be taken without affecting or depleting that variable. Figure 7-5 shows a small circle at the information take-off. This small circle represents, not removal of the content of the symbol, but only the transfer of information about the magnitude of the content. Although the information take-off indicator will be used at the beginning of all information lines, it is critical only where information is being taken about a level variable that is itself a part of the information network. Otherwise, the information take-off indicator is redundant. No depletion type of flow could occur from a rate or auxiliary variable, so only information connections are possible. No information flow, only information take-off, could exist at a level which is not itself information. For example, an information line leaving an inventory can only be an information take-off because the transport flow would be shown by a material-flow line.

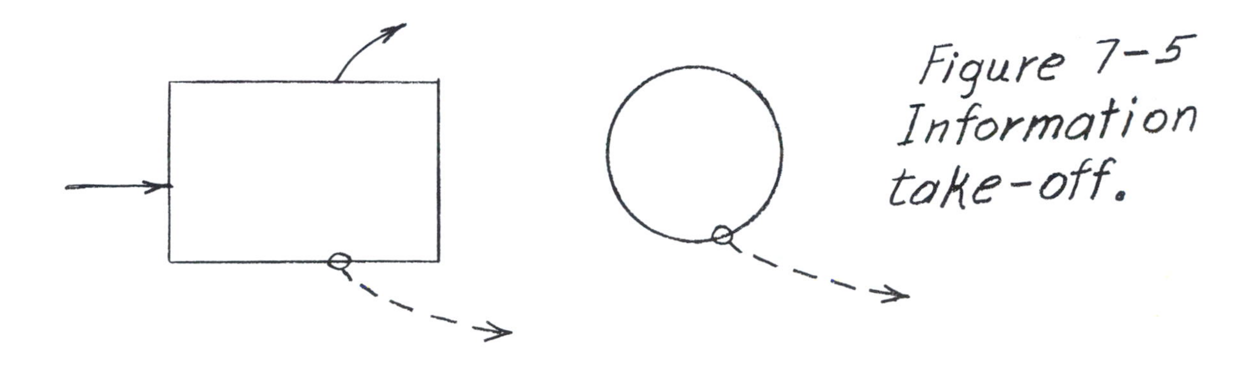

Figure 7-5
Information
take-off.

Parameters (Constants) Parameters are those values which are constant throughout a simulation run. They can, of course, be changed between successive simulations. As in Figure 7-6, the parameter is shown with an underline or overline having an information take-off. The name of the parameter should appear beside its abbreviation. Parameters are always information inputs to rates, either directly or by way of auxiliary equations.

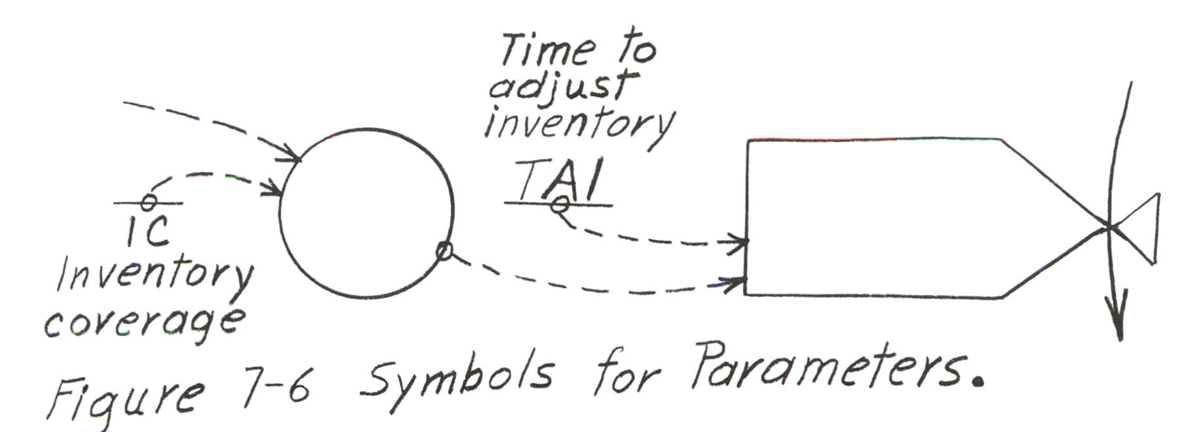

Figure 7-6 Symbols for Parameters.

Sources and Sinks When the source of a flow exerts no influence on the system, the flow is shown as coming from an "infinite" source. An infinite source can not be exhausted. For the purposes of the particular model it will yield any flow that the model equations demand. Figure 7-7 shows such a nonactive source and the converse sink for terminating flow lines when they cross the model boundary.

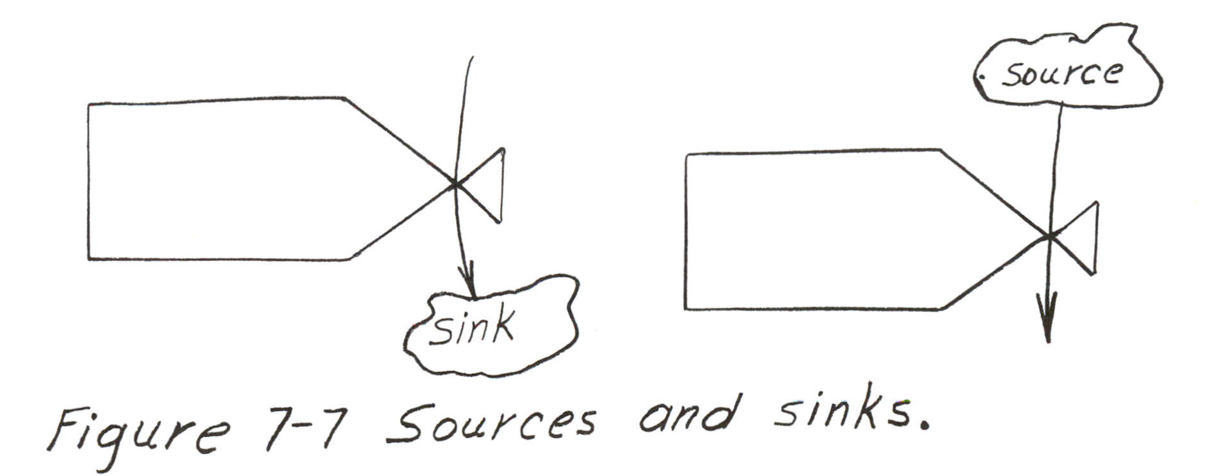

Figure 7-7 Sources and sinks.

<u>Variables from Other Diagrams</u> The flow diagrams of a system may extend over several pages. The flow lines and information transfers that cross from one page to another should be identified as in Figure 7-8 by origin and destination, giving name, abbreviation, equation number, and type of equation.

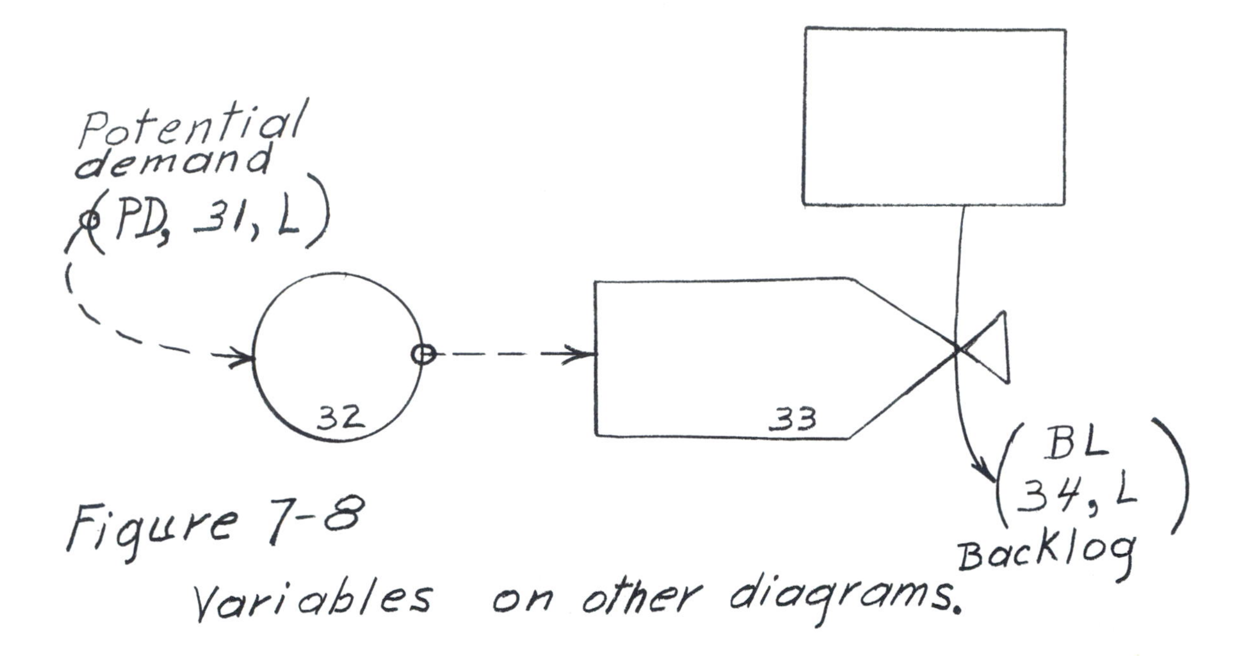

Figure 7-8
Variables on other diagrams.

CHAPTER 8

DYNAMO COMPILER

The DYNAMO compiler is a computer program which accepts the equations for a model of a dynamic feedback system and produces the requested simulation results as numerical tables and graphical plots.*

DYNAMO (for DYNAmic MOdels) has been designed to execute models that follow the structure and equation conventions used in this book. The DYNAMO compiler is widely available and has been adapted to computers of several manufacturers. Some of the following discussion assumes DYNAMO II which is more general than earlier versions. This book does not require the use of a computer and this chapter is not intended as instruction in DYNAMO but only describes the features needed for reading this book.

The DYNAMO compiler accepts a model written in the form of level, rate, and associated equations, using the time postscript conventions as described in Chapter 5. The equations can appear in any sequence. The compiler performs several duties:

1. It checks the equations for logical errors and prints defects. Many errors that can occur in a set of equations can be located because they do not satisfy the logical concepts of a feedback model. For example,

---

*DYNAMO has been designed by the Industrial Dynamics group at the Sloan School of Management, Massachusetts Institute of Technology. For a more complete description see [Reference here to DYNAMO II user's manual when available.]

all variables used on the right side of equations must themselves have a defining equation, no quantity should be defined more than once, no auxiliary equations should form a closed loop without an intervening level, the number of entries in a table should agree with the specification for the table, all special functions should contain the correct "arguments" to define the function, time postscripts should agree with the equation type, proper control instructions should be included, etc.

2. It reorganizes the model in accordance with the structural concepts of a dynamic system, grouping level and rate equations, and arranging the required sequences of those auxiliary equations which depend on one another.

3. It programs the model. That is, the equations with their algebraic notation are converted to detailed computer operating instructions.

4. It executes the step-by-step computation, based on the control instructions that give solution interval and run length, and produces the simulated results of the system represented by the model.

5. It prepares and prints the requested output in tabular and graphical form.

Table 8 is the short but complete model used for the single negative feedback loop in Section 2.2. It shows the format of equations and instructions as delivered to the DYNAMO compiler. Explanatory notes have been inserted.

```
        FILE B47    FIRST-ØRDER NEGATIVE LØØP    08/02/67   0751.7

0.1   *      FIRST-ØRDER NEGATIVE LØØP
0.2   RUN   STD.     FIG. 8A, 8B, 8C, 8D
0.3   NØTE   THE FØLLØWING FIVE LINES GIVE THE ACTIVE EQUATIØNS
0.4   NØTE   IN THE MØDEL.
1     R     ØR.KL=(1/AT)(DI-I.K)
1.1   C     AT=5
1.2   C     DI=6000
2     L     I.K=I.J+(DT)(ØR.JK)
2.1   N     I=1000
2.4   NØTE   THE FØLLØWING ARE SUPPLEMENTARY EQUATIØNS FØR PRINTING
2.5   NØTE   ØNLY, NØT PART ØF THE ACTIVE MØDEL.  THEY SHØW PARTIAL
2.6   NØTE   CØMPUTATIØNS IN THE LEVEL EQUATIØN.
2.7   NØTE
3     S     IE.K=DI-I.K
4     S     CI.K=(DT)(DI-I.K)
4.3   NØTE   THE PRINT INSTRUCTIØN GIVES THE CØLUMN NUMBER, THE VARIABLE
4.4   NØTE   NAME, AND THE SCALE.  IN THE SCALE (0.0), THE FIRST DIGIT
4.5   NØTE   IS THE PØWER ØF 10 BY WHICH THE VALUES ARE TØ BE MULTIPLIED
4.6   NØTE   (10 TØ THE ZERØ PØWER IS 1), AND THE SECØND DIGIT GIVES THE
4.7   NØTE   NUMBER ØF DIGITS TØ BE PRINTED TØ THE RIGHT ØF THE DECIMAL
4.8   NØTE   PØINT.
4.9   PRINT 1)CI(0.0)/2)I(0.0)/3)IE(0.0)/4)ØR(0.0)
5     NØTE   THE PLØT INSTRUCTIØN GIVES THE VARIABLE TØ BE PLØTTED,
5.1   NØTE   THE SYMBØL TØ BE USED ØN THE PLØT, AND THE SCALE.  WHEN
5.2   NØTE   NØ SCALE IS GIVEN DYNAMØ SELECTS A SUITABLE SCALE.
5.3   PLØT  I=I(0,6000)/ØR=Ø(0,1000)
5.4   NØTE   CØNTRØL INSTRUCTIØNS.  DT IS THE SØLUTIØN INTERVAL.  LENGTH
5.5   NØTE   IS IN TIME AND SPECIFIES HØW FAR TØ CARRY ØUT THE CØMPUTATIØN.
5.6   NØTE   PRTPER AND PLTPER GIVE THE INTERVALS BETWEEN PRINTING AND
5.7   NØTE   PLØTTING.
6     C     DT=2
6.1   C     LENGTH=24
6.2   C     PRTPER=2
6.3   C     PLTPER=2
6.6   NØTE   RERUNS WITH CHANGED CØNSTANTS.
6.7   RUN   1
7     C     AT=10
7.1   C     LENGTH=12
7.4   RUN   2
8     C     AT=20
```

B-2096

Table 8. DYNAMO model.

Figure 8a shows the printed and plotted output produced by DYNAMO in response to the control cards and the print and plot instructions given in the main body of the model in Table 8.

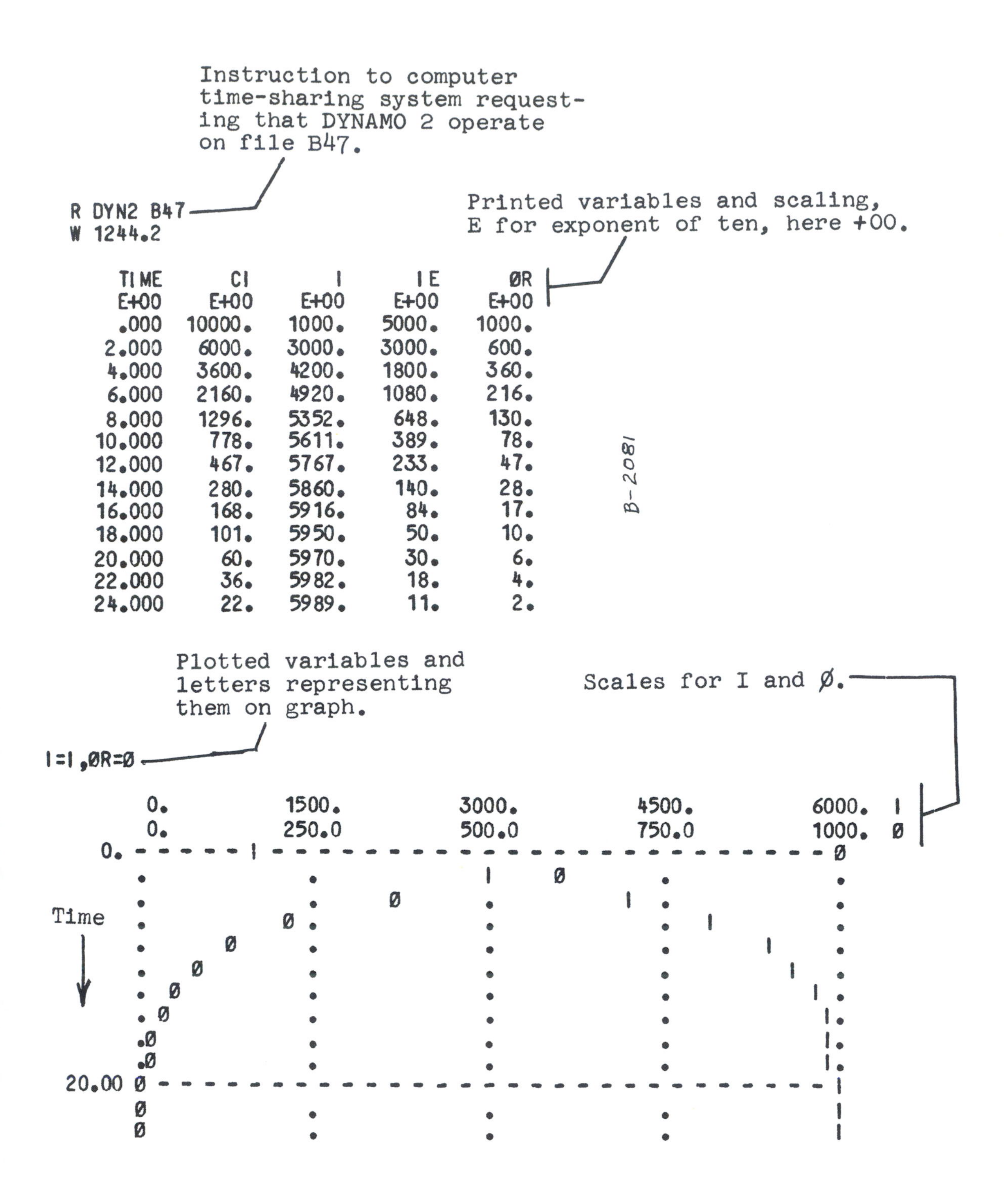

TIME E+00	CI E+00	I E+00	IE E+00	ØR E+00
.000	10000.	1000.	5000.	1000.
2.000	6000.	3000.	3000.	600.
4.000	3600.	4200.	1800.	360.
6.000	2160.	4920.	1080.	216.
8.000	1296.	5352.	648.	130.
10.000	778.	5611.	389.	78.
12.000	467.	5767.	233.	47.
14.000	280.	5860.	140.	28.
16.000	168.	5916.	84.	17.
18.000	101.	5950.	50.	10.
20.000	60.	5970.	30.	6.
22.000	36.	5982.	18.	4.
24.000	22.	5989.	11.	2.

Figure 8a. DYNAMO printed and plotted output.

If repeated simulation runs are desired with changed values of parameters, these can be requested as reruns. The DYNAMO output from Figure 8a continues in Figure 8b where first the parameter changes from the original model are indicated, followed by the requested print and plot.

```
              CHANGES
                         AT    LENGTH
          PRESENT    10.00     12.00
          ØRIGINAL   5.000     24.00


       TIME       CI        I        IE       ØR
       E+00      E+00      E+00      E+00     E+00
       .000    10000.     1000.     5000.     500.
      2.000     8000.     2000.     4000.     400.
      4.000     6400.     2800.     3200.     320.
      6.000     5120.     3440.     2560.     256.
      8.000     4096.     3952.     2048.     205.
     10.000     3277.     4362.     1638.     164.
     12.000     2621.     4689.     1311.     131.


I=I ,ØR=Ø

          0.           1500.          3000.          4500.          6000.  I
          0.           250.0          500.0          750.0          1000.  Ø
     0. - - - - - I - - - - - - - - - Ø - - - - - - - - - - - - - - - - -
          .              .   I   Ø    .              .              .
          .              . Ø      I   .              .              .
          .              Ø            .    I         .              .
          .           Ø  .            .         I    .              .
          .        Ø     .            .              I.             .
          .     Ø        .            .              . I            .
TYPE CHANGES
```


Figure 8b. Rerun with longer adjustment time AT and shorter run length.

DYNAMO was used as part of a time-sharing computer system in preparing the illustrations for this book. With time-sharing the model can be entered through a typewriter input keyboard and model editing and changes can be made. When a simulation run is completed, DYNAMO goes into the rerun mode and will accept altered constants from the typewriter to initiate a rerun using different model parameters.

DYNAMO is a tool for handling dynamic system models. But it is only a tool. By itself it is ineffective unless the model formulation is soundly conceived and properly related to the real-world system that the model represents. The principles discussed in this book and judgment based on experience with system behavior are essential to successful system design. DYNAMO by itself guarantees neither. Except for the less complete model checking and the less efficient use of the investigator's time, other computer compilers can be used to implement the same model concepts.

The following two sections describe some commonly used special functions performed by DYNAMO which will appear later in specifying models. DYNAMO also performs other infrequently used functions which will be explained at the point of use and in addition the user can specify personal special functions which can then be called up in response to a single name. The following descriptions are brief and to be used for future reference. The justification for the desirability and design of the functions, especially those in Section 8.2, is not given here but will emerge in later chapters as the functions are used.

Special functions perform operations more complex than those implied by the algebraic signs. For example, the square root function might appear in an equation as follows:

A.K = B.K + SQRT(C.K)

This equation says that A.K equals B.K plus the square root of C.K. A special function is identifiable on sight and by the DYNAMO compiler because it is not enclosed in the conventional multiply symbols. If SQRT above were a constant multiplied by C.K the equation would be written in one of the following forms where either parentheses or the asterisk indicate multiplication.

A.K = B.K + (SQRT)(C.K)

A.K = B.K + SQRT*C.K

The letter group, not itself enclosed in parentheses, and followed by a left parenthesis indicates a function. Any of the special functions might stand alone in defining a variable or be imbedded in a more elaborate equation.

Each function indicator is followed by parentheses that enclose the "arguments" that give the quantities on which the function is to operate. These arguments will be constants expressed as numbers or by their symbolic name or will be variables with appropriate time postscript. Where appropriate, an argument may include a negative sign. For example, if Q is itself negative, the square root is meaningless because there is no real square root of a negative number, but SQRT(-Q) is possible.

8.1 Functions Without Integration

Several groups of special functions have the nature of rate or auxiliary equations. These groups provide special computational procedures, table interpolation, test inputs, randomness, and logical choices. Being without integration that can be imposed in any flow channel of a model, they introduce no time delays and no periodicity-dependent distortion. They alter only the instantaneous amplitude of signals. Being by nature rate or auxiliary equations, and usually being used as or in auxiliary equations, these special functions without integration will be shown in flow diagrams as a modification of the circle used for auxiliary equations.

The group of computational aids includes the square root, exponential, and logarithmic functions. Flow diagram symbols for these three appear in Figure 8.1a. If the function defines a named variable in a separate numbered equation, as VAR and 8, these can appear as in the SQRT symbol.

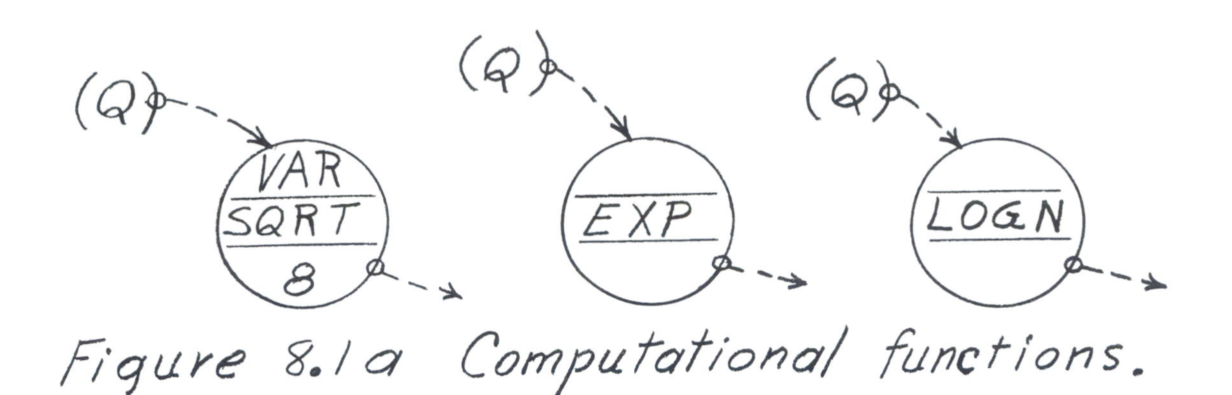

Figure 8.1a Computational functions.

SQRT(Q) means $\sqrt{Q}$ where the function generates the square root of Q. Q must be equal to or greater than zero. Q is the argument, either a constant or a variable with time postscript, on which the function operates.

EXP(Q) means e^Q where e is the base of natural logarithms.

LOGN(Q) means ln Q or the natural logarithm of Q. Q must be greater than zero.

The second class of functions provides for interpolation in a table. The need for interpolation was encountered in Equations 2.5-8 and 2.5-13 with their accompanying Figures 2.5b and 2.5c. Nonlinear relationships appear repeatedly in systems. The interpolation functions locate, by straight-line interpolation, the values intermediate between the points entered in a table. Their flow diagram symbols appear in Figure 8.1b.

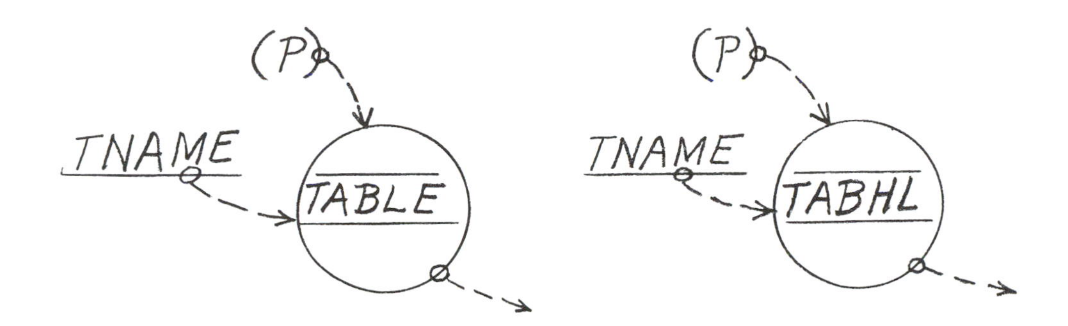

Figure 8.1b Interpolation functions.

TABLE performs linear interpolation between points in a table. It is written

TABLE(TNAME,P.K,N1,N2,N3)

TNAME = E1/E2/------/EM Eq. No., T

TNAME--the name of the table on which the function is to operate
P--the input variable for which the corresponding table entry is to be located
N1--the value of P at which the first table entry is recorded
N2--the value of P for the last table entry
N3--the interval in P between table entries
E1--the numerical value of the table at P=N1
E2--the second table value at P=N1+N3
EM--the last table entry giving the value when P=N2

P is the independent variable used to enter the table and will be a level or auxiliary variable. The table of numerical values itself is known by the symbol appearing at position TNAME. TNAME consists of a series of numerical values, uniformly spaced along the P-axis. The "equation" giving the numerical values is identified by T after the equation number. Figure 8.1c illustrates the TABLE function.

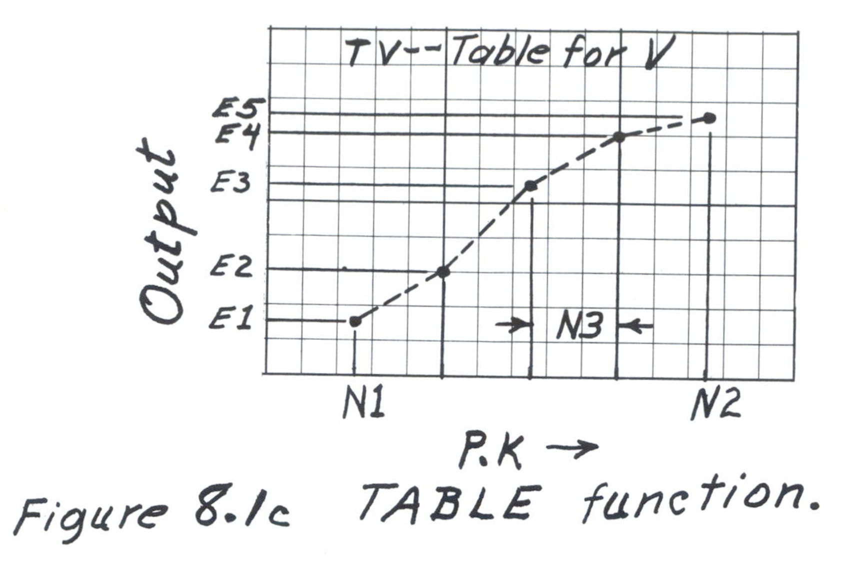

Figure 8.1c TABLE function.

N1 specifies the smallest value of P represented in the table, N2 specifies the largest value of P in the table, N3 is the uniformly spaced interval in P between points in the table. The values E1 through E5 are the numerical values of the table output for the corresponding values of P. The TABLE function assumes straight line segments between the specified points and will find the output corresponding to any value of P within the range N1 to N2. If the value of P becomes smaller than N1 or larger than N2, the TABLE function interprets this as an error and provides an error comment. The table TNAME itself must have exactly the correct number of entries corresponding to N1, N2, and N3. This number will be

$$\text{Number of entries} = M = \frac{N2-N1}{N3} + 1$$

Following these conventions for a TABLE function, Equation 2.5-13 and Figure 2.5c would be written

```
SE.K = TABLE(TSE,DDR.K,0,6,0.5)                                      Eq.8.1-1,A
 TSE = 400/400/390/370/350/320/290/250/210/180/150/120/100   Eq.8.1-1.1,T
```

TABHL (for High-Low extension) is like the TABLE function except that it does not assume an error if P becomes less than N1 or larger than N2. Instead, the last value in the table is extended. E1 is used for all values of P less than N1 and EM is used for all values of P larger than N2. The TABHL function is similar in form to TABLE and is written TABHL(TNAME,P.K,N1,N2,N3).

A group of functions consisting of STEP, RAMP, SIN, and COS are used primarily as test inputs. These provide excitation to which the model system reacts and yields useful information about the system dynamic behavior. The flow diagram symbols appear in Figure 8.1d.

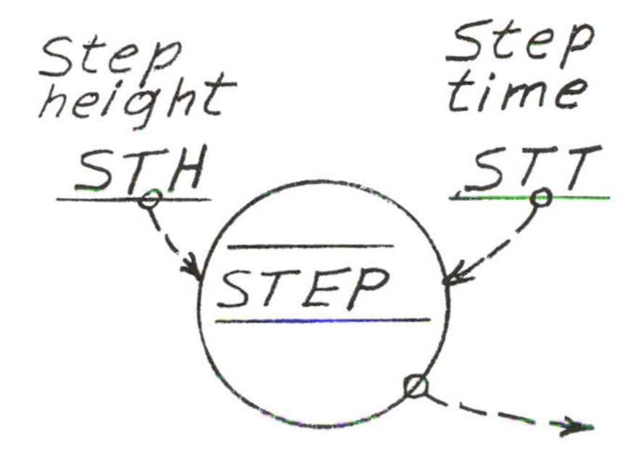

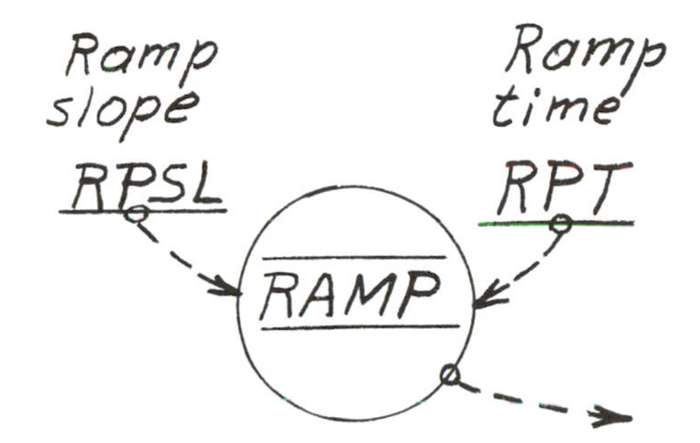

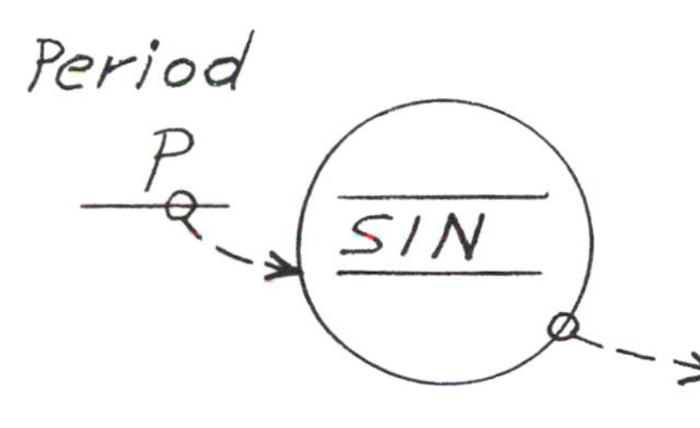

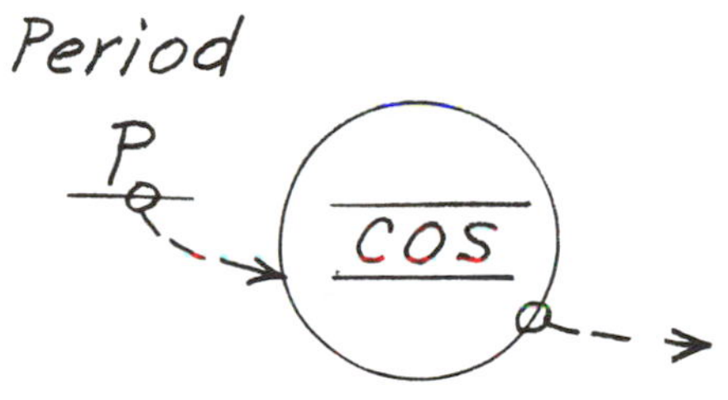

Figure 8.1d Test functions.

STEP changes to a specified value from zero at a specified time. It is written

STEP(STH,STT)

STH--step height, value of step after time STT
STT--time at which step changes from zero to STH

The STEP function is frequently used as a shock excitation to determine the dynamics of system recovery. Figure 8.1e illustrates a step, ramp, sine, and cosine. Table 8.1 shows the DYNAMO statements that produce Figure 8.1e.

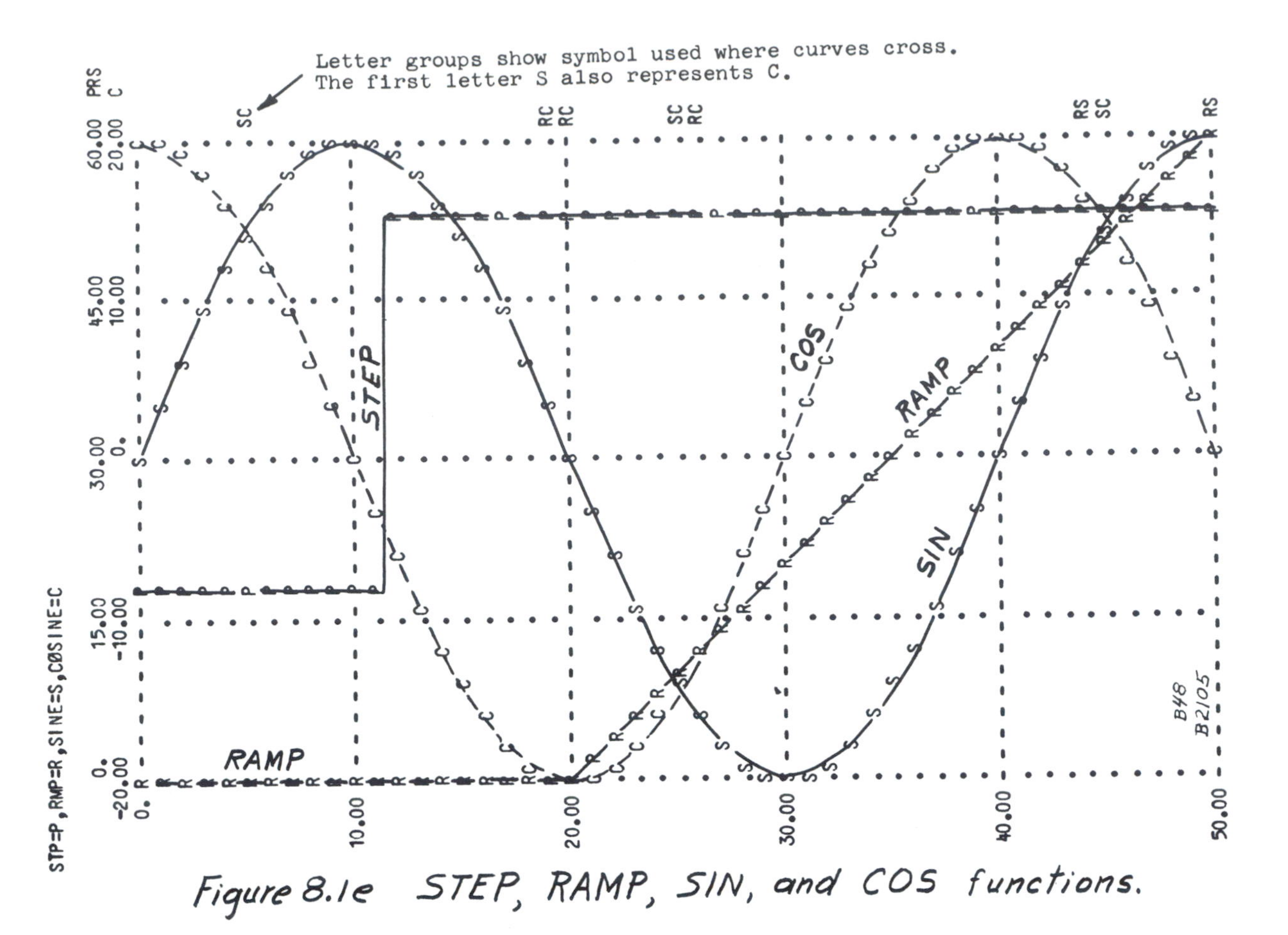

Figure 8.1e STEP, RAMP, SIN, and COS functions.

```
                FILE B48     TEST FUNCTIØNS                    08/02/67   0801.4

     0.1    *        TEST FUNCTIØNS
     0.2    RUN   STD.       TEXT TABLE 8.1, FIG 8.1E
     1      A     STP.K=18+STEP(HIGH,WHEN)
     1.1    C     HIGH=35
     1.2    C     WHEN=12
     2      A     RMP.K=RAMP(2,20)
     3      A     SINE.K=30+(30)*SIN(6.283*TIME.K/40)
     4      A     CØSINE.K=(20)*CØS(6.283*TIME.K/CPER)
     4.1    C     CPER=40
     4.4    PLØT  STP=P,RMP=R,SINE=S(0,60)/CØSINE=C
     5      C     DT=.5
     5.1    C     LENGTH=50
     5.2    C     PLTPER=1
     5.3    C     PRTPER=0
```

B2099

Table 8.1 Model to produce Figure 8.1e.

RAMP produces a variable of constant specified slope beginning at zero value at a specified time. It is written

RAMP(RPSL,RPT)

RPSL--Ramp slope (units/time)
RPT--Ramp time at which the slope starts (time)

SIN generates a sine (sinusoidal) fluctuation having a unit amplitude and a specified period. The expression for a sine in terms of period P and the independent variable time t is ordinarily written sine$\left(\frac{2\pi}{P} t\right)$ where in the DYNAMO format this becomes

SIN(6.283*TIME.K/P)

TIME--Model time supplied as a variable by DYNAMO
P--Period of the sine fluctuation (time units)

COS in a like manner generates a cosine fluctuation of unit amplitude and specified period. The cosine curve is of course the same shape as the sine but occurs a quarter-period earlier in time. The COS function is written

COS(6.283*TIME.K/P)

Both the SIN and COS functions can be used as computational functions when the arguments inside the parentheses are replaced with other variables or constants.

The functions NOISE, NORMRN, and CPONSE generate random sequences. Decision processes contain uncertainty (that is, random) components caused by processes that are not described by the known policy descriptions governing the decisions. Much of the behavior of actual systems reflects the manner in which the system responds to the random sequences that impinge on it. To study and to reproduce the effects of random behavior, generators of sequences of random numbers are needed. But a single source does not suffice because there can be several statistical specifications of randomness for different purposes, so DYNAMO contains three "noise" generators having different characteristics. Noise is a term arising in communications engineering to refer to random signals. It implies an unpredictable, meaningless signal. The noise generators in DYNAMO are the "pseudo-random" type meaning that they operate by a numerical process that is repeatable but yet produces a sequence of numbers that pass the same tests for randomness as do numbers generated by natural chance processes. The flow diagram symbols for the noise sources appear in Figure 8.1f.

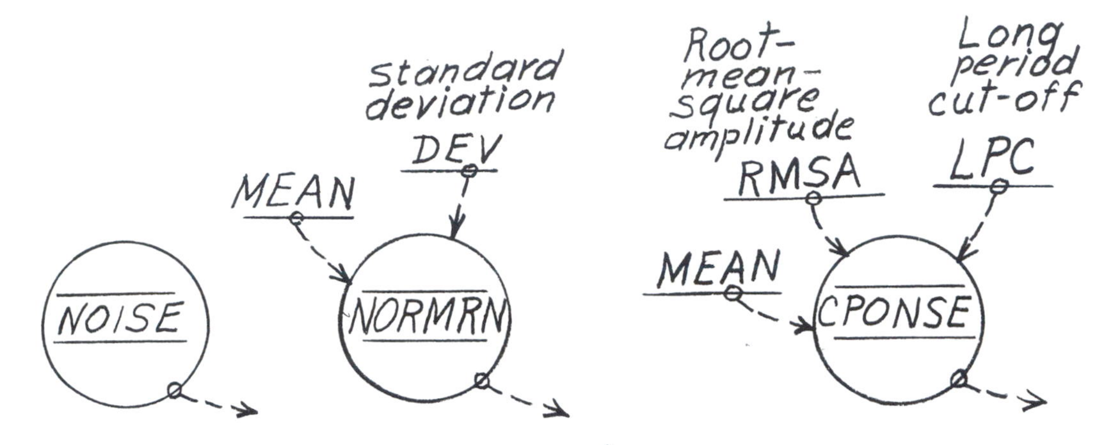

Figure 8.1f Noise generators.

NOISE generates random numbers uniformly distributed between -0.5 and 0.5. The function is designated by

NOISE()

There are no arguments necessary to specify the operation of NOISE but the empty parentheses are needed to indicate that the preceding letter group is a function and not the name of a constant.

NORMRN generates random numbers normally distributed with a specified mean and standard deviation.

NORMRN(MEAN,DEV)

MEAN--Mean value of the random numbers
DEV--Standard deviation of the normal distribution

A sequence from the NORMRN noise generator is shown in Figure 8.1g.

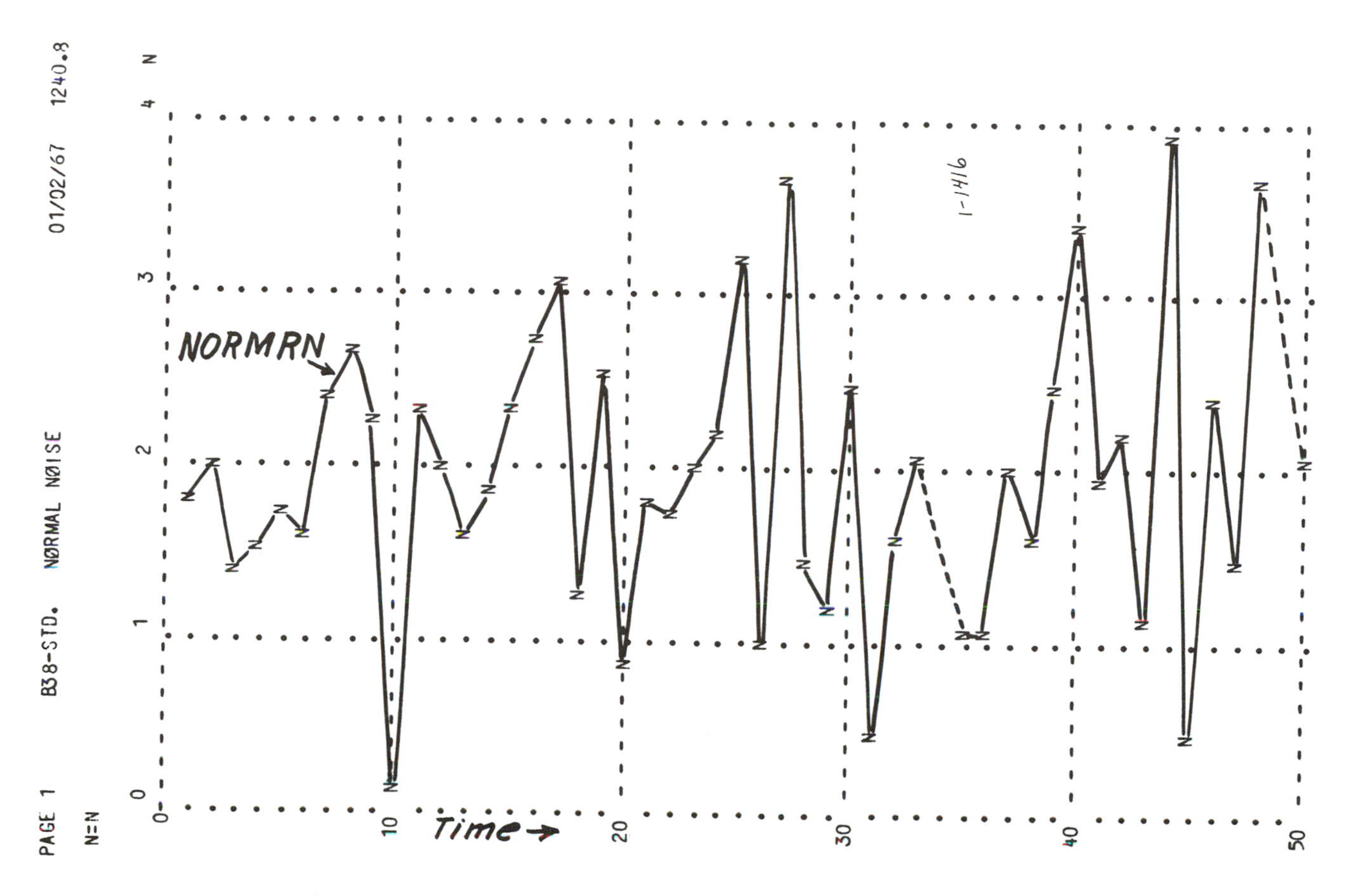

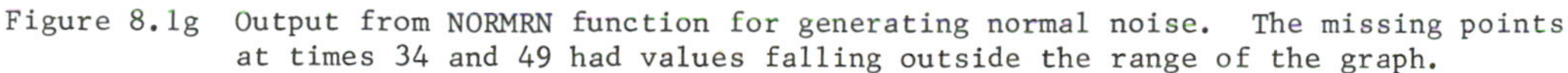
Figure 8.1g Output from NORMRN function for generating normal noise. The missing points at times 34 and 49 had values falling outside the range of the graph.

CPONSE generates constant-power-per-octave noise. "Octave" here means a band of periodicities bounded by periods that are a factor of two apart. In a sequence of random numbers as generated by NOISE or NORMRN, the short-period fluctuation is very strong. They are essentially "white noise" as the term is used by the engineer. Such noise has an energy in any octave that is twice the energy in the next longer-period octave. But for many purposes it is desirable to have a random signal with equal power in each octave. However, the probable amplitude of such a signal rises with the length of the periodicities it contains and often an upper limit on signal period is desirable. The CPONSE generator accepts three arguments to specify the mean value, the root-mean-square amplitude, and the long period cutoff. The details of how it operates are explained in Appendix . The function is written

CPONSE(MEAN,RMSA,LPC)

MEAN--mean value of noise signal
RMSA--Root-mean-square amplitude
LPC--Long period cutoff (time)

Figure 8.1h shows a short sequence from the CPONSE generator.

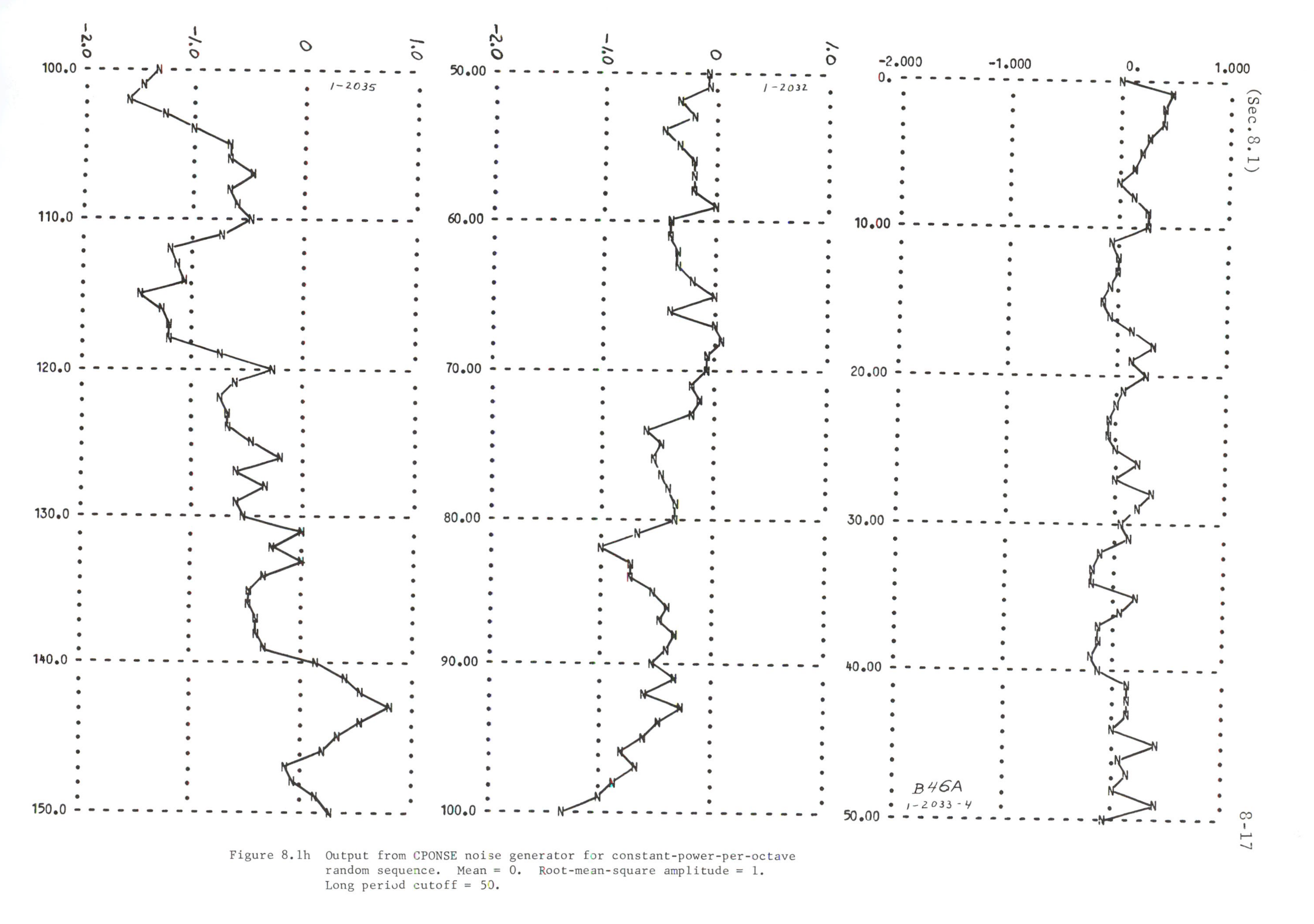

Figure 8.1h Output from CPONSE noise generator for constant-power-per-octave random sequence. Mean = 0. Root-mean-square amplitude = 1. Long period cutoff = 50.

The last group of functions in this section perform logical operations--MAX, MIN, SAMPLE, CLIP, SWITCH. Figure 8.1i gives the flow diagram symbols.

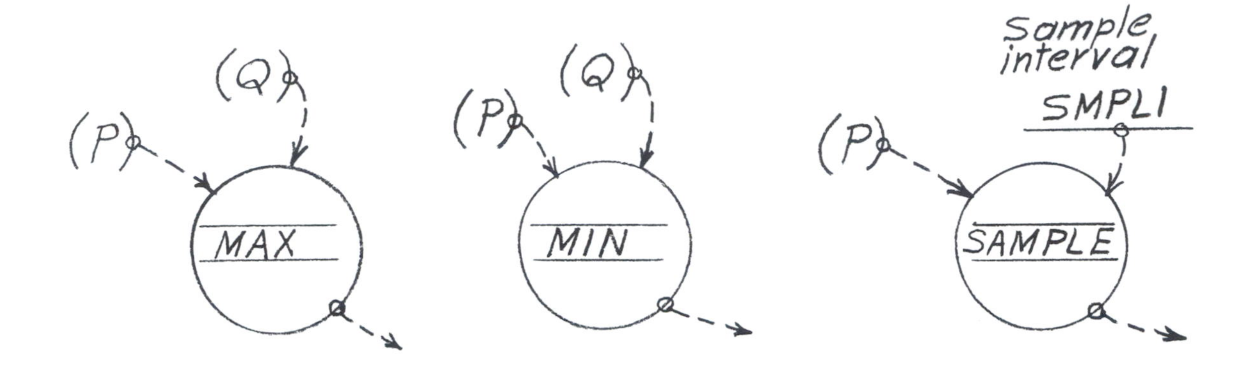

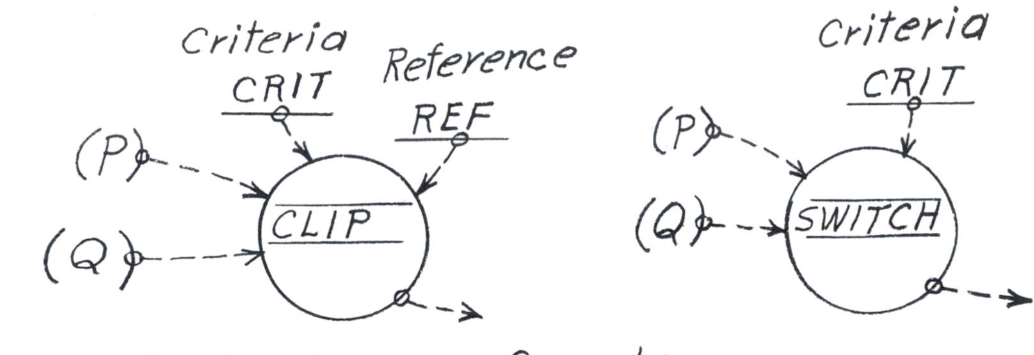

Figure 8.1i Logic functions.

MAX selects the greater from two inputs.

MAX(P,Q)

MIN selects the lesser from two inputs.

MIN(P,Q)

SAMPLE is set equal to P at uniformly spaced sampling intervals SMPLI and holds the value until the next sample is taken.

SAMPLE(P,SMPLI)

CLIP makes a choice between two quantities P and Q on the basis of the relative values of two (the same or other) quantities, the criteria CRIT and the reference REF.

CLIP(P,Q,CRIT,REF)

$$\text{CLIP} = \text{P if CRIT} \geq \text{REF}$$

$$\text{CLIP} = \text{Q if CRIT} < \text{REF}$$

SWITCH is similar but makes a choice on the basis of the criteria being zero or non-zero.

SWITCH(P,Q,CRIT)

$$\text{SWITCH} = \text{P if CRIT} = 0$$

$$\text{SWITCH} = \text{Q if CRIT} \neq 0$$

8.2 Functions Containing Integration

The functions SMOOTH, DLINF1, DLINF3, and DELAY3 contain integration. They normally are inserted in the path of a flow channel or information channel in a model. Because they include level equations that integrate, these functions change the time-shape of quantities moving between their inputs and outputs. These four functions are assemblies of elementary level and rate equations and have been established as special functions because the particular groupings are used so often.

These delay-producing functions in DYNAMO actually cause the corresponding elementary equations to be generated. The created variables within the added equations are given special names that are not allowable as normal model abbreviations, hence they can not interfere with names selected in the model itself. In the following paragraphs, the created variables start with the $ sign followed by L or R to indicate level or rate and then a serial number to identify the particular variable.

Internally, these functions contain level equations for which initial conditions are necessary. DYNAMO automatically generates initial conditions to define an internal steady-state condition that matches the initial value of the input variable to the function.

Because these functions contain integration, the rectangle for the level variable is used as the basic flow diagram symbol. Figure 8.2a shows the symbols to be used. The symbol shows the name of the function, the name of the output variable as specified in the system equations, the kind of output variable, and the name of the delay or averaging parameter of the function.

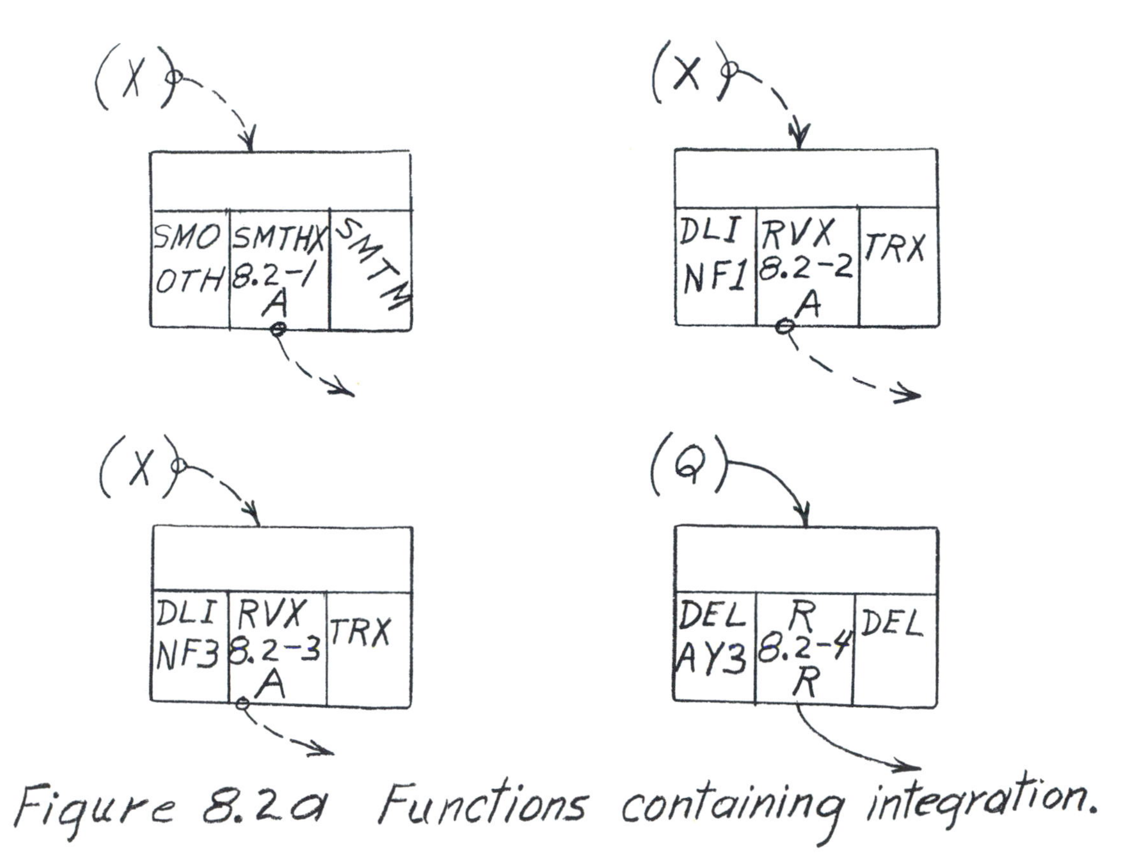

Figure 8.2a Functions containing integration.

SMOOTH produces a first-order exponential average of a physical (non-information, that is, conserved) rate of flow. It might be specified in a model in equation form as:

SMTHX.K = SMOOTH(X.JK,SMTM)   Eq. 8.2-1, A

SMTHX--Smoothed (averaged) value of X (same units/time as X)
SMOOTH--Function indication
X--variable rate to be smoothed (units/time)
SMTM--Smoothing time (time units)

Internally, the SMOOTH function will generate and execute the following equations which appear in flow diagram Figure 8.2b.

$L1.K = $L1.J + (DT)(X.JK-$R1.JK)   L
$L1 = (X)(SMTM)   N
SMTHX.K = $L1.K/SMTM   A
$R1.KL = SMTHX.K   R

$L1--Created level (units)
X--The rate being observed or smoothed (units/time)
$R1--Created rate (units/time)
SMTHX--Smoothed value of X, output of the function (same measure as X)
SMTM--Smoothing time (time)

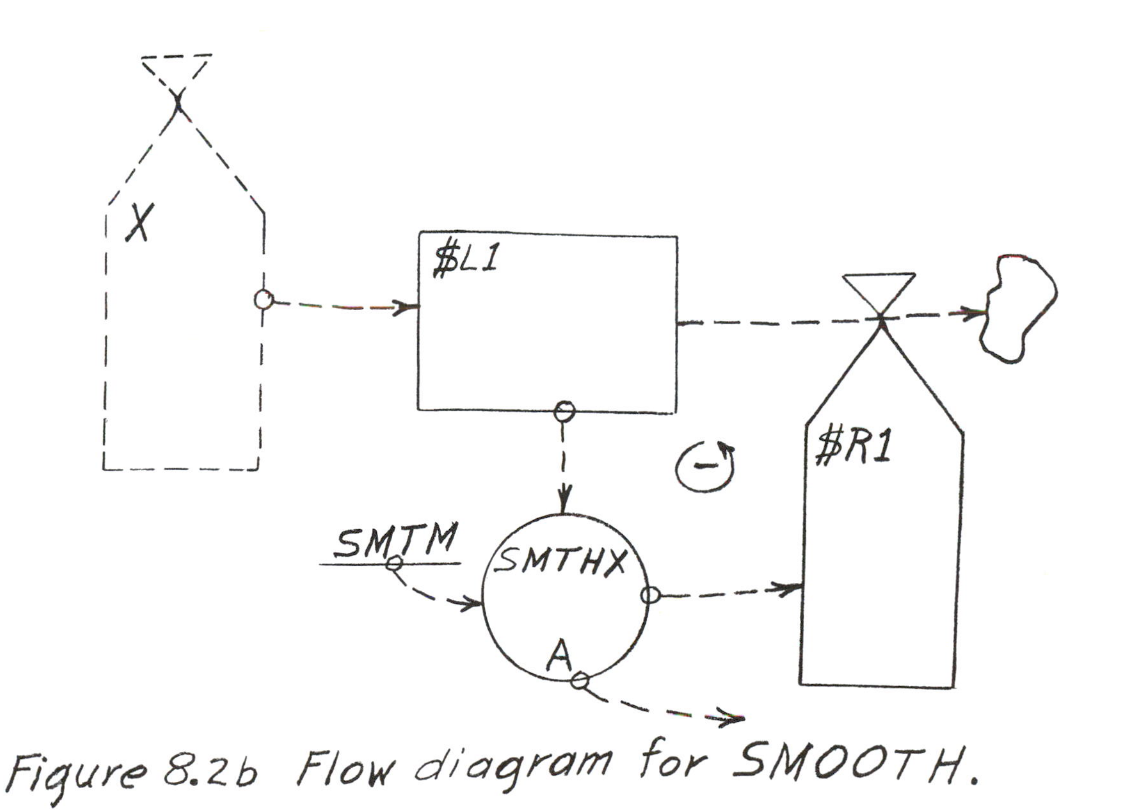

Figure 8.2b Flow diagram for SMOOTH.

The SMOOTH function describes a simple first-order negative-feedback loop having an exponential behavior similar to that already seen in Sec.2.2. Later chapters will further examine this averaging process.

DLINF1 is used in an information channel to produce a first-order exponential delay. It represents the process of a gradual, delayed adjustment of recognized information moving toward the value being supplied by a source. It is used to generate a delayed awareness of a changing situation. DLINF1 performs the same functions as Equations 2.5-11 and 2.5-12 for delivery delay recognized in Sec.2.5. The model equation:

RVX.K = DLINF1(X.K,TRX) Eq.8.2-2, A

RVX--Recognized value of X (same measure as X)
X--Level or auxiliary variable whose value is delayed
TRX--Time to recognize X (time units)

will cause the following to be generated and executed as shown in Figure 8.2c

\$R1.KL = (X.K-\$L1.K)/TRX	R
\$L1.K = \$L1.J + (DT)(\$R1.JK)	L
\$L1 = X	N
RVX.K = \$L1.K	A

The first two equations define internally generated variables for the rate and level that describe a first-order delay. The third

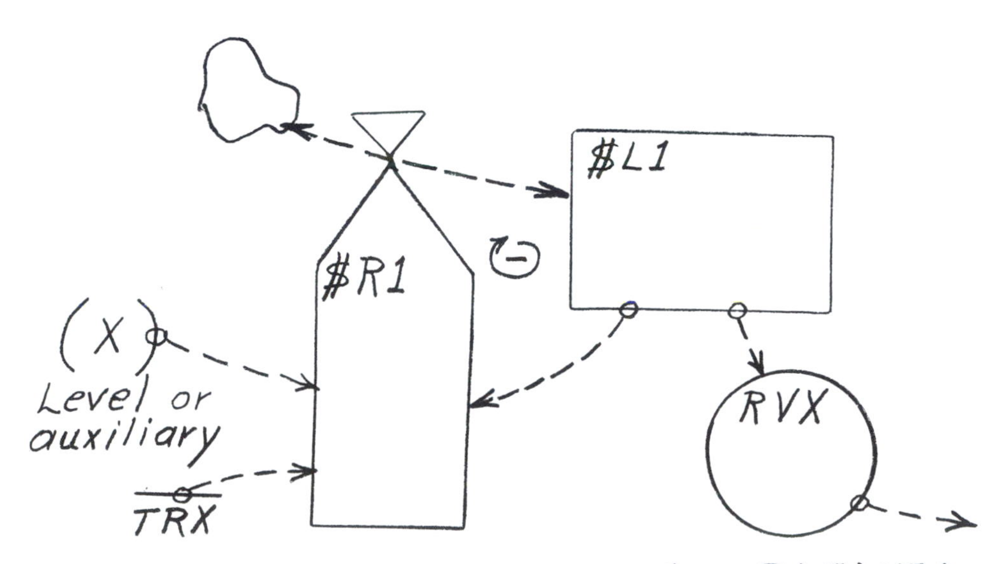

Figure 8.2c Flow diagram for DLINF1.

equation renames the result to that specified by the model equation. Being a first-order, negative loop, this function also has a simple exponential response to changes at its input.

The remaining two functions--DLINF3 and DELAY3--each are cascades of three first-order feedback loops to produce third-order exponential delays.

DLINF3 is placed in an information channel, as is DLINF1. But DLINF3 produces a different time-shape in its output. The output responds to the input more slowly at first than DLINF1 but then catches up. The result is closer to a "pipeline" delay or "pure" delay in which the input would be exactly reproduced at the output except for a time delay. The difference between the responses of DLINF1 and DLINF3, both with a delay of 15 time units, is shown in Figure 8.2d where the input to each is the step function.

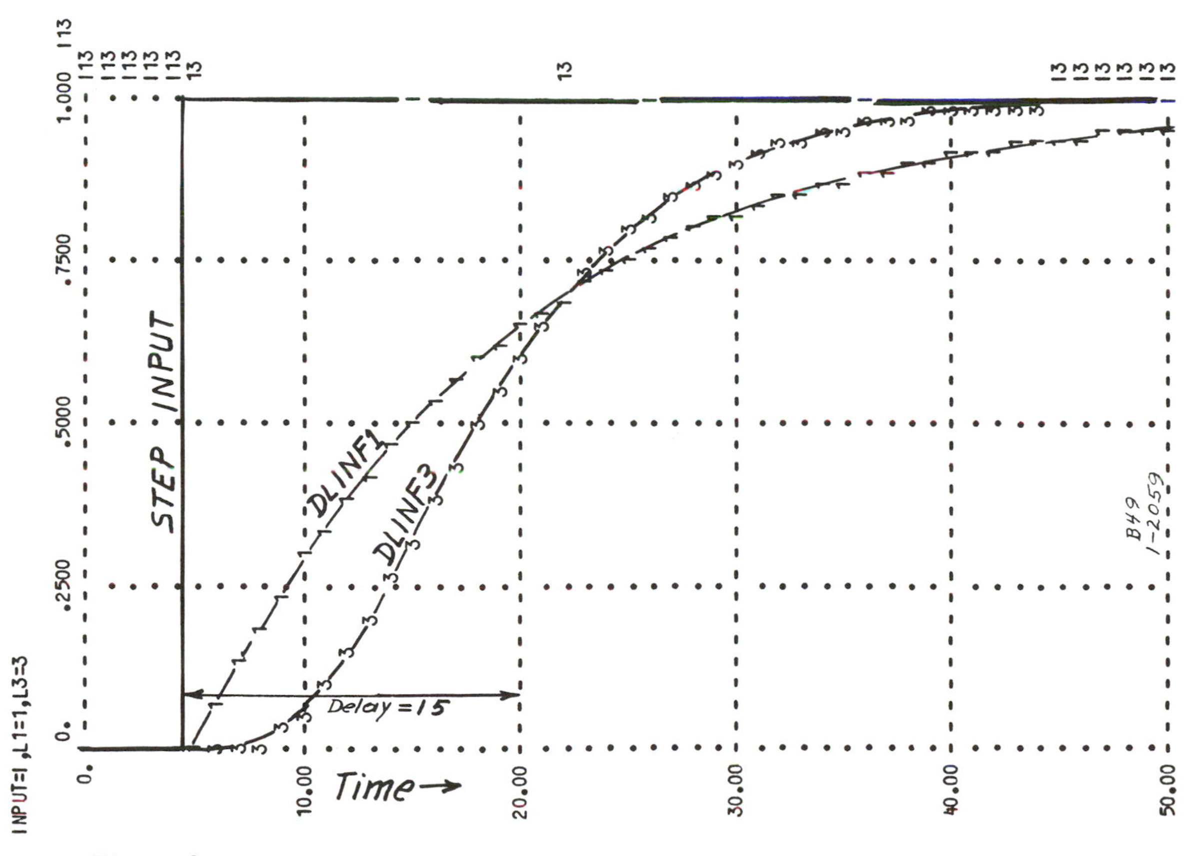

Figure 8.2d. DLINF1 and DLINF3 response to a step input with delay of 15 time units.

DLINF3 might appear in an equation as

RVX.K = DLINF3(X.K,TRX) Eq. 8.2-3, A

with the same meanings as for DLINF1. Figure 8.2e shows the flow diagram for the following equations that DYNAMO creates to produce the third-order exponential delay in the information stream.

```
$R1.KL = (X.K-$L1.K)/(TRX/3)       R
 $L1.K = $L1.J + (DT)($R1.JK)      L
   $L1 = X                         N
$R2.KL = ($L1.K-$L2.K)/(TRX/3)     R
 $L2.K = $L2.J + (DT)($R2.JK)      L
   $L2 = X                         N
$R3.KL = ($L2.K-$L3.K)/(TRX/3)     R
 $L3.K = $L3.J + (DT)($R3.JK)      L
   $L3 = X                         N
 RVX.K = $L3.K                     A
```

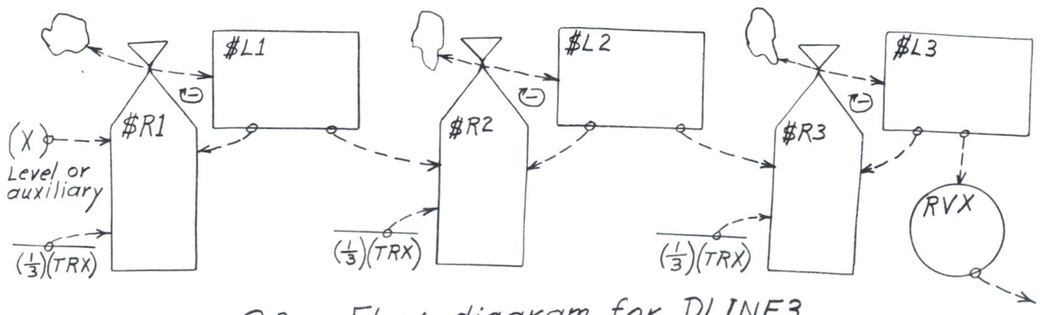

Figure 8.2e Flow diagram for DLINF3.

These equations and the flow diagram show three sections like that for DLINF1 except that in each section the delay is one-third of the total.

DELAY3 is similar to DLINF3. It is also three cascaded first-order exponential feedback loops. Therefore, it has the same dynamic response. It differs in being arranged for insertion in a flow channel that transports a quantity being moved from one point to another. As such, it receives an input rate of flow and delivers an output rate of flow. DELAY3 is used to create a delay in the transmission of a quantity from the input to the output. As shown in Figure 8.2f, DELAY3 differs from DLINF3 by having the rate equations after the levels and arranged to move the flow from one

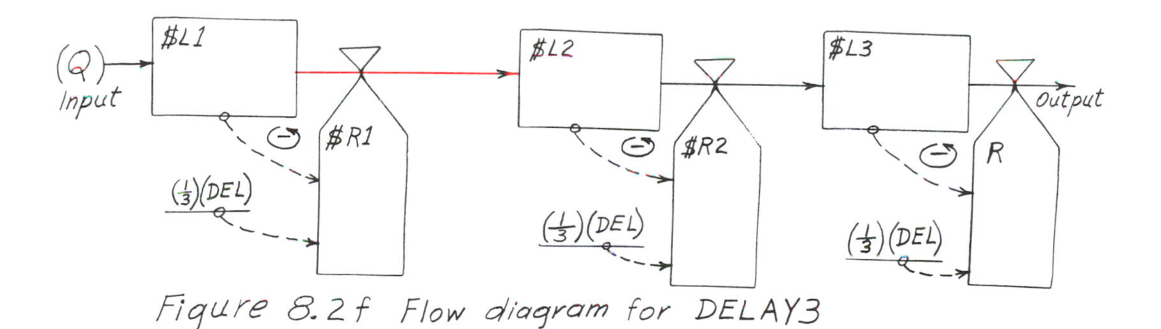

Figure 8.2f Flow diagram for DELAY3

level to the next rather than to adjust the value of a following level to that of the preceding. DELAY3 could be written in an equation as:

R.KL = DELAY3(Q.JK,DEL) Eq. 8.2-4, R

R--Rate output from the delay (units/time)
Q--Rate input to the delay (units/time)
DEL--Delay between Q and R (time units)

In response to the preceding function request, DYNAMO would create a set of equations that operate as follows:

```
$L1.K = $L1.J + (DT)(Q.JK-$R1.JK)        L
  $L1 = (Q)(DEL/3)                       N
$R1.KL = $L1.K/(DEL/3)                   R
$L2.K = $L2.J + (DT)($R1.JK-$R2.JK)      L
  $L2 = (Q)(DEL/3)                       N
$R2.KL = $L2.K/(DEL/3)                   R
$L3.K = $L3.J + (DT)($R2.JK-R.JK)        L
  $L3 = (Q)(DEL/3)                       N
 R.KL = $L3.K/(DEL/3)                    R
```

In response to a step input at Q, the output of DELAY3 at R would have the same shape as seen for DLINF3 in Figure 8.2d. Figure 8.2g shows how DELAY3 responds to a ramp input. DLINF3 would do the same. As an example of using the DYNAMO functions, Table 8.2a is the complete set of instructions needed to generate Figure 8.2g.

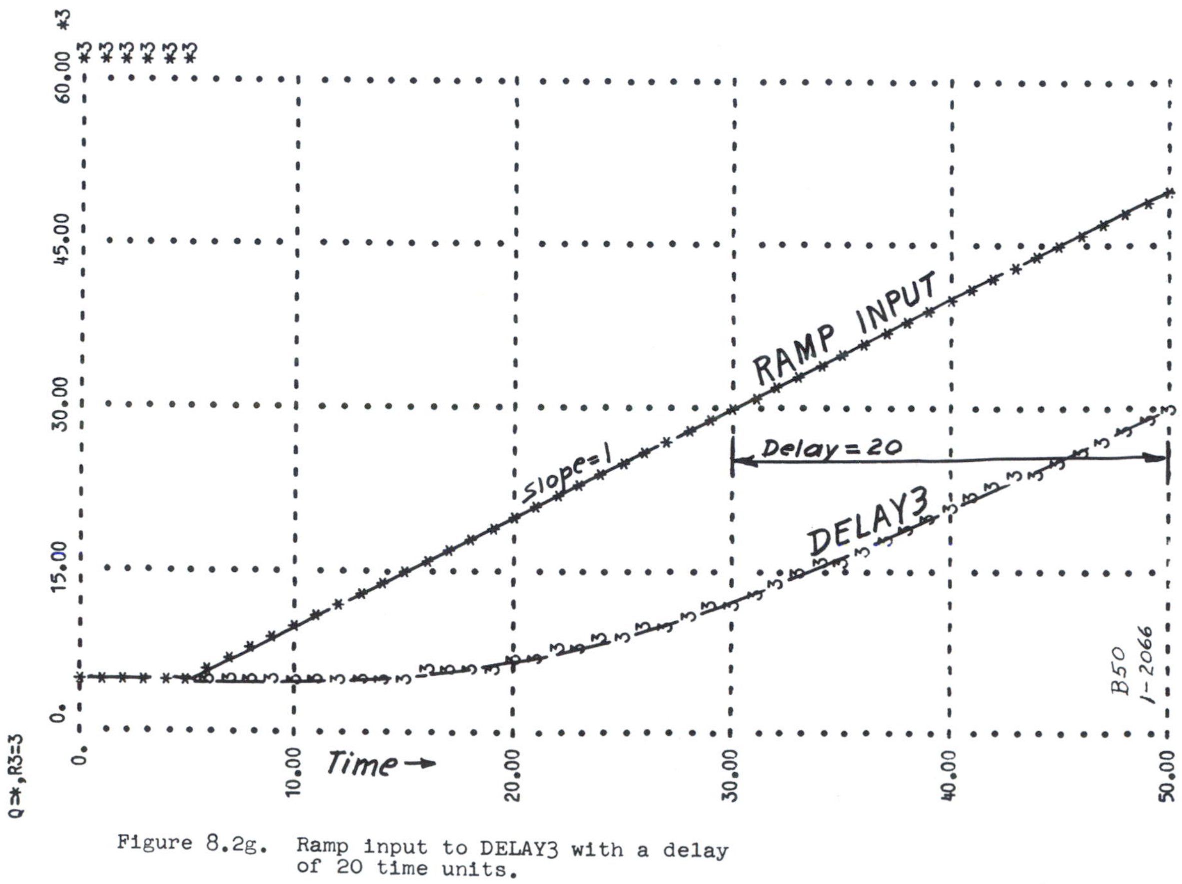

Figure 8.2g. Ramp input to DELAY3 with a delay of 20 time units.

```
            FILE B50     RAMP AND DELAY3                   08/01/67   0921.1

0.1    *        RAMP AND DELAY3
0.2    RUN     STD.     FØR TEXT FIG. 8.2G, TABLE 8.2A
1      R       Q.KL=QI+RAMP(RS,RT)
1.1    C       QI=5       INITIAL VALUE
1.2    C       RS=1       RAMP SLØPE
1.3    C       RT=5       RAMP TIME
2      R       R3.KL=DELAY3(Q.JK,DEL)
2.1    C       DEL=20     DELAY
2.4    PLØT    Q=*,R3=3(0,60)
3      C       DT=1
3.1    C       LENGTH=50
3.2    C       PLTPER=1
3.3    C       PRTPER=0
```

1-2068

Table 8.2a. DYNAMO model for Figure 8.2g.

The dynamic behavior of exponential delays is illustrated in Figure 8.2h where a sinusoid is generated as input for two DELAY3 functions, one having a delay equal to one-fifth of the sine period, the other being equal to the sine period. Note that the longer delay suppresses the fluctuation more sharply. The output of each delay has a lower amplitude than the input and is delayed in time. The longer delay produces the greater effect.

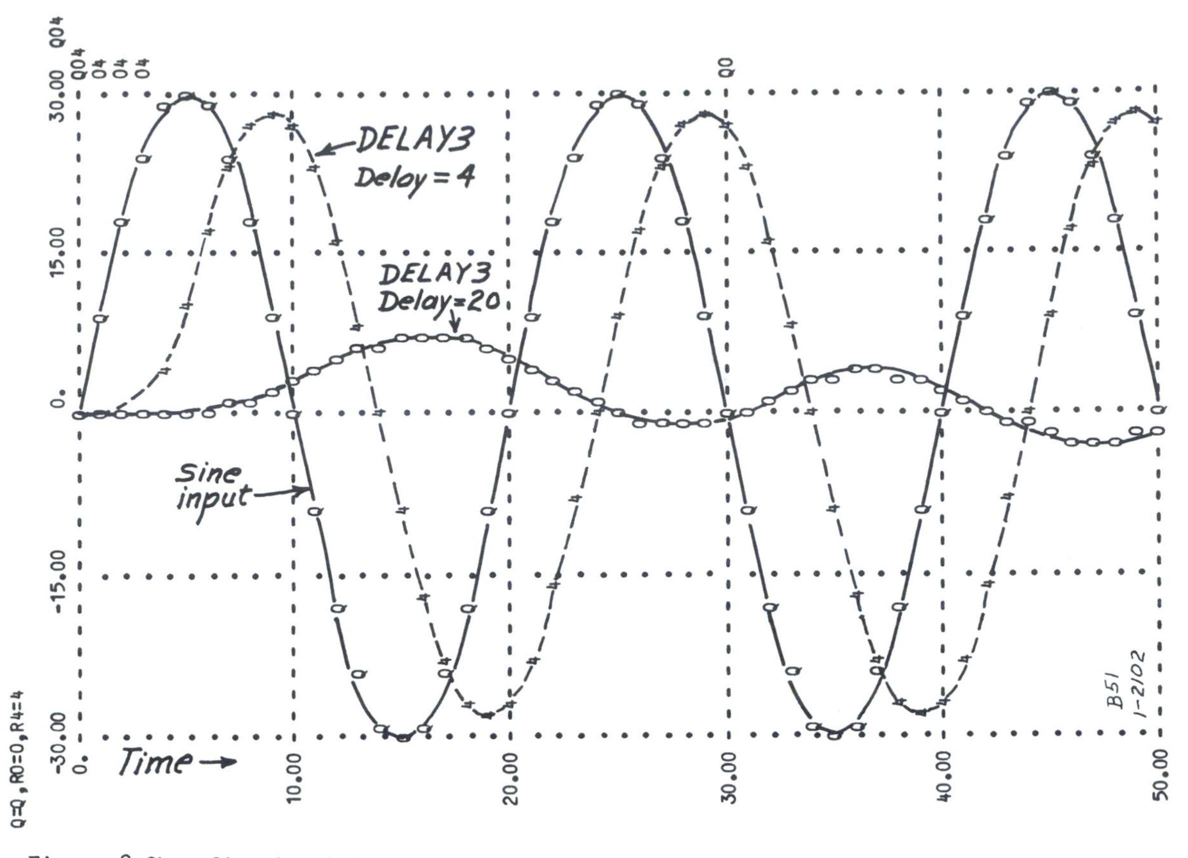

Figure 8.2h. Sine input to two DELAY3 functions having delays of 4 and 20 time units.

Table 8.2b gives the DYNAMO functions to generate Figure 8.2h.

```
          FILE B51     SINUSØID AND DELAY3               08/02/67   0811.7

0.1    *        SINUSØID AND DELAY3
0.2    RUN     STD.      FIGURE 8.2H, TABLE  8.2B
1      R       Q.KL=30*SIN(6.283*TIME.K/SP)
1.1    C       SP=20
2      R       R0.KL=DELAY3(Q.JK,D20)
2.1    C       D20=20
3      R       R4.KL=DELAY3(Q.JK,D4)
3.1    C       D4=4
3.4    PLØT    Q=Q,R0=0,R4=4(-30,30)
4      C       DT=1
4.1    C       LENGTH=50
4.2    C       PLTPER=1
4.3    C       PRTPER=0
```

1-2101

Table 8.2b. DYNAMO instructions to produce Figure 8.2h.

CHAPTER 9

INFORMATION LINKS

In the structure of a system, information links connecting levels to rates have a quite different character from the flow streams between levels. The information links show the information sources on which the rates depend. The information links leading to a rate equation do not affect the source levels from which they come, but the flows that are controlled by the rate equation do cause the levels to change.

The rate equations control the movement of a quantity to or from a level. The quantity within the level is "conserved," that is, it does not change except as controlled by the flows. A flow transports a quantity from one level to another (or to or from a source or sink). One level is reduced while the other is increased.

But information about a level can be taken as an input to a rate equation without affecting the source level. Information links from a level to a rate do not carry away the content of the level from which they emanate. Information is not a "conserved" flow. Information does not disappear when used. Information can be duplicated without destroying or depleting the information.*

---

*In uncommon situations the act of gathering information is recognized as influencing the condition of the source. If one inquires about a person's attitude, he may cause a thought sequence that alters the attitude. A sample of a chemical taken for analysis alters the amount of chemical remaining. In small-particle physics, the position or velocity of a particle may be changed by the acts necessary to measure position or velocity. But these are not exceptions to the above statements about the conservation of the content of a level and the non-conserved nature of information. Where measuring interacts with the quantity measured, there are several simultaneous processes. Through one process, information is obtained about a level. Through a related, and perhaps inseparable, process a rate of flow to or from the level is created.

Figure 9a shows a fragment of a system to illustrate the difference between information links and flow rates. The rate R1 is a physical flow from a source into the level L1. The rate R2 moves the same kind

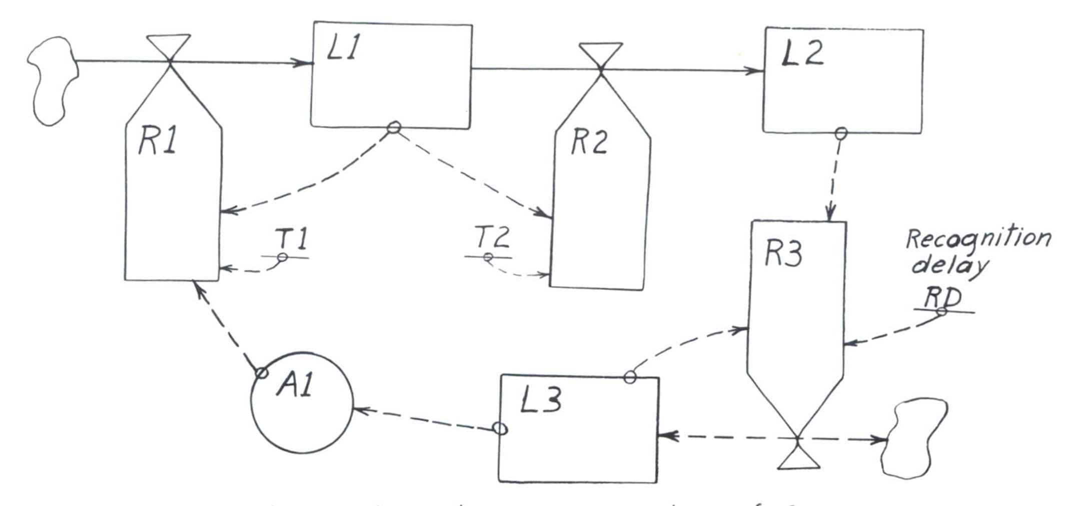

Figure 9a. Information links and rates of flow.

of item as in rate R1 from level L1 to level L2. The levels L1 and L2 must have the same units of measure because the content of one flows into the other. The dotted lines from L1 to R1 and from L1 to R2 are information links. Information about the magnitude of L1 is used as a basis for determining the rates R1 and R2 but these links do not themselves represent a flow that will deplete L1. The level L3 is within the information network. As with other levels, it is changed only by its associated flow R3. The dotted information line from L3 to the source-sink represents the flow that causes L3 to change and is different in nature from the dotted links starting at L2 and L3 which terminate in R3. These latter two links show that R3 depends on the values of both L2 and L3 but these links, shown as starting at the small information take-off circles, do not affect the source levels. The flow between L3 and the source-sink is controlled by R3 and does cause change in the level L3. Only information links are inputs to rate equations. Auxiliary equations as at A1 are subdivided parts of the rate equations and can only exist in the information links.

These distinctions between information links and flows are simple, yet confusion on these points often leads to difficulties in model construction. The points already made are sufficiently important to repeat as principles of model construction.

* *
* *

Principle 9-1. <u>Auxiliary variables only in the information links.</u>

An auxiliary variable is a subdivision of a rate equation and must lie in an information link that connects a level to the rate.

* *
* *

All level equations integrate the in and out flow rates. A level is changed only by a related flow. We can speak of a level as a "conserved" quantity. A level can only be increased by bringing in more of the quantity it contains from elsewhere or decreased by taking away the quantity. In the study of physics, the laws of the conservation of matter, energy, and momentum are clearly recognized. But the concept of conservation is much more widely applicable. There is conservation of money as it is moved between different bank accounts. There is conservation of people as they are transferred between assignments. There is conservation of reputation in the sense that it changes only as the processes of decay or as favorable and unfavorable acts cause change. Any level, whether in a physical or informational subsystem, provides system continuity; it changes only as the related rates of flow cause it to change.

* *
* *

Principle 9-2. <u>Levels exist in conservative subsystems.</u>

All levels are "conserved" quantities. They can be changed only by moving the contents between levels (or to or from a source or sink).

* *
* *

Within a conservative subsystem, all levels contain the same kind of item, the units of measure are identical. The associated rates of flow are in identical terms of items per time unit.

* *
* *

Principle 9-3. Same units within conservative subsystem.

Within any subsystem of conserved flows, all levels have the same units of measure; all rates are measured in those same units divided by time.

* *
* *

Information links do not represent flows from the content of the source level. Information is not conserved. It can be used without being depleted.

* *
* *

Principle 9-4. Information not a conservative flow.

Information is not depleted by use. It is not subject to the conservation laws. Information can be transmitted to another point without destroying its existence at the source.

* *
* *

The connections from levels to rates are always through information links that do not themselves affect the source levels.

* *
* *

Principle 9-5. Information links connect levels to rates.

Information links connect levels to the control of rates. Information links are the only inputs to rate equations.

* *
* *

A policy (rate equation) governing a rate of flow can be responsive only to the available information at the particular point in a system. Very often we tend to overlook the distortions in the information network that occur between the "true" levels and the apparent values of those levels. Information can be delayed. It can be disturbed by random error. It may be biased so that it consistently indicates a displacement from the "true" value. It can be distorted to produce errors that depend on the time-shape of the information stream itself. And it is subject to "cross talk" whereby the information shifts in apparent definition or source. All these processes occur within the information linkages. In principle, "true" information is never available at a policy point in a system; although as a practical matter, we can often overlook the discrepancy between true and apparent values. Apparent values of information arrive at the policy point, not directly from the "true" level but instead from an intervening auxiliary equation or information level. An auxiliary equation can insert a simple algebraic variation into the information stream, such as the addition of bias or random error. An intervening information level can introduce time-dependent distortion.

* *
* *

Principle 9-6. <u>Decisions</u> (<u>rates</u>) <u>only based on available information</u>.

Only apparent or available information can influence a decision. "True" system levels are often altered by processes within the information network before they become available at a decision point.

* *
* *

It follows from Principles 9-2 and 9-5 that the information links form the connecting tissue in a system between the different conserved subsystems. For example in Figure 9b, men move only between levels of men, inventories move only between levels of inventory units. But information links connect the number of men in the personnel system to the manufacturing rate (inventory units/month) in the inventory system to determine transfer rate between raw material and finished inventory.

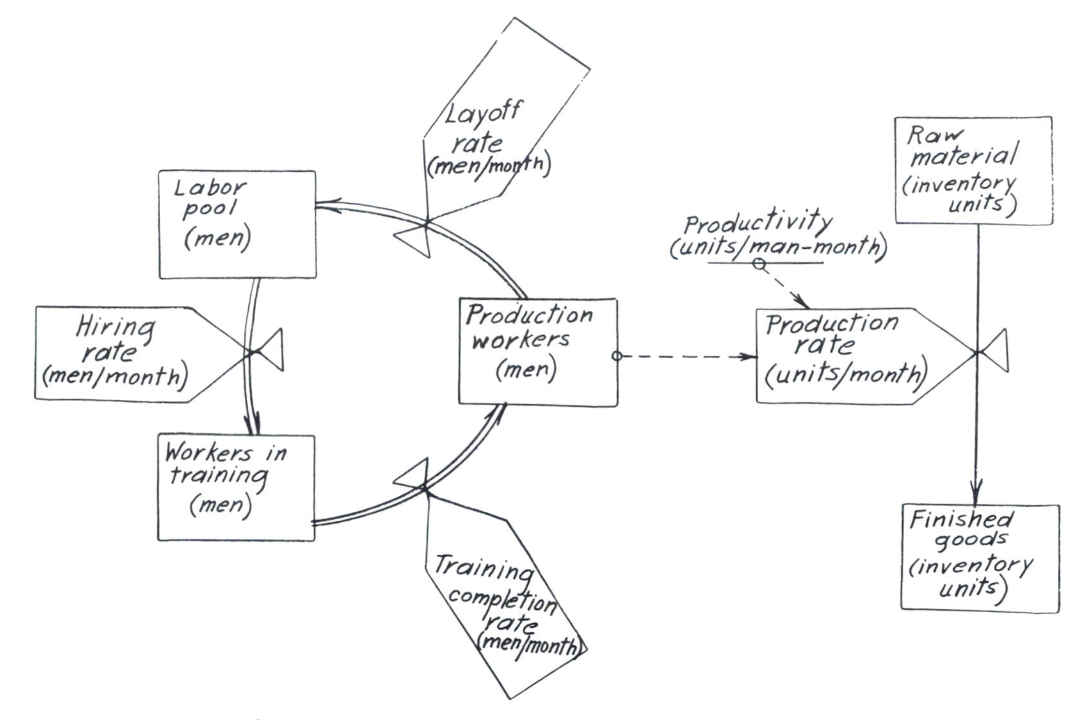

Figure 9b. Information link between conservative subsystems.

* *
* *

Principle 9-7. Information as connecting tissue of systems.

Only information links can connect between conservative subsystems. Information about levels in one subsystem can control rates of flow in a different subsystem.

* *
* *

Conversion coefficients usually accompany the information links in the rate equations. The separate conservative subsystems are usually measured in different units. Furthermore, there is usually a time-measure difference between the units of a rate and those of the level that supplies an information link. In Figure 9b, both conversions are provided by the productivity coefficient that relates men in the worker pool to the production rate.

(men)(units/man-month) = units/month

Here the productivity coefficient converts men to units (units/man) and a level to a rate (1/month).

* *
* *

Principle 9-8. <u>Conversion coefficients exist only in the information links.</u>

Information links usually require conversion coefficients to change units of measure between different conservative subsystems or to create a 1/time conversion of a level to a rate. Conversion coefficients exist only in the information linkages between the levels and the rates of flow.

* *
* *

Conversion coefficients should never be inserted in a system model merely to balance the dimensions of measure. Each coefficient should have a clear and identifiable real-life meaning. It should have a numerical value which can be derived from directly related observations on the system. Conversion coefficients should be more than correlation relationships derived statistically from time-series data taken from the actual system; the coefficients should describe specific processes within the system.

* *
* *

Principle 9-9. Conversion coefficients identifiable within real system.

Conversion coefficients are not inserted merely to correct the units of measure nor are they abstract values derived only from a statistical analysis. They should be relatable to the actual processes in the real system and should have numerical values which can be deduced from observation of the associated system levels.

* *
* *

CHAPTER 10

INTEGRATION

Chapters 4 and 5 described the structure of a feedback loop as a sequence of alternating levels and rates. Levels accumulate the rates of flow. The level equations, being integrations, involve the relentless forward movement of time.

Dynamic behavior arises from the process of integration. Integration can produce a variable with a time-shape and time-position different from those of the input rate. Only with a clear understanding of integration and its influence on the time-shape of a flow rate can one understand the behavior of feedback loops, constructed as they are of alternating integrations and flows.

10.1 Integrating a Constant

To see how integration can produce a variable with a time-shape different from that of the input flow rate, consider the influence of a series of integrations on a constant flow rate. Figure 10.1a is a flow diagram for the following equations where levels are measured in units, and rates are measured in units/time:

```
R1.KL = C                          R
    C = 1                          C
 L1.K = L1.J + (DT)(R1.JK)         L
   L1 = 0                          N
R2.KL = L1.K/A2                    R
   A2 = 1                          C
 L2.K = L2.J + (DT)(R2.JK)         L
   L2 = 0                          N
R3.KL = L2.K/A3                    R
   A3 = 1                          C
 L3.K = L3.J + (DT)(R3.JK)         L
   L3 = 0                          N
```

C--a constant flow to the first level (units/time)

A2, A3--constants relating levels to flow rates (units per time per unit = 1/time)

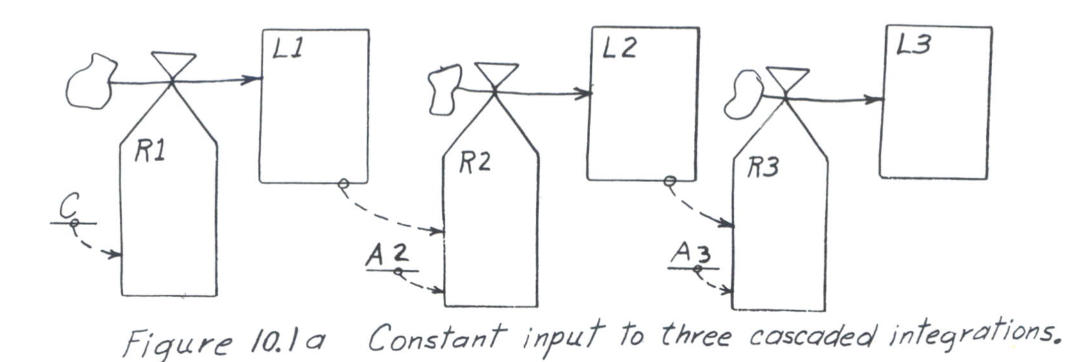

Figure 10.1a Constant input to three cascaded integrations.

In the equations and the flow diagram, a constant flow rate is accumulated in the first level. The second rate is proportional to the first level and is integrated in the second level. The third rate is proportional to the second level and is integrated in the third level. Figure 10.1b shows the time-shapes of these four quantities, the constant input rate C, which equals

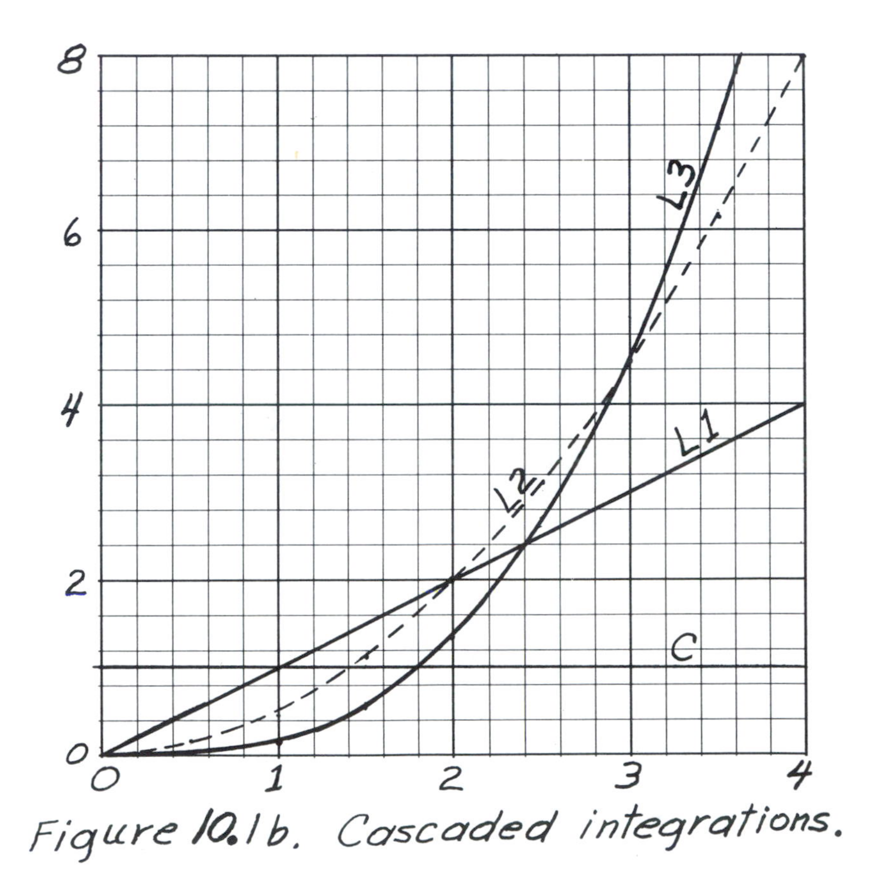

Figure 10.1b. Cascaded integrations.

1 unit/time, and the corresponding values of the three cascaded levels L1, L2, and L3. As one knows from calculus, integrating a constant C yields a result proportional to time:

$$\int C dt = Ct$$

which is a straight line whose value changes with the slope C as time moves forward. In Figure 10.1b, the first integral appears as the value of L1.

In turn, the integral of Ct is

$$\int Ct dt = \frac{1}{2} Ct^2$$

which is a parabola as plotted for L2 in Figure 10.1b. When the parabola is in turn integrated,

$$\int \frac{1}{2} Ct^2 dt = \frac{1}{2} C \int t^2 dt = \frac{1}{2} C \left(\frac{1}{3} t^3 \right) = \frac{C}{6} t^3$$

which is the cubic curve L3 in Figure 10.1b. The figure shows, by comparing the curves for C, L1, L2, and L3, how integration can alter the time-shape of a variable.

The changes in time-shapes in Figure 10.1b are produced by the open-loop system of Figure 10.1a. There is no connection from the output of the integrations back to the input. If there were such a connection to make a feedback loop, the output would then become the input, and the output shape would necessarily be the same as the input shape. Is there some shape of curve for which the shape is not altered by integration? If so, that time-shape would be capable of recirculating in a closed loop that contains integration.

A family of curves does exist, the shapes of which are not altered by integration. A feedback loop generates one or more members of that family. The exponential family contains the simple exponential functions and also the complex exponentials known more commonly as the sine and cosine functions.

10.2 Integration Creates the Exponential

All positive feedback loops and the first-order, negative-feedback loop generate a time response of simple exponential shape. Higher-order negative loops can generate sinusoidal behavior. This section considers how the exponential is generated by the process of integration in both positive and negative first-order loops and in higher-order positive loops.

The simple exponential function is $e^{t/T}$ where e = 2.718......, the base of natural logarithms. In the exponent, t is time and T is the "time constant" of exponential change. The time constant T is measured in units of time so that the exponent t/T is dimensionless.

Figure 10.2a shows the exponential function for the case where the exponent is positive:

$$\text{Value} = e^{t/T}$$

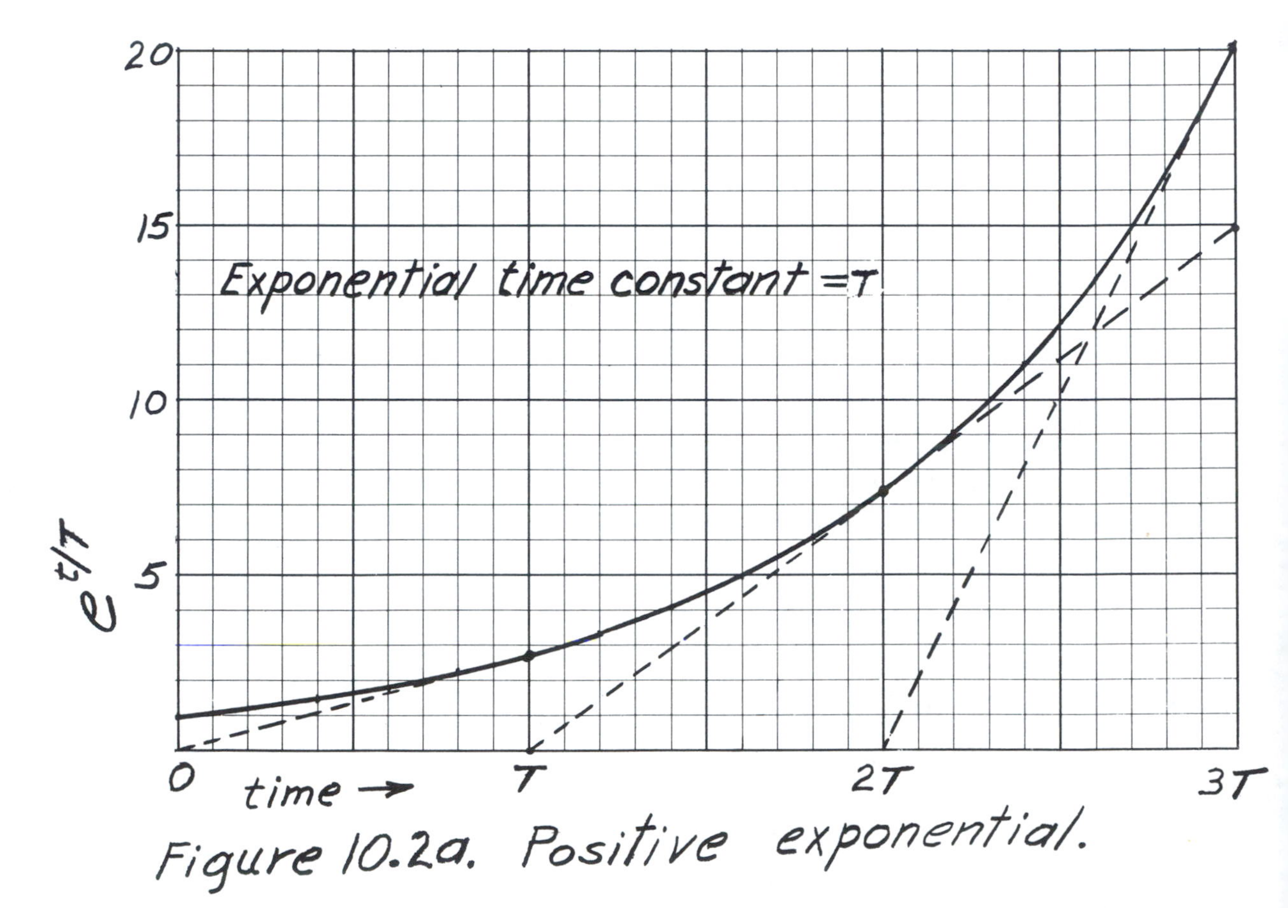

Figure 10.2a. Positive exponential.

The exponent is positive in the positive-feedback loop. The value of the function increases with time at an ever increasing rate. At any point, the slope is such that, if the slope were continued, the value would double in one time constant, that is in the time T. But actually the curvature is upward so that in one time constant the value of the curve increases by the multiple, e = 2.718.... Actual doubling occurs in the time for $e^{t/T} = 2$ or $t/T = .7$, approximately, or $t = .7T$. That is, the value doubles in about 0.7 of a time constant. In Figure 10.2a, one notes that the slope from any point, when extended backward in time, intercepts the unstable equilibrium value (here zero) at a time which is one time constant in the past.

Figure 10.2b shows the exponential function when the exponent is negative, representing the goal-seeking, negative-feedback loop.

$$\text{Value} = e^{-t/T}$$

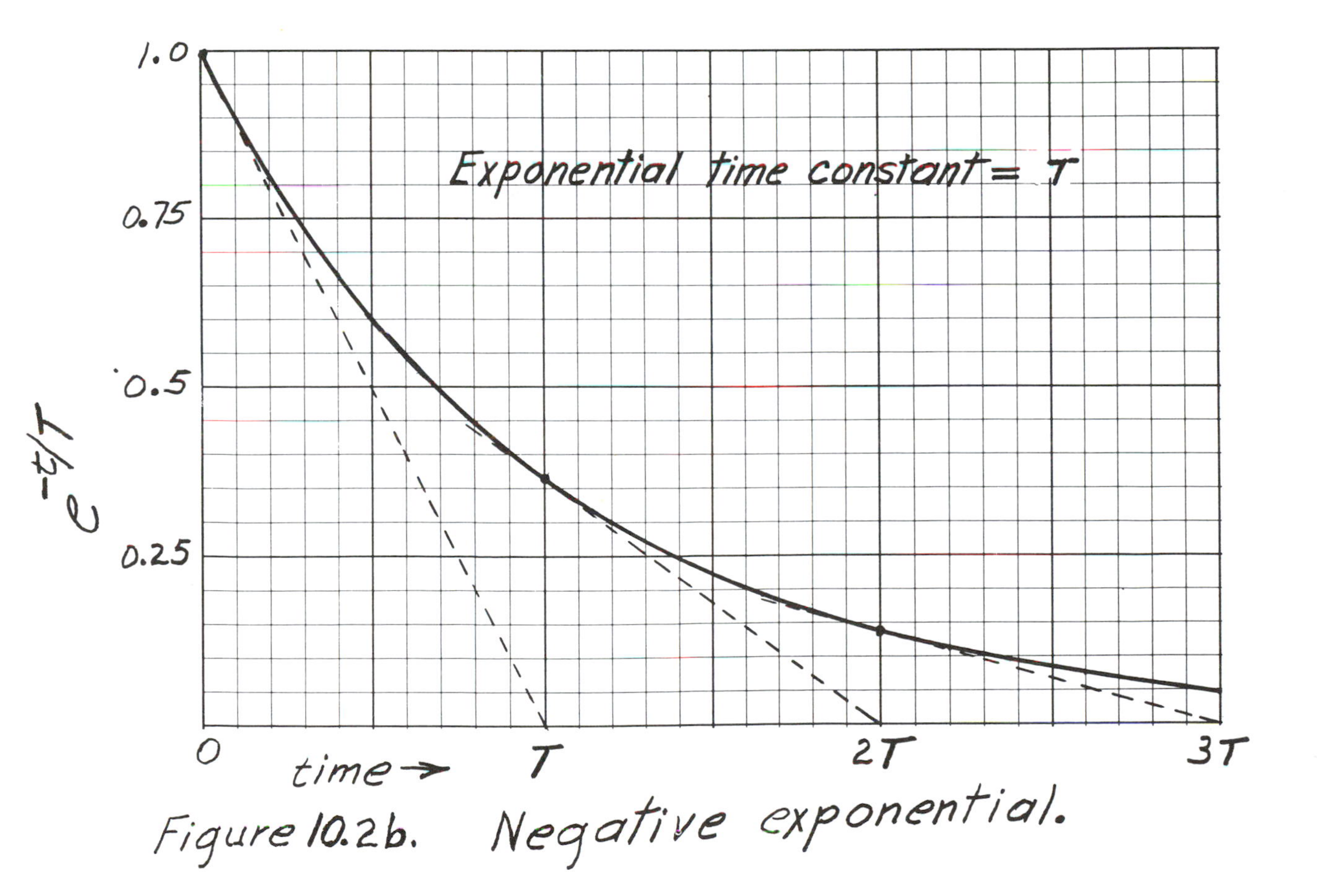

Figure 10.2b. Negative exponential.

Here the curve is declining toward zero such that the extended slope forward in time from any point will intersect the equilibrium value (here zero) in one time constant. But because the curve is becoming less steep, the value actually falls to $1/e = .37....$ of its initial value in one time constant.

Unlike the change in shape seen in comparing the powers of t in Figure 9.1b, the exponential is not changed in shape by integration. Only a constant multiplier term can appear as a result of integration:

$$\int e^{t/T} dt = Te^{t/T}$$

Likewise, for the negative exponential,

$$\int e^{-t/T} dt = -Te^{-t/T}$$

The integral of the exponential equals the reciprocal of the time constant multiplied by the exponential itself. The exponential is, therefore, a function which, when integrated, yields itself except for a multiplier term.

The structure of the first-order feedback loop forces a dynamic response in which the level and the rate variables have the same time shape. That is, the integrated rate has the same shape as the rate itself except for a multiplier coefficient. The structure of the loop forces such a relationship because the output of the level variable becomes its own input.

Consider the simple feedback loop of Figure 10.2c which is described by the following equations:

R.KL = L.K/A Eq. 10.2-1, R

L.K = L.J + (DT)(R.JK) Eq. 10.2-2, L

L = N Eq. 10.2-2.1, N

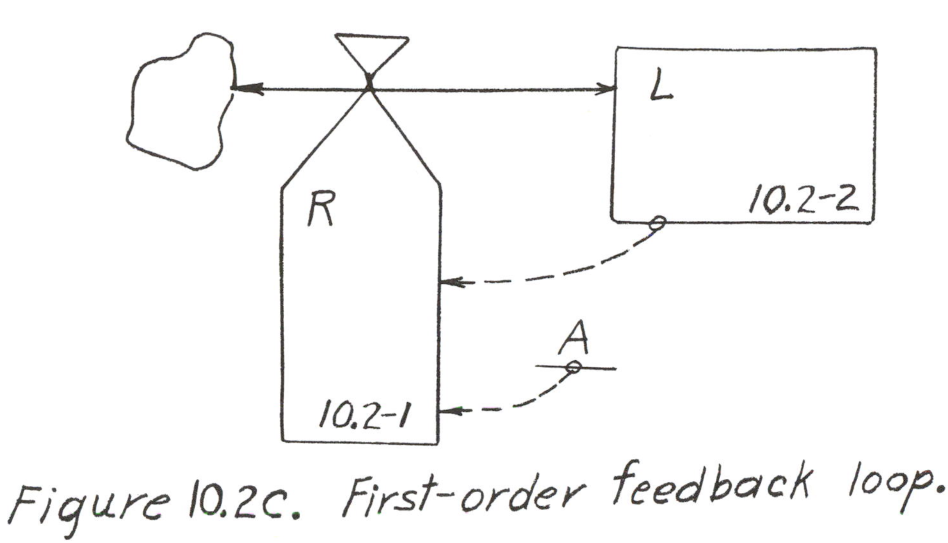

Figure 10.2c. First-order feedback loop.

By tracing the loop in Figure 10.2c, it is clear that L produces R which produces L. The time shape of the dynamic behavior of L must be such that when L is divided by A to produce R and when R is integrated it must produce the same L. Only the exponential function satisfies this requirement of retaining its same shape when integrated. The analytic solution to Equations 10.2-1 through 10.2-2.1, given in the form of continuous calculus notation rather than in difference equations, is

$$L = Ne^{t/T} \qquad \text{Eq. 10.2-3}$$

Although this is known to be the solution to the first-order feedback loop, the following steps show that the exponential solution in Equation 10.2-3 meets the closed loop requirement that the level variable produce itself when divided by A (in Equation 10.2-1) and integrated (in Equation 10.2-2). In Equation 10.2-3, if t=0, $e^{t/T}=1$. Then L = N which is its correct initial value at t = 0. The initial condition N is therefore the correct coefficient in the solution. From Equations 10.2-1 and 10.2-3,

$$R = \left(\frac{L}{A}\right) = \left(\frac{N}{A}\right)\left(e^{t/T}\right) \qquad \text{Eq. 10.2-4}$$

using the continuous notation of calculus. The level L is its initial value plus the integral of R (as indicated by the level Equation 10.2-2) so

$$L = N + \int_0^t R\,dt = N + \frac{N}{A}\int_0^t e^{t/T}dt$$

$$= N + \frac{N}{A}(T)\left[e^{t/T}\right]_0^t$$

$$= N + \frac{N}{A}(T)\left(e^{t/T} - 1\right)$$

$$= \left[N - \frac{N}{A}(T)\right] + \frac{N}{A}(T)e^{t/T} \qquad \text{Eq. 10.2-5}$$

Now Equations 10.2-5 and 10.2-3 both define L and can only be correct if they are identical. To be identical regardless of the value of T, the coefficients of all corresponding terms in t must be equal. The constant term, which does not exist in Equation 10.2-3 , must equal zero in Equation 10.2-5 so,

$$N - \frac{N}{A}(T) = 0$$

$$T = A \qquad \text{Eq. 10.2-6}$$

From this we learn that the time constant T of the exponential must be the reciprocal of the multiplier (1/A) which multiplies the level variable to produce the rate in the first-order feedback loop.

The coefficients of the $e^{t/T}$ terms in Equations 10.2-5 and 10.2-3 must be equal,

$$\frac{N}{A}(T) = N$$

which is true when substituting T = A from Equation 10.2-6.

Substituting the value of T = A into Equation 10.2-3, the solution for L in terms of the parameter A is

$$L = Ne^{t/A} \qquad \text{Eq. 10.2-7}$$

So, having started by assuming a solution for L as given in Eq.10.2-3, Equation 10.2-7, after traversing the loop, arrives at the same function for L. In other words, the assumed solution is a possible solution and is consistent with the loop structure which imposes the condition that the input to the level is the output divided by A.

In Equations 10.2-1 through 10.2-7, the time constant T = A could be either positive or negative to produce responses from the feedback loop as in Figures 10.2a or 10.2b. These are the positive and negative feedback systems as also discussed in Sections 2.4 and 2.2. Figure 2.2c shows negative exponential behavior moving upward toward a goal or equilibrium value. Figure 10.2b is also a negative exponential but starting above and descending toward its equilibrium value.

* *
* *

Principle 10.2-1 <u>Exponential behavior of first-order loop</u>.

The first-order feedback loop always exhibits an exponential time-shape. For positive feedback, the positive exponential diverges from an equilibrium value. For negative feedback, the exponential converges to the equilibrium.

* *
* *

* *
* *

Principle 10.2-2 <u>Time constant of first-order loop relates level to rate</u>.

The exponential time constant of a first-order loop is the reciprocal of the multiplier that defines the rate in terms of the level.

* *
* *

In the same manner as for the first-order loop, the exponential growth of a second-order, positive-feedback loop will now be examined.

The system is illustrated in Figure 10.2d with two levels and two rates. The levels are simple integrations, that is, neither is imbedded in a first-order feedback loop around itself as, for example, was the goods-on-order level GO in Figure 2.3a. The equations for this second-order system (two levels) are as follows where A1 and A2 are both positive. (If either were negative, the loop would be the negative feedback case which will be discussed in the next section):

R1.KL = L2.K/A1	Eq. 10.2-8, R
L1.K = L1.J + (DT)(R1.JK)	Eq. 10.2-9, L
L1 = N1	Eq. 10.2-9.1, N
R2.KL = L1.K/A2	Eq. 10.2-10, R
L2.K = L2.J + (DT)(R2.JK)	Eq. 10.2-11, L
L2 = N2	Eq. 10.2-11.1, N

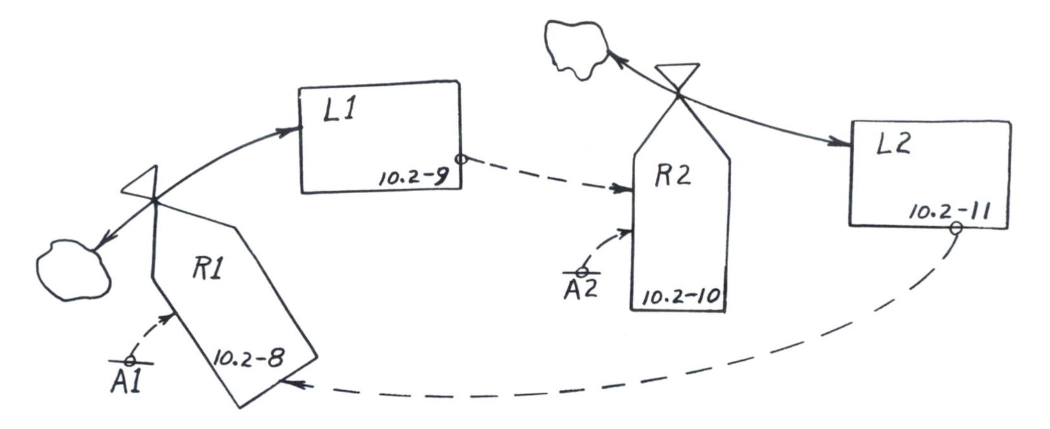

Figure 10.2d Second-order loop without first-order loops around the individual integrations.

The following analysis demonstrates that the exponential is a dynamic response that conforms to the structure of the second-order, positive-feedback loop. The nature of the dynamic time pattern for the level L1 is assumed to be an exponential, which is known to be true although the direct derivation lies outside the scope of this book. Using the assumed exponential, the consequences are traced around the loop to yield a second expression for L1. Because the two expressions for L1 must be identical, being the same point in the system, their corresponding coefficients can be equated and evaluated. This will determine how the initial conditions must be related to one another and how the exponential time constant depends on the values of A1 and A2. Also, because the square-root of the product of A1 and A2 appears, we will see that A1 and A2 must both be positive (or both negative) to make the exponential time constant a real number. (If one were negative, it would indicate the sinusoidal solution in the next section.)

So first assume, again using continuous calculus notation, that

$$L1 = (N1)e^{t/T} \qquad \text{Eq. 10.2-12}$$

If the dynamic solution is of this form, it is clear that the coefficient of the exponential must be N1 which, when t = 0, gives $e^{t/T} = 1$ and makes L1 equal its initial value of N1. Combining Equations 10.2-10 and 10.2-12

$$R2 = \frac{L1}{A2} = \frac{N1}{A2}e^{t/T} \qquad \text{Eq. 10.2-13}$$

Integrating R2 as specified by Equation 10.2-11,

$$\begin{aligned} L2 &= (N2) + \int_0^t (R2)dt \\ &= (N2) + \left[\frac{N1}{A2}(T)e^{t/T}\right]_0^t \\ &= (N2) + \frac{N1}{A2}(T)\left(e^{t/T} - 1\right) \\ &= (N2) - \frac{N1}{A2}(T) + \frac{N1}{A2}(T)e^{t/T} \qquad \text{Eq. 10.2-14} \end{aligned}$$

Dividing L2 by A1 to produce R1 as indicated by Equation 10.2-8,

$$R1 = \frac{N2}{A1} - \frac{N1}{(A1)(A2)}(T) + \frac{N1}{(A1)(A2)}(T)e^{t/T} \qquad \text{Eq.10.2-15}$$

R1, being the rate of flow into L1, must be integrated to produce L1,

$$L1 = (N1) + \int_0^t (R1)dt$$

$$= (N1) + \left[\frac{N2}{A1} - \frac{N1}{(A1)(A2)}(T)\right] t + \frac{N1}{(A1)(A2)}\left(T^2\right)\left(e^{t/T} - 1\right)$$

$$= \left[N1 - \frac{N1}{(A1)(A2)}\left(T^2\right)\right]$$

$$+ \left[\frac{N2}{A1} - \frac{N1}{(A1)(A2)}(T)\right] t$$

$$+ \left[\frac{N1}{(A1)(A2)}\left(T^2\right)\right] e^{t/T} \qquad \text{Eq. 10.2-16}$$

Now, Equations 10.2-16 and 10.2-12 must be identical for all values of t if the exponential in Equation 10.2-12 is a correct statement for L1. For the identity to be true, the three coefficients of terms in t in Equation 10.2-16 must each equal the corresponding coefficient in Equation 10.2-12 (in which the first two coefficients are zero, there being only the exponential term). Equating the first coefficient in Equation 10.2-16 to its corresponding zero value in Equation 10.2-12,

$$N1 - \frac{N1}{(A1)(A2)}\left(T^2\right) = 0$$

$$T^2 = (A1)(A2)$$

$$T = \sqrt{(A1)(A2)} \qquad \text{Eq. 10.2-17}$$

The positive square root* gives the time constant of exponential growth in terms of the two parameters A1 and A2 in the rate Equations 10.2-8 and 10.2-10. So that the time constant will be real, it is evident that the coefficients A1 and A2 must have the same algebraic sign.

Now from the coefficient of t in Equation 10.2-16, equating it to the zero coefficient of the non-existent term in Equation 10.2-12,

$$\frac{N2}{A1} - \frac{N1}{(A1)(A2)}(T) = 0$$

$$N2 = \frac{(N1)(T)}{A2}$$

Substituting the value of T from Equation 10.2-17,

$$N2 = (N1)\sqrt{\frac{A1}{A2}} \qquad \text{Eq. 10.2-18}$$

This expression shows how the initial value N2 for L2 must be related to the initial value N1 for L1 if the pure exponential shape of L1 is to appear immediately at time = 0. Were N2 of some different value, it would not affect the long-term existence of the exponential growth time constant as given in Equation 10.2-17, but it would generate some additional transient terms in the solution which would require time to subside before the pure exponential growth would dominate.

The coefficients of the $e^{t/T}$ term in Equations 10.2-16 and 10.2-12 must also be equal so,

$$\frac{N1}{(A1)(A2)}(T)^2 = N1$$

---

*Note that $T = -\sqrt{(A1)(A2)}$ is also a solution. This gives a negative time constant which in Equation 10.2-18 requires a negative value of the initial condition N2. For this special case, the negative initial value N2 for L2 produces a reduction toward zero of the positive L1 and the positive initial value of L1 produces a reduction toward zero of the negative L2, bringing the system to the unstable equilibrium point.

Using the value of T from Equation 10.2-17, the above expression is true without the need to define any new values or relationships.

Equation 10.2-17 can be substituted into Equation 10.2-12, giving the solution in terms of the originally stated parameters,

$$L1 = (N1)e^{t/\sqrt{(A1)(A2)}}$$

For this to be the complete statement of dynamic change without the presence of transient terms, the initial value for L2 must be as in Equation 10.2-18. The progression indicated by Equations 10.2-6 and 10.2-17 continues to the general expression for the time constant of a higher-order positive feedback loop,

$$T = \sqrt[n]{(A1)(A2)....(An)}$$

for a loop containing pure integrations without minor loops.

* *
* *

Principle 10.2-3 <u>Higher-order, positive-feedback loops show exponential growth.</u>

Positive feedback loops of nth order exhibit simple exponential growth (ignoring possible initial transients). The time constant $T = \sqrt[n]{(A1)(A2)....(An)}$ when the levels are simple integrations (no minor loops) and the A's are the reciprocals of the multipliers relating levels to succeeding rates.

* *
* *

In this section we have seen how the exponential time-shape is the natural behavior produced by a first-order feedback loop wherein an integration is back-coupled to itself so that the value of the level controls the rate that produces the level. For the positive feedback loop, where an increasing value of the level produces an increasing rate, the loop exhibits positive exponential growth. Where the sense of feedback is negative, that is, an increasing value of the level produces a reduction in the rate of inflow, the loop exhibits negative exponential approach to a goal.

For the second-order, positive-feedback loop, the time-response is also a positive exponential growth with a time constant related to the integration coefficients by $T = \sqrt{(A1)(A2)}$. Because the time constant would become imaginary, containing $\sqrt{-1}$ if one of the parameters were negative, the behavior of the second-order, negative-feedback loop containing only simple integrations must be other than a simple exponential shape.

10.3 Integration Creates the Sinusoid

Figure 10.2d and Equations 10.2-9 through 10.2-11.1 can be used to also represent a second-order negative feedback loop that does not have minor first-order loops. But Equation 10.2-8 must be changed to introduce the sign reversal that produces the negative sense in the feedback connection,

R1.KL = – L2.K/A1 Eq. 10.3-1, R

For the second-order negative loop, without subloops around the separate integrations, it is known that the time-response is sinusoidal. Such is suggested by Equation 10.2-17 where a negative value for A1 leads to an imaginary time constant containing $\sqrt{-1}$. The sine and cosine functions can be written in terms of exponentials with imaginary time constants.

For a simple situation where L1 has an initial value N1 and the initial value of L2 is N2 = 0, the time-shape of L1 will be a simple cosine curve starting at t = 0. Figure 10.3a shows the behavior of this system when A1 = 5 and A2 = 8.1 (chosen as will be seen later to produce a period of fluctuation of 40 time units as shown by the figure).

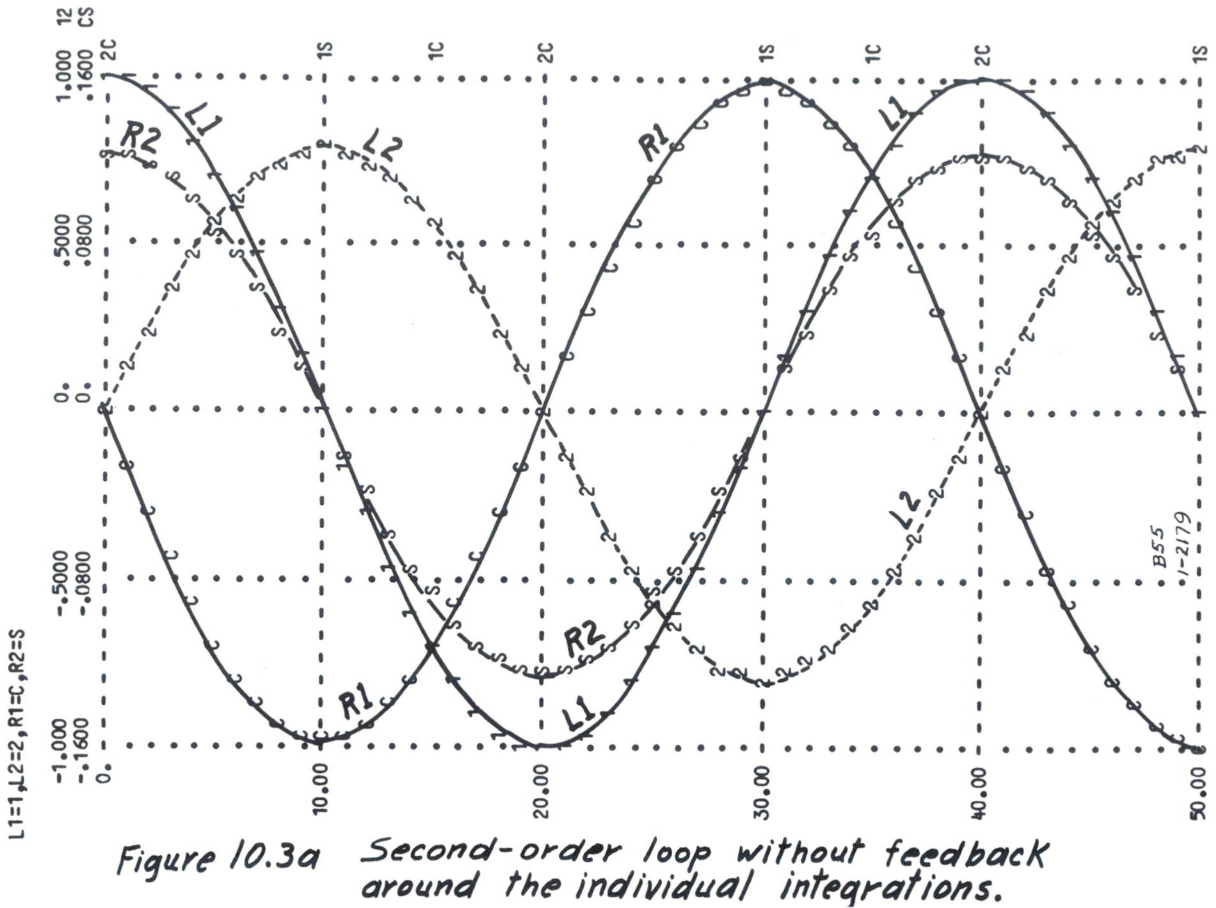

Figure 10.3a Second-order loop without feedback around the individual integrations.

Table 10.3 gives the DYNAMO compiler statements to produce the figure.

```
               FILE B55   LØØP--TWØ SIMPLE INTEGRATIØNS   10/05/67   2008.6

0.1    *       LØØP--TWØ SIMPLE INTEGRATIØNS
1      R       R1.KL=-L2.K/A1
1.1    C       A1=5
2      L       L1.K=L1.J+(DT)(R1.JK)
2.1    N       L1=1
3      R       R2.KL=L1.K/A2
3.1    C       A2=8.1
4      L       L2.K=L2.J+(DT)(R2.JK)
4.1    N       L2=0
4.4    PLØT    L1=1,L2=2(-1,1)/R1=C,R2=S(-.16,.16)
5      C       DT=.01
5.1    C       LENGTH=50
5.2    C       PLTPER=1
```

1-2188

Table 10.3 Model for Figure 10.3a.

In Figure 10.3a, note that L1 is the cosine curve. The rate R2 is proportional to L1 and is also a cosine. The level L2 is the integral of R2 and so lags a quarter period behind the rate R2 and appears as a sine curve. The rate R1 is the negative of and proportional to the level L2. The level L1 is the integral of R1 and lags a quarter period after R1. Traversing the loop from L1, there is one full period of phase shift before returning to L1. Two quarter periods of lag occur in the two integrations; and the negative sign of A1 inverts the phase between L2 and R1 and is equivalent to another half period of phase shift.

To verify the cosine time-shape for L1, the cosine will be assumed, following the procedure of the previous section. The consequences of that assumption will be traced around the loop, and then the coefficients of similar terms will be equated to establish the parameter and initial condition values which are necessary if the closed loop structure is to be valid for the cosine as the dynamic behavior of L1.

We start by assuming that

$$L1 = (N1) \cos \frac{2\pi}{P} t \qquad \text{Eq. 10.3-2}$$

where t is time and P is the period of the cosine function. The initial condition N1 is the proper coefficient of the cosine as is evident by noting that at t = 0 the cosine of zero is one and L1 = N1 as it should. From Equations 10.2-10 and 10.3-2, and using calculus notation,

$$R2 = \frac{L1}{A2} = \frac{N1}{A2} \cos \frac{2\pi}{P} t$$

Equation 10.2-11 states that L2 is the integral of R2,

$$L2 = (N2) + \int_0^t (R2)dt$$

$$= (N2) + \left(\frac{N1}{A2}\right)\left(\frac{P}{2\pi}\right) \sin \frac{2\pi}{P} t \Bigg|_0^t$$

$$= (N2) + \left(\frac{N1}{A2}\right)\left(\frac{P}{2\pi}\right) \sin \frac{2\pi}{P} t$$

From Equation 10.3-1,

$$R1 = \frac{-L2}{A1} = \frac{-N2}{A1} - \frac{N1}{(A1)(A2)}\left(\frac{P}{2\pi}\right)\sin\frac{2\pi}{P}t$$

Substituting according to Equation 10.2-9,

$$L1 = (N1) + \int_{o}^{t}(R1)dt$$

$$= (N1) - \frac{N2}{A1}t - \frac{N1}{(A1)(A2)}\left(\frac{P}{2\pi}\right)\left(\frac{-P}{2\pi}\right)\cos\frac{2\pi}{P}t\Big|_{o}^{t}$$

$$= (N1) - \frac{N2}{A1}t + \frac{N1}{(A1)(A2)}\left(\frac{P}{2\pi}\right)^{2}\left(\cos\frac{2\pi}{P}t - 1\right)$$

$$= (N1) - \frac{N1}{(A1)(A2)}\left(\frac{P}{2\pi}\right)^{2} - \frac{N2}{A1}t + \frac{N1}{(A1)(A2)}\left(\frac{P}{2\pi}\right)^{2}\cos\frac{2\pi}{P}t \qquad \text{Eq. 10.3-3}$$

The coefficients of corresponding terms in t must be identical if Equation 10.3-3 is to be identical to Equation 10.3-2. So for the coefficient not involving t, which is zero in Equation 10.3-2,

$$(N1) - \frac{N1}{(A1)(A2)}\left(\frac{P}{2\pi}\right)^{2} = 0$$

$$\left(\frac{P}{2\pi}\right)^{2} = (A1)(A2)$$

$$P = 2\pi\sqrt{(A1)(A2)} \qquad \text{Eq. 10.3-4}$$

This equation gives the period in terms of the two parameters A1 and A2. The relationship between Equations 10.3-4 and 10.2-17 is of interest. From Equation 10.3-4, $P/2\pi$, which is the time for one radian of angular rotation in the sinusoidal behavior of the negative loop, has the same value as the time constant of response in the positive loop given in Equation 10.2-17.

The coefficient of t in Equation 10.3-3 must be zero,

$$-\frac{N2}{A1} = 0$$

$$N2 = 0$$

which must be the initial condition of L2 if the simple cosine is to be the time-shape of L1. Were N2 not zero, a sustained fluctuation would still be the system behavior but the cosine curve for L1 would have an amplitude depending on both N1 and N2 and would have a phase shift with respect to t = 0.

Equating the coefficients of $\cos \frac{2\pi}{P}t$ in Equations 10.3-3 and 10.3-2,

$$\frac{N1}{(A1)(A2)} \left(\frac{P}{2\pi}\right)^2 = N1$$

which is true by the relationship in Equation 10.3-4.

The simple (having no minor loops around single integrations), second-order, negative-feedback loop oscillates in a sustained sinusoidal manner if it is disturbed from equilibrium. The period of the oscillation increases as the "coupling time-constants" (the A's in Equation 10.3-4) become longer. Often the rate equations, for example Equations 10.2-10 and 10.3-1, are written, not with the A's in the denominator but with the reciprocals $\left(\frac{1}{A1}, \frac{1}{A2}\right)$ in the numerator and when so done the coefficient is often called a "gain" or "amplification" coefficient. If G1 = 1/A1 and G2 = 1/A2, then the period

$$P = \frac{2\pi}{\sqrt{(G1)(G2)}}$$

In terms of gain coefficients, the period of oscillation decreases as the gain increases.

* *
* *

Principle 10.3-1 Sinusoidal oscillation in simple, second-order, negative loop.

The second-order, negative loop with no minor loops oscillates as a sustained sinusoid with a period $P = 2\pi \sqrt{(A1)(A2)}$ where the A's are the coupling time-constants or the reciprocals of the gain multipliers that relate levels to succeeding rates.

* *
* *

WORKBOOK FOR PRINCIPLES OF SYSTEMS

Chapters W1 through W10

by

Jay W. Forrester
Professor of Management
Massachusetts Institute of Technology

Cambridge
Massachusetts

CHAPTER W1

SYSTEMS

W1.1 The Ubiquity of Systems

In this book we examine the nature of systems whether they be systems of physical parts or systems of people. Social systems representing interaction between people include the family, small groups, business enterprises, countries, national economies, and international relationships. Physical systems surround us in modern technology. Some of our most important systems involve the interaction between social and physical elements. The reader may be surprised to find a common theory that bridges from the control of earth satellites to individual psychology and to international power politics. Yet it is in such a sweep of generality that one discovers the challenge and excitement of system principles.

Until recently the "laws" governing the way systems change, grow, and fluctuate have been wrapped obscurely in the mathematics of differential equations and Laplace transforms. Such mathematics is so difficult that it is accessible only to a skilled mathematician. Even with the forefront of mathematical knowledge being applied to systems theory, mathematics is so weak when confronted by the important systems questions, that it is inadequate for dealing with the realities of most of the world's problems. But in the available mathematical theory of systems are to be found principles that can be used to guide work with systems that are so complex that they lie beyond the capability of formal mathematical analysis.

This workbook and the accompanying text attempt to release system theory from confinement within mathematics and to bring it as living ideas into everyday affairs. To do this requires the building of a broad foundation of concepts that may be new to the reader. Creating this foundation requires many small building blocks of separate ideas which must be gradually cemented together.

The experiences of earlier students have shown that learning about systems behavior is not a spectator sport, one must participate and learn by doing.

This workbook provides exercises which are an essential accompaniment to the text. After each section of the text, the reader should complete the corresponding section of the workbook before continuing with the next section of text.

In the following exercises the reader should cover the answer with a large card or folded paper until he has followed the instructions and completed the blanks. Fill the spaces in writing. In later examples using graphs, fill in your answers on the page before comparing with the answer provided.

1. An assembly of parts that operate together for a common purpose is called a _______________.
   A watch is a _________________ that indicates time.
   * * * * *
   * * * * *
   system, system

2. A governor and the engine to which it is coupled form a _______________ to deliver power at constant engine speed.
   * * * * *
   * * * * *
   system

3. A lawn mower is an assembly of parts that form a ___________ for cutting grass.
   * * * * *
   * * * * *
   system

4. A system may include people as well as physical components.
   A research laboratory is a ______________ of ____________ for developing new products.
   * * * * *
   * * * * *
   system, people

5. Accounting is a ________________ of records and data transfer procedures that produce part of the input information on which managerial decisions are based.

   * * * * *

   * * * * *

   system

W1.2 System Principles as the Structure of Knowledge

Understanding is easiest to acquire in those areas where knowledge has been structured and organized. Contrast the increase in pace that has occurred during the last century in a curriculum of science study compared to the little change between past and present in a curriculum of liberal arts.

A liberal arts education has been essentially the same over the last century. Each student starts at the same point, traverses the same literature, and attempts to distill for himself a structure, or rationale, that relates the past to the present and future. The structuring is left to the student. The structure is not itself the substance of the material being taught but instead the content deals with specific situations and incidents.

But, by contrast, scientific education deals with the generalizations, or structure, that relate the variables of the physical processes. Before the advent of modern science, the technical arts were learned by apprenticeship through a process of retracing the learning stages followed by one's predecessors. But today science at the high school level covers the same material that a generation earlier lay at the frontier of new scientific discovery. This accelerated access is possible because of the scientific concentration on exposing the systematic structural relationships that underlie the observations of physical behavior.

To illustrate what we mean by the structuring of scientific knowledge, consider the following elementary electric circuit for current flowing through a resistance. Here the voltage E applied to the terminals causes the current I to flow through the resistance R.

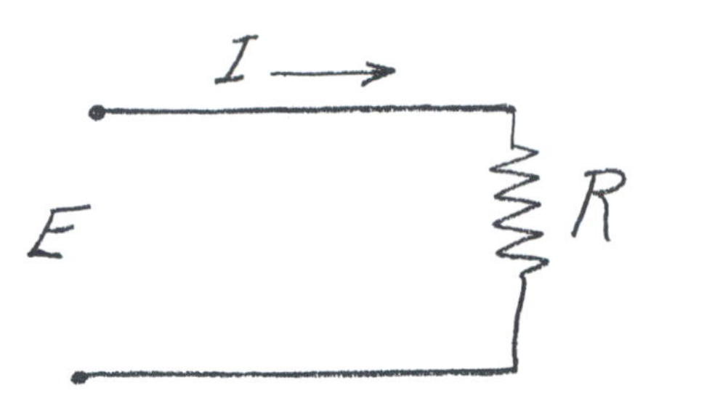

Fig. W1.2 Current through a resistance.

Suppose we wish to know the relationship between E, I, and R. One way would be to tabulate numerous measurements in tables and look up needed values:

E	I	R
5	2.5	2
4	8	.5
17	8	2.125

But, of course, such a table can be fully replaced by the equation E=IR showing that voltage equals the product of current times resistance. The knowledge of this structure (or equation) allows one to grasp the relationship of the variables far more quickly than from a table of experimental results.

1. A map shows the ______________ of relationships between features of the terrain.

   * * * * *

   * * * * *

   structure

2. The photographic relationship

   $$\text{Exposure} = \frac{(\text{constant})(\text{light intensity})(\text{shutter opening})}{(\text{f-number})^2}$$

   conveys the ________________ or relationships necessary to maintain a constant film exposure.

   * * * * *

   * * * * *

   structure

3. The ________________ provided by laws serves as a guide to permissible behavior that does not encroach on the rights of others.

   * * * * *

   * * * * *

   structure

4. But no one knows the entire legal code as set down in the statute books. Instead, a guiding moral ____________ conveying the fundamental nature of right and wrong helps us to estimate approximately what the law must be.

   * * * * *

   * * * * *

   structure

W1.3 Systems--Open and Feedback

This book treats the behavior of feedback systems. A feedback system is one in which an action is influenced by the consequences of previous action. The identification of the action determines the scope of the pertinent system because the system boundaries should include the significant channels whereby the consequences of an action are brought back to influence future action.

Any feedback system can be a component of a more comprehensive feedback system. As the purpose and goals of a system are broadened, the system becomes more extensive and incorporates subsystems each of which has its own narrower purpose and goal. The structure of goals and related feedback loops guide the components as they interact with one another.

Section 1.3 of the text describes a watch as an open system. That is true in terms of the goal of showing the correct time. But, as a mechanical system having the goal of merely keeping the balance wheel oscillating, it is a feedback system. At the proper point in each cycle, the escapement releases another pulse of energy to drive the balance wheel. From this strictly mechanical viewpoint, we see a feedback system in which the balance wheel controls the release of spring power to drive the balance wheel. The watch is not a feedback system in terms of a time-keeping goal because the watch by itself has no measure or awareness of whether or not it keeps the correct time. Relative to the proper time, the watch is an open system to which the input is winding the spring and the output is a position of the hands, irrespective of whether they show true time.

A factory further illustrates how the system formulation depends on the viewpoint. To a person dealing with the broad scope of a company and its market, a factory may be but an open-system component in a larger feedback loop. The factory might be represented as nothing more than a simple delay between the receipt and the shipment of orders. But from the viewpoint of the factory manager it is a complex interaction of many subsystems involving scheduling, purchase of materials, investment in machines, authority, and morale. Each subsystem has local goals and together they form a feedback system that tries to satisfy its goal of matching output to demand.

1. From the viewpoint of biology and psychology, a person is a complex _______________ system that adjusts actions toward achieving goals.

   * * * * *

   * * * * *

   feedback

2. But from the viewpoint of reordering to maintain inventory in a warehouse, a person might be seen as a simple _________ system component of a larger feedback system that encompasses the mail, the supplier, and the delivery of goods.

   * * * * *

   * * * * *

   open

3. Temperature regulation of the human body is a _____________ system that reacts to the temperature that is being controlled.

   * * * * *

   * * * * *

   feedback

4. A student adjusts the time devoted to study in accordance with his past learning rate and grades and as such he is part of a ________________ system that adjusts toward a desired academic performance.

   * * * * *

   * * * * *

   feedback

5. Electrical and mechanical feedback systems are used in the control of many manufacturing processes. A chemical reactor combined with a thermometer to measure tank temperature and an amplifier and control valve to admit steam to the reactor comprise a ______________ system for the control of reactor temperature.

   * * * * *

   * * * * *

   feedback

6. Feedback systems are found as the basis of control in all of the biological processes. One can keep his eyes fixed on an object as he moves his body because his eyes, brain, and muscles form a _______________ system that moves the muscles of the eyes and neck to prevent deviation of the eyes from the object.

   * * * * *

   * * * * *

   feedback

7. The human body is a complex structure of _______________ loops that cooperate to sustain life. However, the individual is but a component in the larger society. From the viewpoint of society or a business, the individual alone is an ______________ system. In the larger setting, he is a component that is coupled to the social system by the guiding forces and information sources that surround him.

   * * * * *

   * * * * *

   feedback, open

8. A research department for the purpose of designing better products is by itself an __________ system until it is connected to its environment by information channels that tell it customer needs and by the reaction of customers to past products that have been developed.

   * * * * *

   * * * * *

   open

9. Control of an inventory in a warehouse would be an __________ system if the content of the inventory were not used as the basis for reorder of supplies. The inventory becomes part of a ____________ system when the state of the inventory is used to guide orders that supply the inventory.

   * * * * *

   * * * * *

   open, feedback

10. In terms of the profit goal of a company, a manufacturing process (is/is not) a feedback system. In terms of the goal of making items to match the flow of customer orders, the manufacturing process (is/is not) a feedback system.

    * * * * *

    * * * * *

    is not, is

W1.4 The Feedback Loop

The feedback loop is the universal structure surrounding any decision-making process. The reader should have a clear picture of the way decision, action, system level*, and available information are related.

---

*"Level" here means condition or state.

(Sec.W1.4)

1. The parts of a feedback system form a structure shaped as a (chain/loop).

   * * * * *

   * * * * *

   loop

2. The ______________ arrangement of a ____________ system brings the result of past actions back to guide present decisions.

   * * * * *

   * * * * *

   loop, feedback

3. The "action" represents the flow of something (energy, goods, money, heat, water, work, thought, etc.) that is controlled by the ______________.

   * * * * *

   * * * * *

   decision

4. The ______________ is based on the available information and controls the ______________.

   * * * * *

   * * * * *

   decision, action

5. The ______________ alters the ____________ of the system.

   * * * * *

   * * * * *

   action, level

6. The ____________ of the system is the true condition of the system and is the source of information about the system.

   * * * * *

   * * * * *

   level

7. The level of the system is the present condition of the system. The level of the system results from an accumulation of all of the past actions. The present level of the system (is/is not) determined by the present action stream.

   * * * * *

   * * * * *

   is not

8. The level of the system accumulates the history of all past ___________________.

   * * * * *

   * * * * *

   actions

9. Knowing only the present decision and action will tell us the level of the system (true or false?).

   * * * * *

   * * * * *

   false (because the level depends on what has resulted from the accumulation of past actions)

10. Slow response in gathering data or mistakes can cause the ________________ to be inaccurate in reflecting the __________________ of the system.

    * * * * *

    * * * * *

    information, true level

11. The decisions form a stream through time that is dependent on the available information about the system. Repeated__________of any particular type act through the system to change the _________________ on which they are based.

    * * * * *

    * * * * *

    decisions, information

12. The recurring ________________ stream governing the fraction of time spent in recreation instead of in work changes as our available ___________________ about our situation, health, income, and attitude changes.

    The information about the system changes as a result of past _____________ governing ____________ that have changed the levels of the system of which we are a part.

    * * * * *

    * * * * *

    decision, information, decisions, actions

CHAPTER W2

PREVIEW OF FEEDBACK DYNAMICS

W2.1 Diversity of Behavior

1. The amount of heat in a thermometer determines its temperature, the greater the heat content the higher the temperature. The rate at which heat flows in depends on the difference in temperature between the thermometer and its surroundings. As the temperature of the thermometer approaches the temperature of the surroundings, heat flow gradually declines. Which curve in Figure 2.1 of the text would best describe how the temperature of the thermometer changes with time after it is immersed in a hot liquid? ____________________

   * * * * *

   * * * * *

   Curve A

2. The position of a pendulum, which is displaced and allowed to swing, is best described by Curve ___________ in Figure 2.1 of the text.

   * * * * *

   * * * * *

   B

3. A student learning a new subject or skill finds that, as his knowledge increases, it for a time becomes easier and easier to acquire still more knowledge. He learns at an ever faster pace till he begins to approach his maximum capability in the subject, then loss of interest or discouragement because he is not making the same rapid progress as earlier may reduce motivation. Which curve in Figure 2.1 of the text describes this process? ____________

   * * * * *

   * * * * *

   Curve D. Early accomplishment makes a greater rate of progress possible until the student approaches a limit to his progress.

4. Automobile A is overtaken and passed by automobile B. Automobile A then catches up with B and tries to follow at a fixed distance behind B. Which curve in Figure 2.1 of the text best describes the velocity of automobile A? ________________

   * * * * *

   * * * * *

   Curve B. The velocity of automobile A must rise until it exceeds the final value to close the gap between the two automobiles. It is not Curve D because, in the accelerating phase, velocity does not itself produce still more velocity to produce an upward curving velocity line.

5. In the industrialization of a society, capital equipment can be used to produce still more capital equipment. Which curve of Figure 2.1 in the text would best describe the amount of capital equipment versus time? ________________

   * * * * *

   * * * * *

   Curve C. The accumulated result of past action (the available capital equipment) determines the present action (construction rate of more equipment). The more equipment there is, the faster is the accumulation.

W2.2 First-Order, Negative-Feedback Loop

At this point it is important to begin developing personal familiarity with the way in which a system can be described by flow diagrams and equations. In addition to practice in system formulation, hand computing of system behavior will give the student an opportunity to observe how the variables in a system interact. While the hand computation and curve plotting may at times seem tedious, it is strongly recommended to the extent of the exercises in this workbook. In this section a simple feedback system for water level control will be used as a basis for exercises in flow diagrams, consistency of measure, time graphing, the accumulation process represented by a level variable, and computation of successive levels of a system.

1. Consider the simple mechanism of a toilet tank float and valve as in the figure.

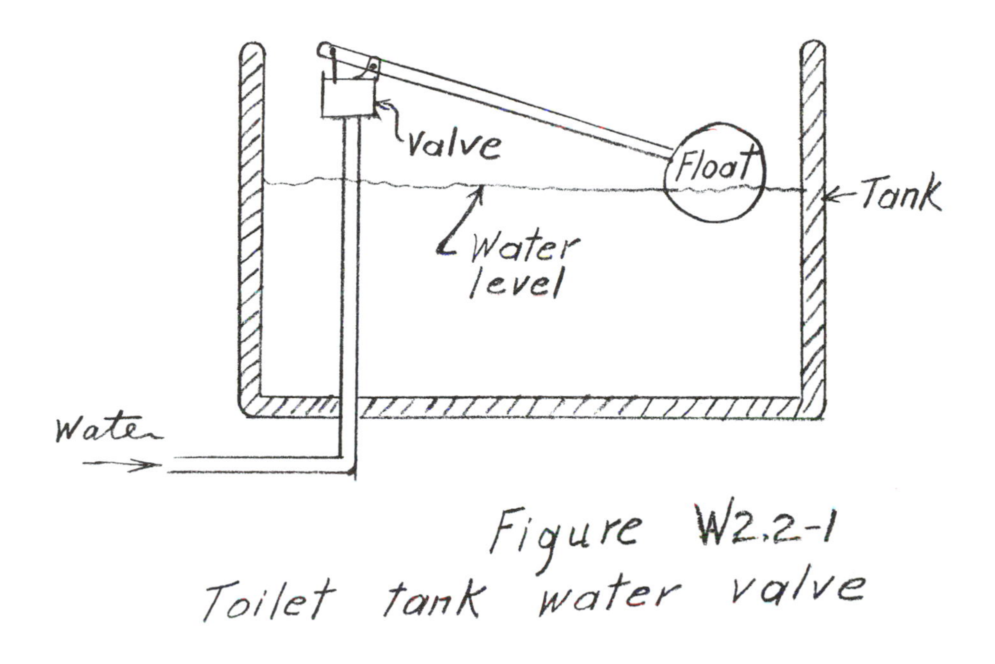

Figure W2.2-1
Toilet tank water valve

As the float drops it turns on the valve to admit water. The rising water level causes the float to rise and this in turn gradually shuts off the water. Because the valve controls the water which controls the valve, this is a ____________ system.

* * * * *

* * * * *

feedback

2. Referring to text Figure 1.4a, what in Figure W2.2-1 corresponds to the level of the system? ____________________

   * * * * *

   * * * * *

   Water level. (If the water level is known, everything else about the system can be deduced.)

3. What is the action variable in the water tank system? ___________

   * * * * *

   * * * * *

   Water flow rate. (It is the water flow which causes the water level to change.)

4. The water flow rate depends on the ______________ ______________.

   * * * * *

   * * * * *

   Water level

5. In this system, do we need to recognize information about the water level as being different from the water level? ___________

   * * * * *

   * * * * *

   No. (Here the float is mechanically coupled to the valve. The time necessary for the float to adjust the valve is very short compared to the filling time of the tank.)

6. However, in many physical and managerial systems, the responses are so slow that one can not make the simplification of using the true ___________ of the system as a direct input to the decisions.

   * * * * *

   * * * * *

   level

7. Sketch and label the flow diagram of the water system following the symbols in text Figure 1.4a.

Figure W2.2-7

* * * * *

* * * * *

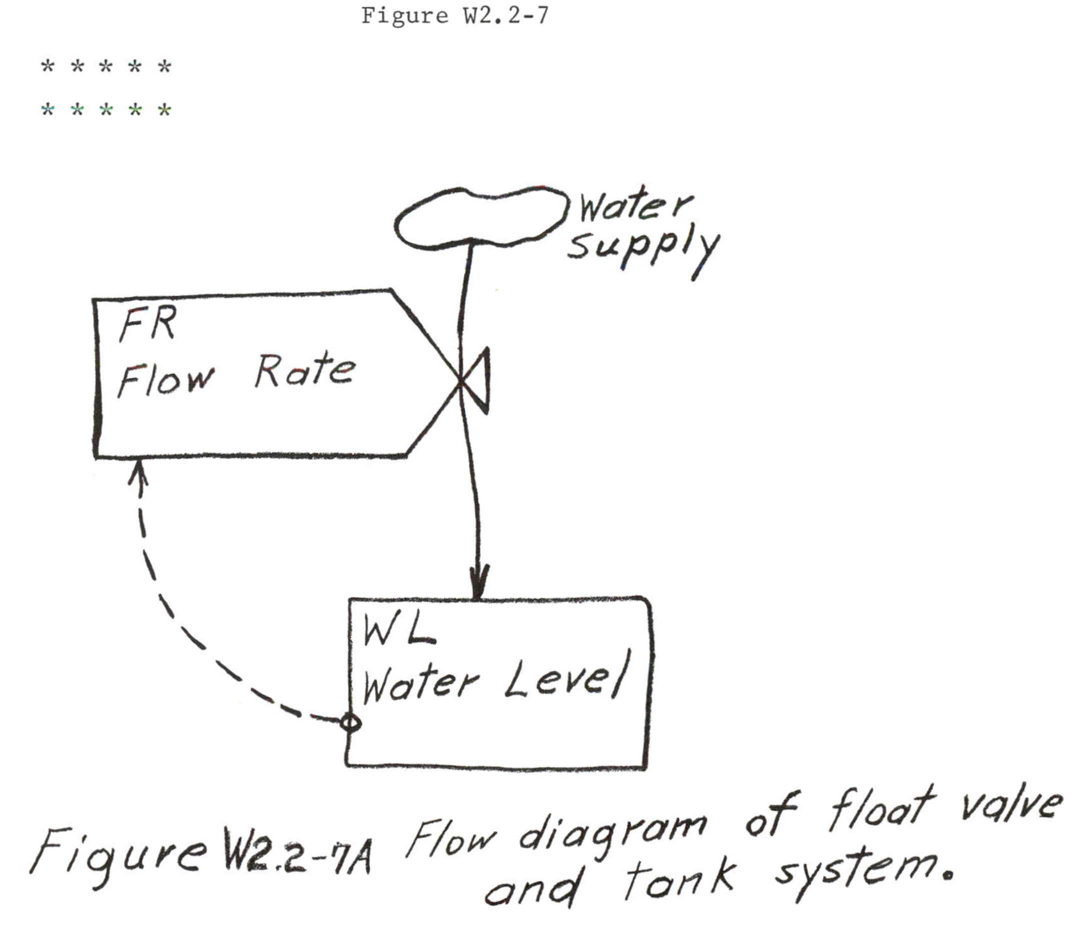

Figure W2.2-7A Flow diagram of float valve and tank system.

8. Before studying the feedback character of the preceding float-valve system, we will examine the simpler system in the following flow diagram which is obtained by removing the float and then hand-controlling the valve.

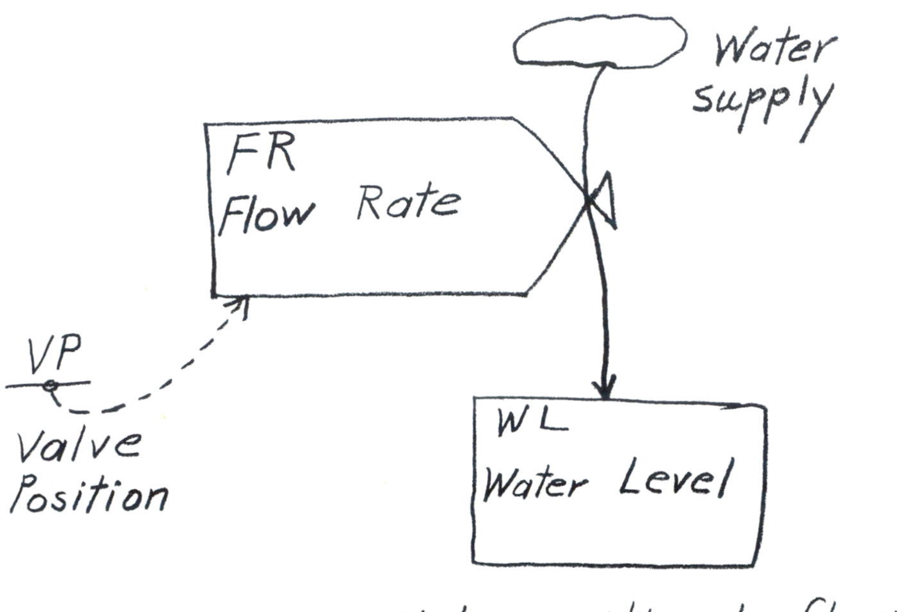

Figure W2.2-8 Valve without float

This figure shows an ____________________ system.

* * * * *

* * * * *

open (Because the feedback loop is broken. The water level no longer controls the valve.)

9. In the open system with hand controlled valve, suppose that the valve position VP is set so that the flow rate FR is 0.1 gallon per second. In ten seconds there will be __________ __________ of water in the tank.

   * * * * *

   * * * * *

   one gallon

10. At the end of 30 seconds there will be __________ __________ of water in the tank.

    * * * * *

    * * * * *

    three gallons

11. How much water would be in the tank at the end of 45 seconds?

    ____

    * * * * *

    * * * * *

    4.5 gallons (Note that the answer is not correct unless the "units of measure," that is, "gallons," are given.)

12. The units of measure must always accompany any numerical value to define the quantity. The units of measure of water flow rate are __________ __________ __________. The water level is measured in __________.

    * * * * *

    * * * * *

    gallons per second (or gallons/second), gallons

13. In dealing with dynamic systems it is important to see the system from several viewpoints--as a physical arrangement (see Figure W2.2-1), as a flow diagram (see Figure W2.2-8), and as a set of actions and consequences that can be shown graphically through time. On the following figure, plot the water flow rate of 0.1 gallon per second.

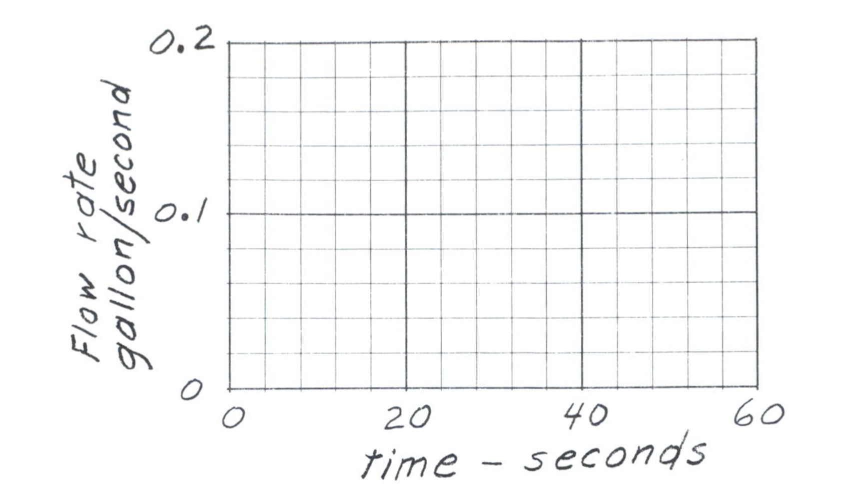

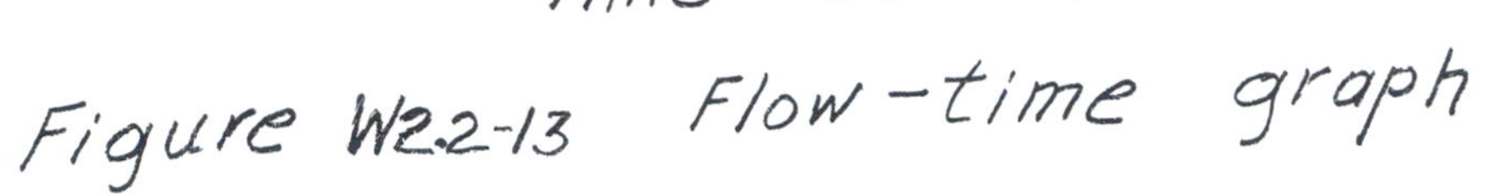

Figure W2.2-13 Flow-time graph

* * * * *

* * * * *

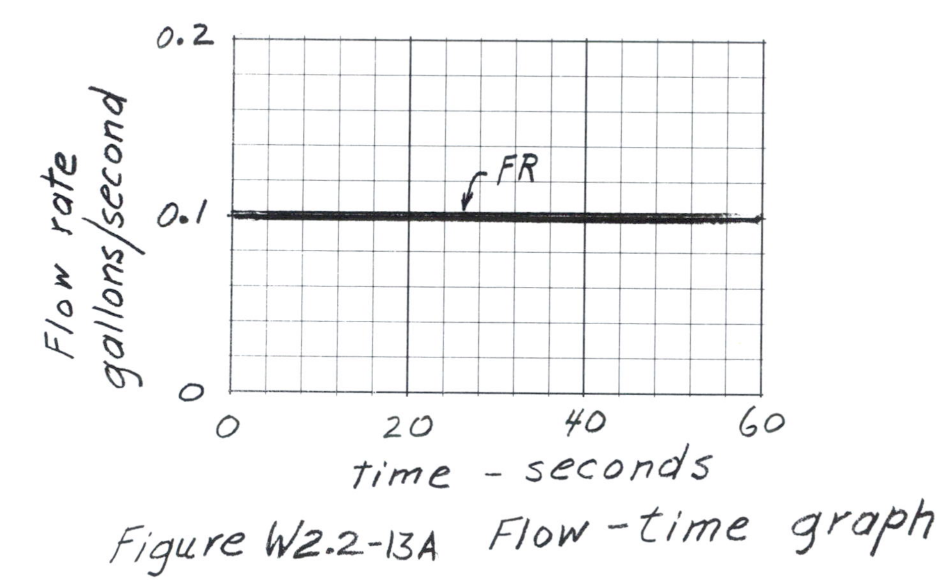

Figure W2.2-13A Flow-time graph

Note that a constant flow rate of 0.1 gallon per second is a horizontal line on the graph at the corresponding vertical position. As one moves horizontally along the graph in time, he finds that the flow rate is at all times 0.1 gallon per second.

14. Plot the curves for flow rates of 0.04 and 0.16 gallon per second.

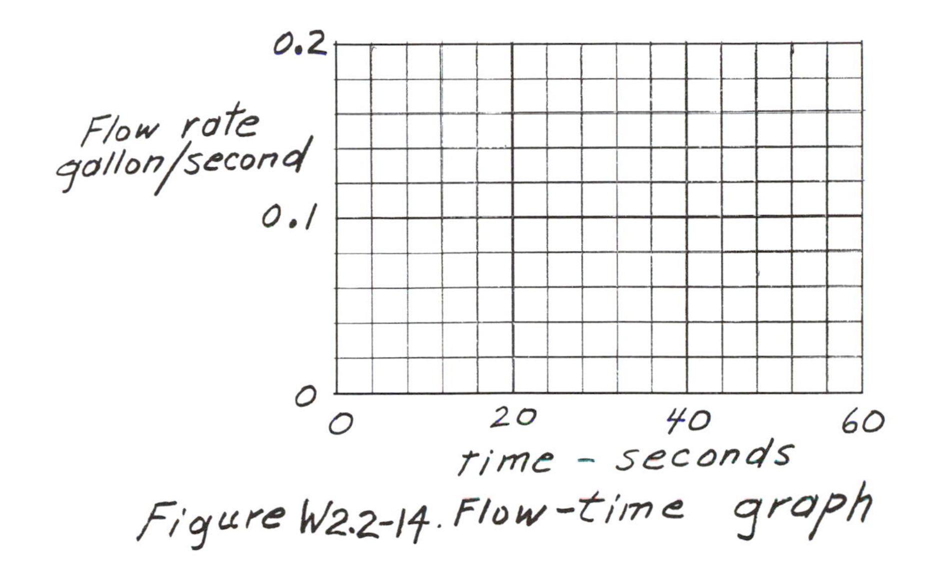

Figure W2.2-14. Flow-time graph

* * * * *
* * * * *

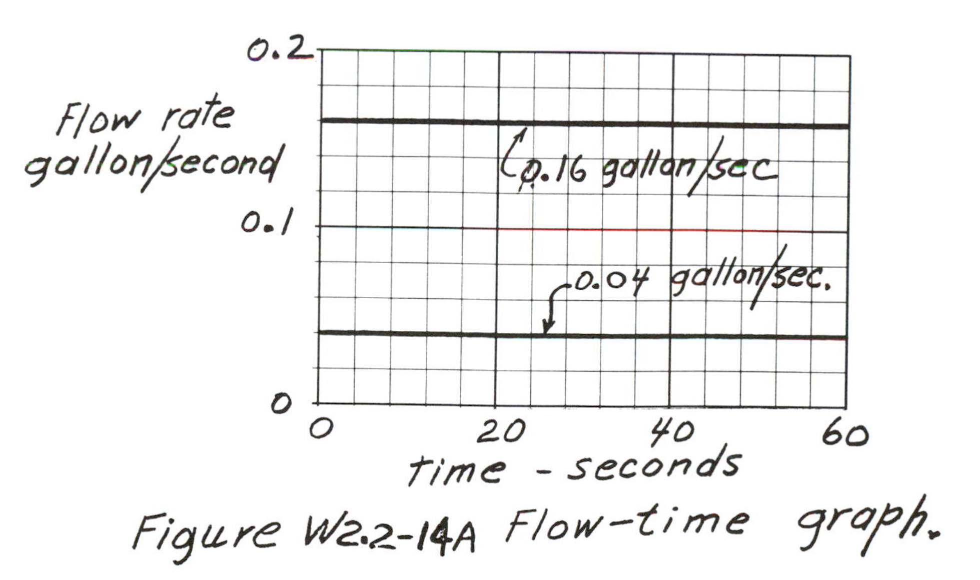

Figure W2.2-14A Flow-time graph.

15. If the flow rate FR is 0.04 gallon per second, and if the tank is empty at time = 0, the amount of water is

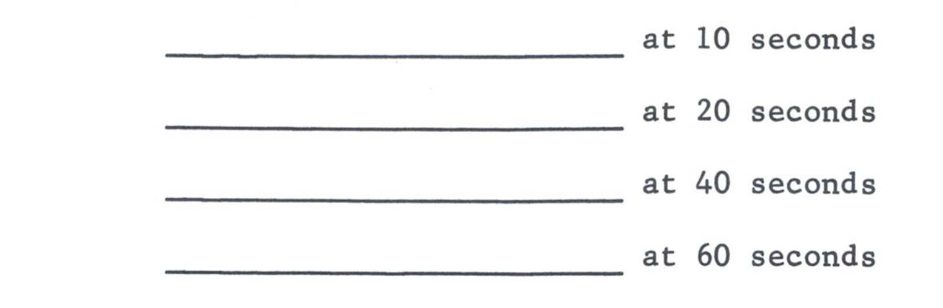
_______________ at 10 seconds

_______________ at 20 seconds

_______________ at 40 seconds

_______________ at 60 seconds

* * * * *

* * * * *

0.4 gallon, 0.8 gallon, 1.6 gallons, 2.4 gallons

16. On this graph plot the amount of water in the tank for every point in time, starting with an empty tank and a flow rate of 0.04 gallon per second.

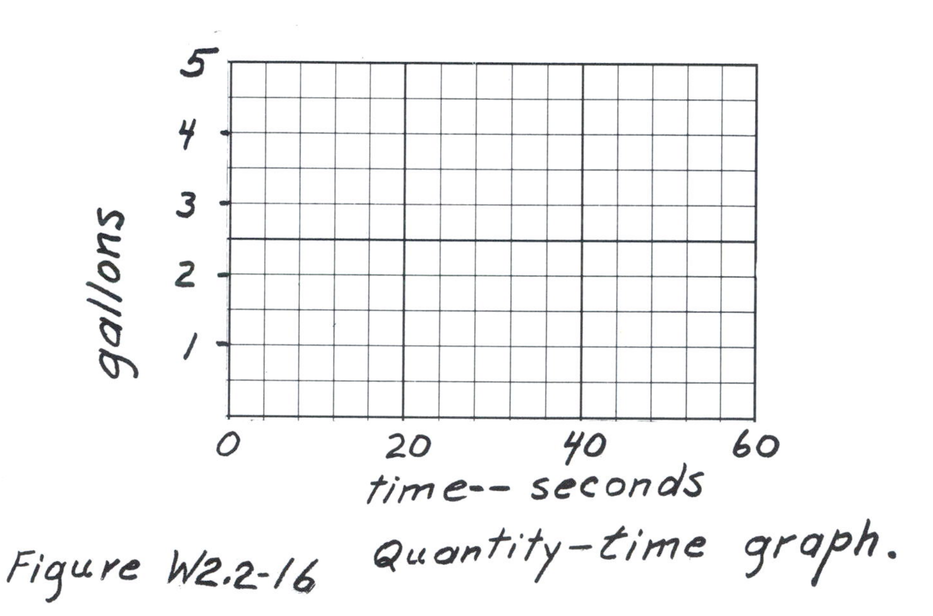

Figure W2.2-16 Quantity-time graph.

* * * * *

* * * * *

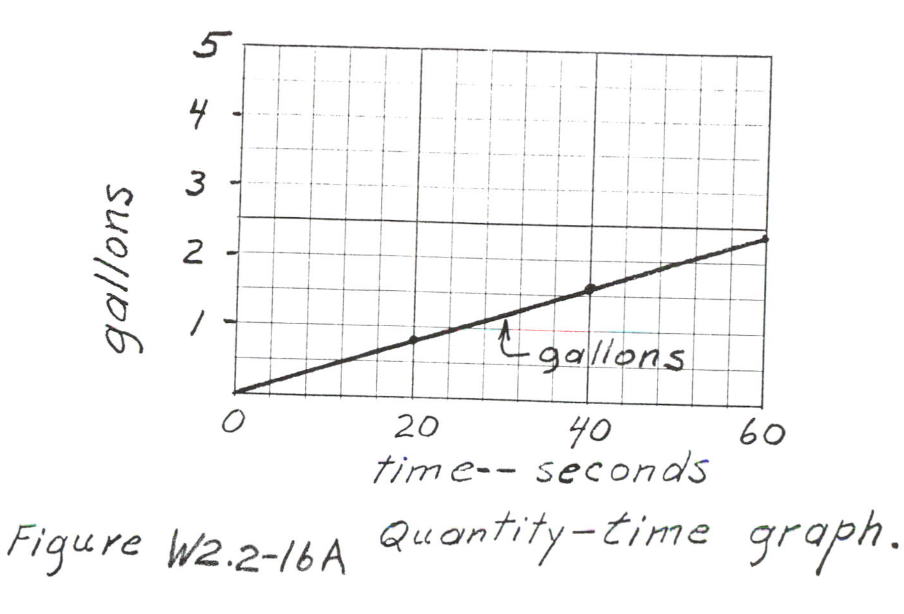

Figure W2.2-16A Quantity-time graph.

17. On this graph, plot the water-time relationships showing the amount of water in the tank for a flow rate of 1/12 gallon per second, for 0.2 gallon per second, and for 1/30 gallon per second. Assume the tank starts empty.

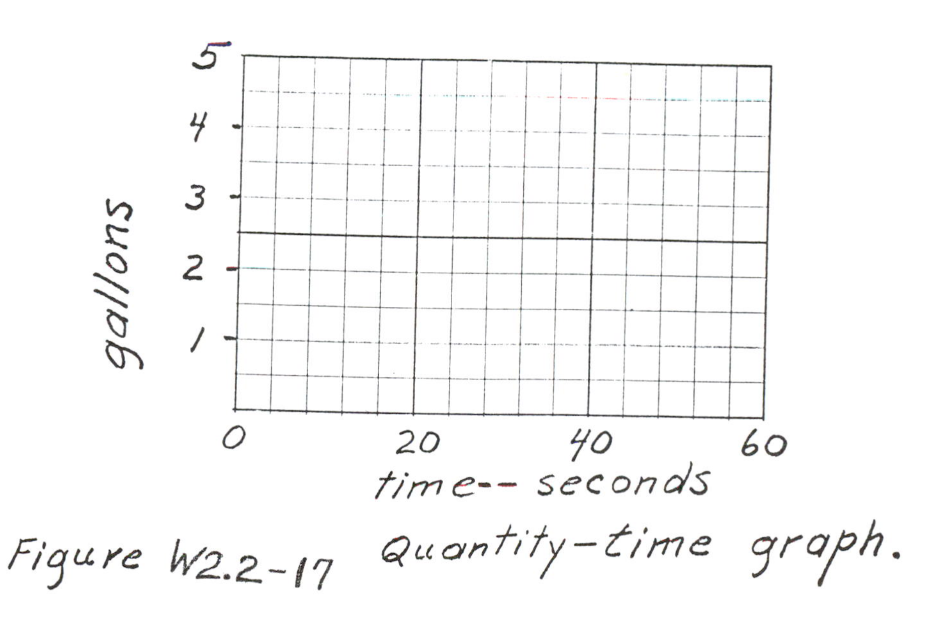

Figure W2.2-17 Quantity-time graph.

* * * * *

* * * * *

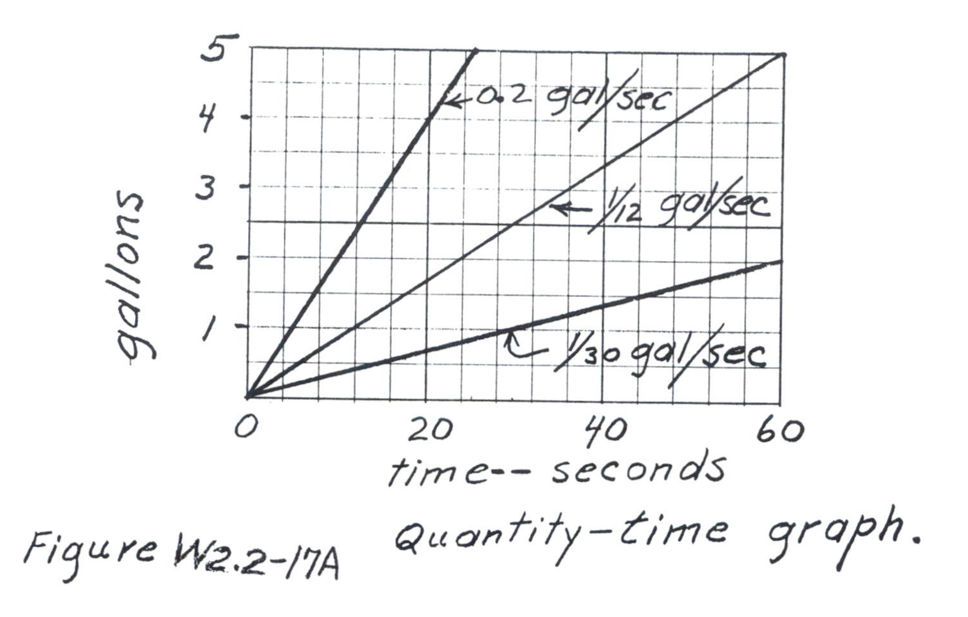

Figure W2.2-17A Quantity-time graph.

18. The "slope" of a line is defined as the ratio of its vertical change divided by the corresponding horizontal distance as shown below.

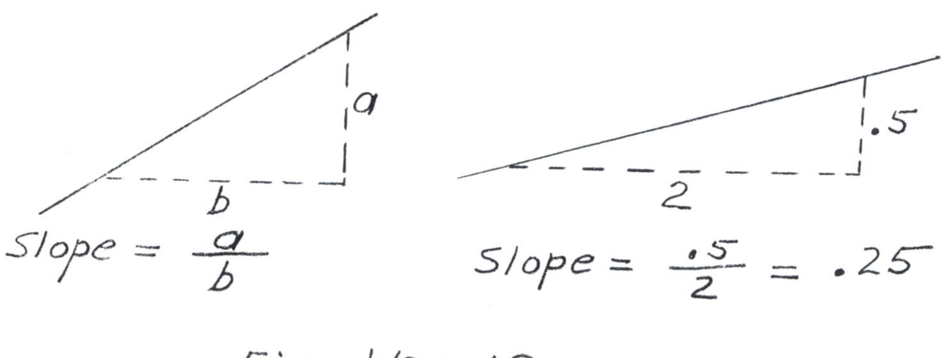

Fig. W2.2-18

(Sec.W2.2)

Draw lines starting at 0, 0 with the slope of 1/10, 1/12, 0.2, and 5/40.

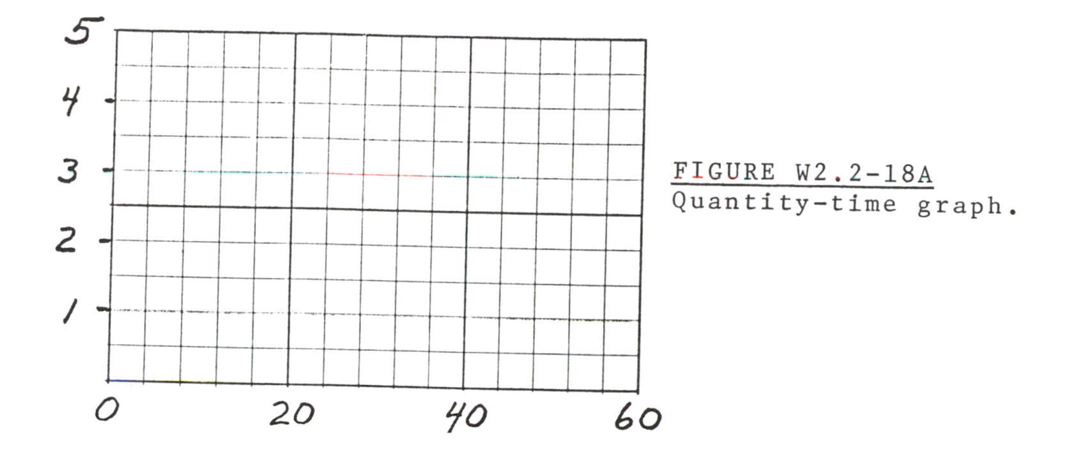

FIGURE W2.2-18A
Quantity-time graph.

* * * * *
* * * * *

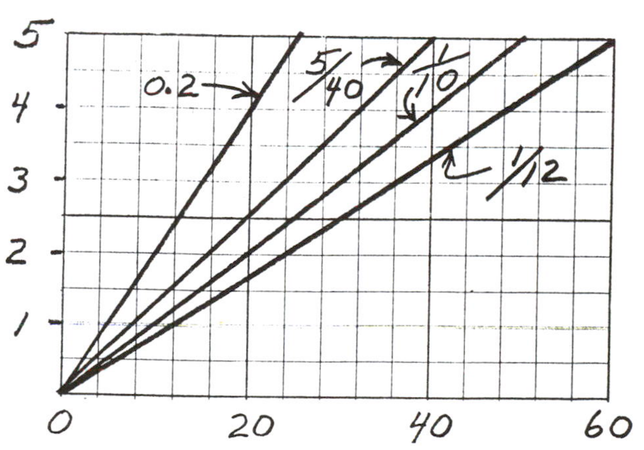

FIGURE W2.2-18B
Quantity-time graph.

19. Refer to Figure W2.2-17A. What is the slope of the line for 1/12 gallon per second? For 0.2 gallon per second?

    * * * * *

    * * * * *

    1/12 gallon/second, 0.2 gallon/second

    (Note that a slope has units of measure as a ratio of the units of the two axes of the graph.)

20. How is the water flow rate FR through the valve related to the slope of the line showing water level through time?

    ______________________.

    * * * * *

    * * * * *

    directly (or proportionally, or the same)

21. On the following graph note that there are two vertical scales. Using the water flow scale, show FR = 0.05 gallon/second. Using the water level scale show the gallons in the tank WL at all points in time, assuming the tank is empty at time = 0.

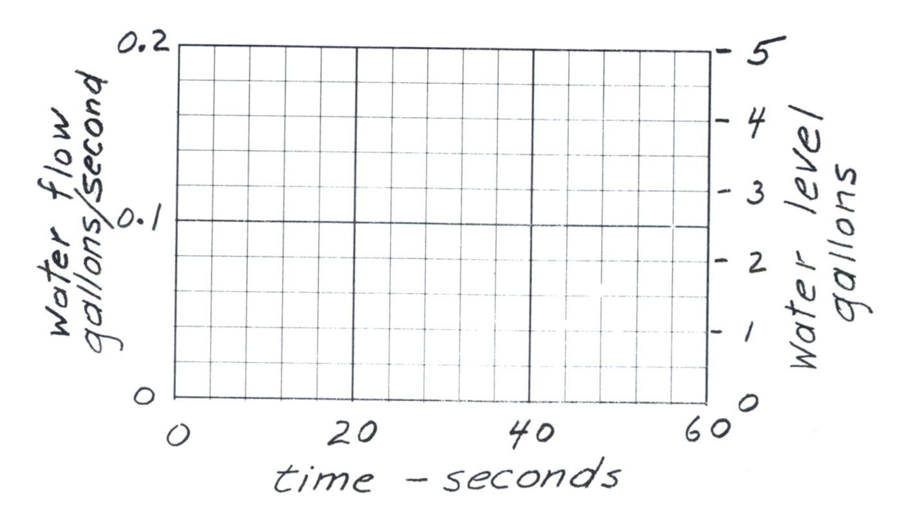

Figure W2.2-21

* * * * *

* * * * *

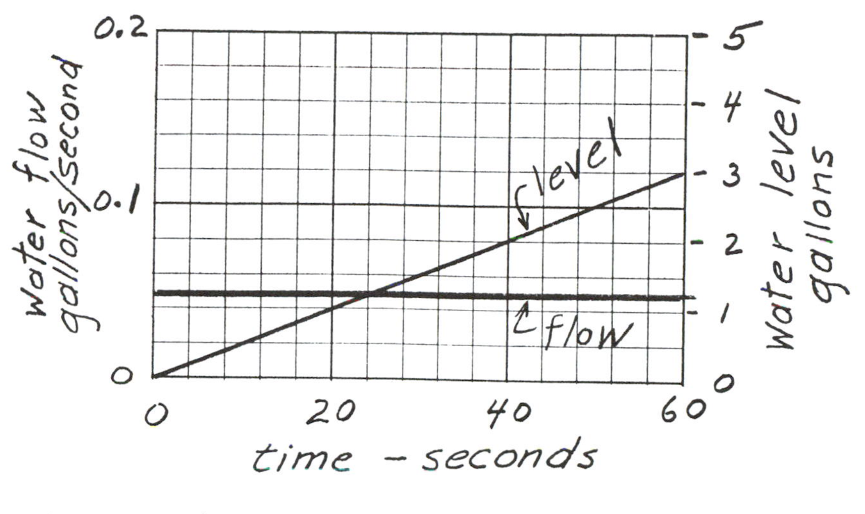

Figure W2.2-21A

22. Show on the graph a water flow rate of 0.1 gallons per second and the corresponding water level, assuming the tank starts empty.

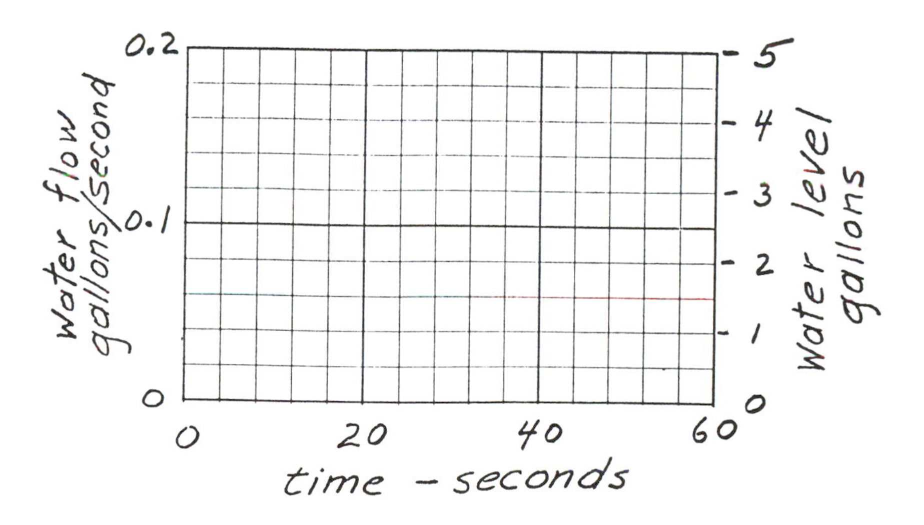

Figure W2.2-22

* * * * *

* * * * *

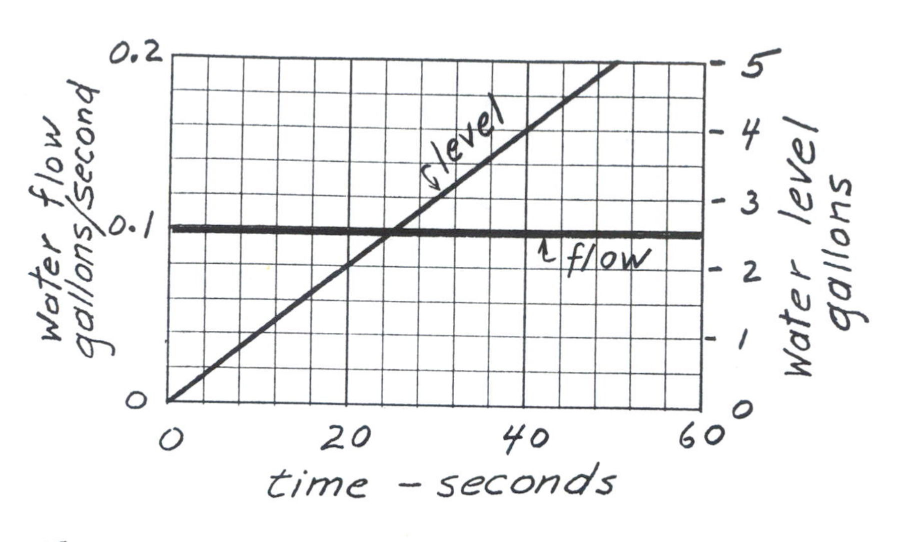

Figure W2.2-22A

23. On the graph show a flow rate FR of 0.1 gallon/second from time = 0 to time = 30 seconds and then a flow rate of 0.05 gallon/second from time = 30 seconds to time = 60 seconds. Plot the corresponding water level WL, starting with an empty tank.

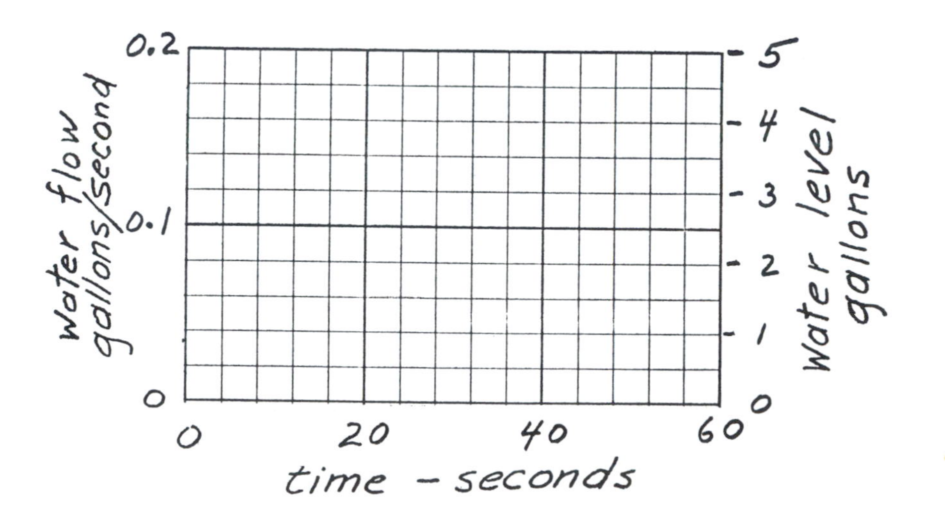

Figure W2.2-23

* * * * *

* * * * *

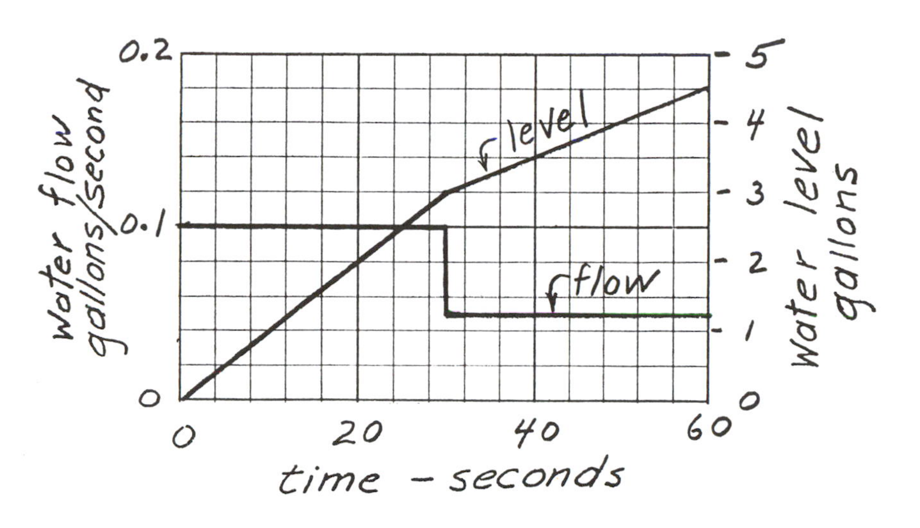

Figure W2.2-23A

24. On the graph, show a flow rate FR of 0.1 gallon per second for 20 seconds, then a rate of 0.075 gallon per second for the interval from 20 to 40 seconds and a rate of 0.05 gallon per second for the last 20 seconds. Starting with WL = 0, plot the water level during the 60 seconds.

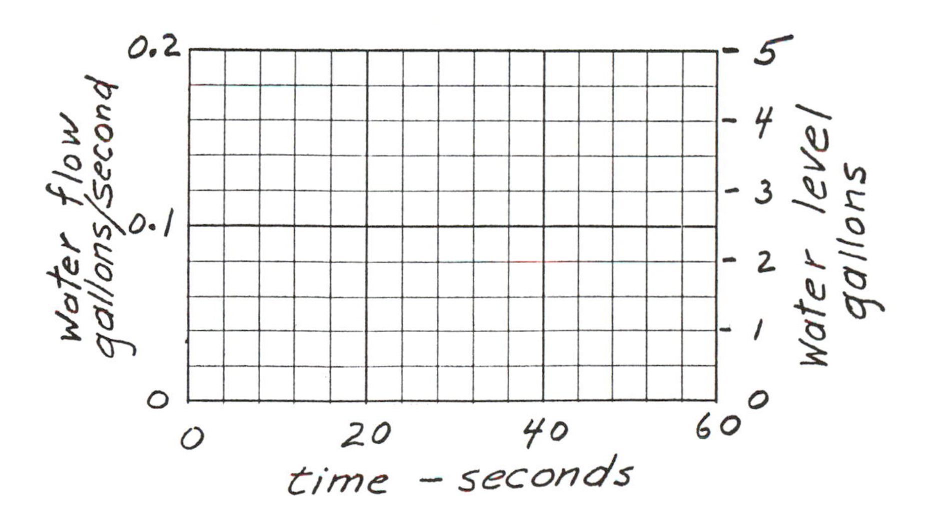

Figure W2.2-24

* * * * *

* * * * *

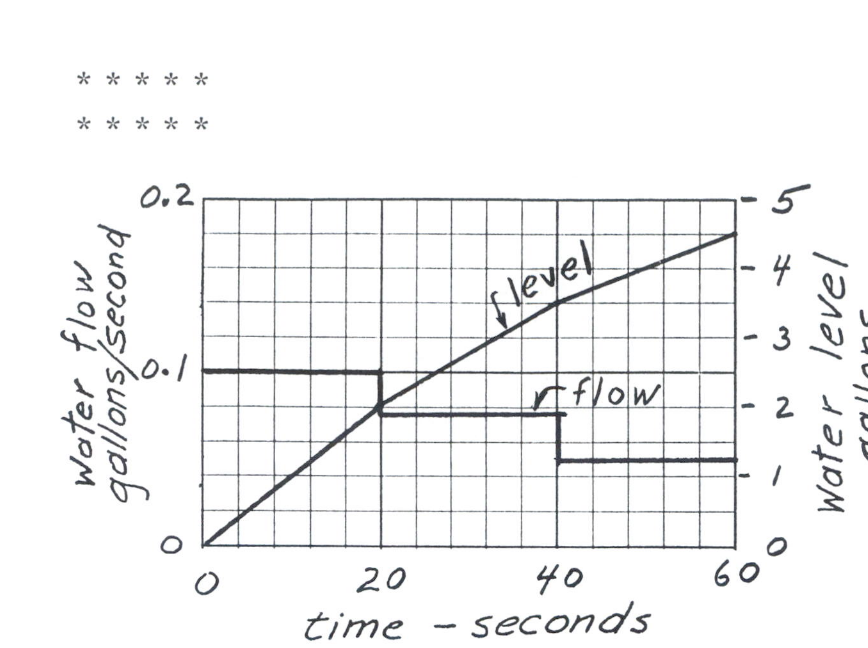

Figure W2.2 -24A

25. Take particular note of how the shapes of the flow rate and the water level are related in Frames 21 through 24. The amplitude (height) of the flow rate determines the __________ of the water level line.

* * * * *

* * * * *

slope

26. In each of the preceding frames asking for a graph of water level versus time (versus means against, that is, water level on one axis against time on the other axis) we have stated that the tank was empty before time = 0. Was this information necessary to permit plotting the line showing the amount of water in the tank? _______________.

* * * * *

* * * * *

yes. (otherwise we do not know the amount of water from which the flow rate FR begins the filling)

(Sec.W2.2)

27. The value of a level variable at the start of operation is called the "initial condition." Show below a water flow rate of 0.05 gallon/second. Starting with 2 gallons in the tank, plot the curve of water versus time. Two gallons is the ________ ________.

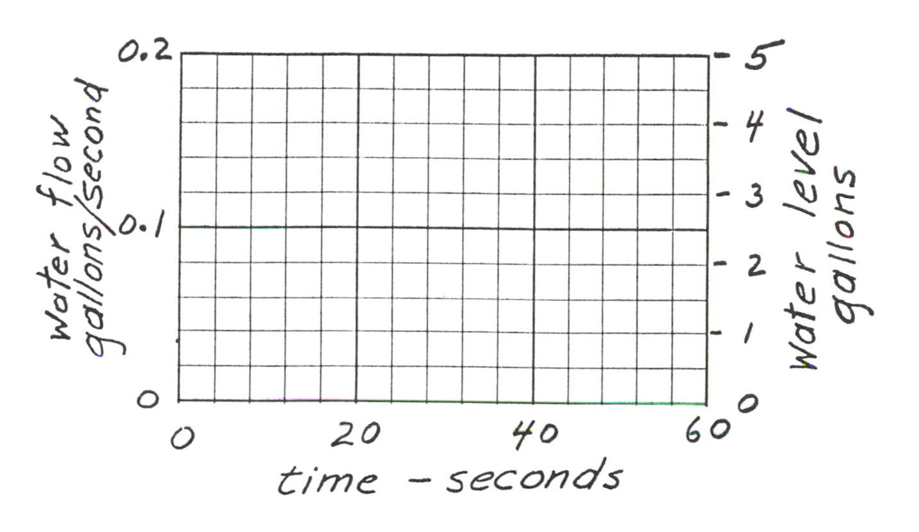

Figure W2.2-27

* * * * *

* * * * *

initial condition

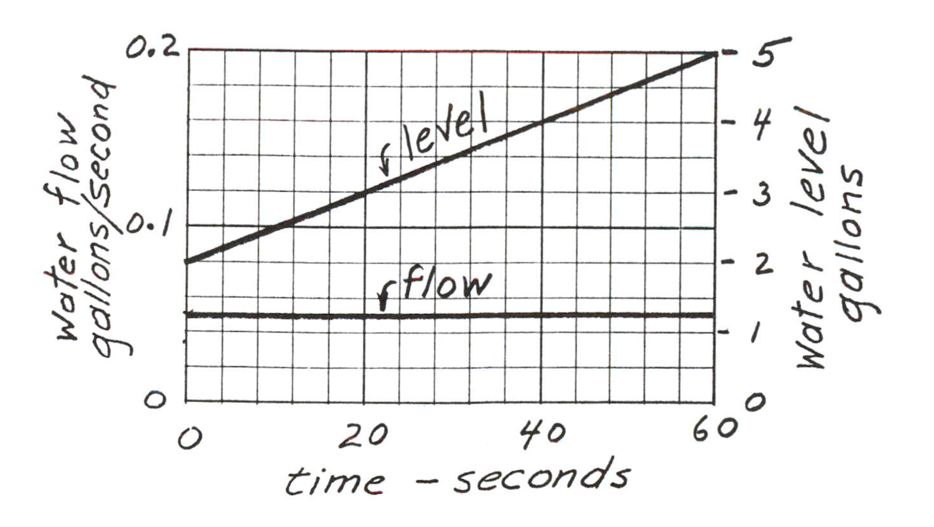

Figure W2.2-27A

28. Compare the slope of the water level line in Figure W2.2-27A with the slope in Figure W2.2-21A. Does the slope depend on the initial condition level of water in the tank?

* * * * *

* * * * *

No. (In each the slope is 1/20 gallon/second. It depends only on the flow rate of water being added.)

29. If we know the initial condition of water level immediately before the start of the time region of interest (that is, before time = 0) does it matter what the water flow rate FR was immediately before the time region of interest?

* * * * *

* * * * *

No. (The level of the system is fully described by the water level. No other information from the past is needed. Whatever the past flow rate FR, its effect is represented in the water level.)

30. If the flow rate of water is zero, the slope of the water level versus time graph will be ____________________.

* * * * *

* * * * *

zero (or horizontal)

31. If the water flow rate were to vary while filling the tank, the greatest slope of the water level versus time curve would occur when the flow rate is at its _______________ value.

* * * * *

* * * * *

maximum (or greatest)

(Sec.W2.2)

32. The change in water level over any specified time interval is found by multiplying the flow rate by the length of the interval. In this figure

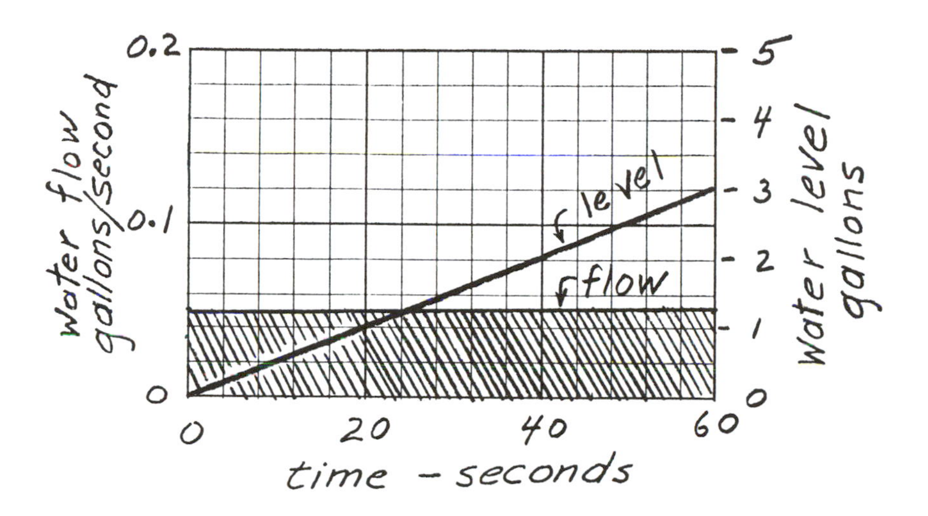

Figure W2.2-32

the final water level is __________ gallons. The change in water level equals the product of flow rate multiplied by the time interval. This product also corresponds to the __________________ under the flow rate line.

* * * * *

* * * * *

3, area. Note that the dimensions of measurement are correct, (gallons/second)(seconds) = gallons.

33. What is the area under the flow rate line in Figure W2.2-23A? ______________. Does this equal the final water level? _________.

* * * * *

* * * * *

4.5 gallons (note that the dimensions, gallons, must be given), yes

34. What is the area under the 0 to 20 second section of the flow rate line in the figure below?

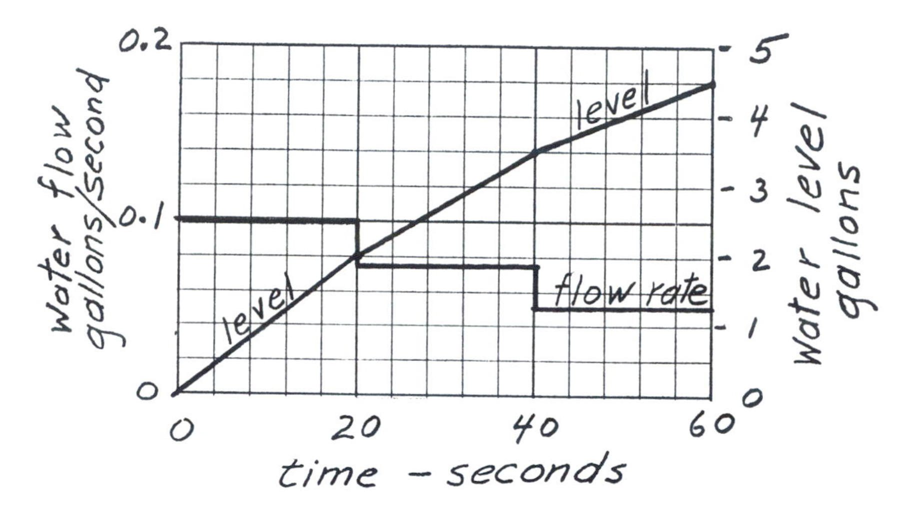

Figure W2.2-34

Does this equal the water level at 20 seconds? ___________. What is the area under the 20 to 40 second section of the flow rate? ____________________. Does this equal the change in water level between 20 and 40 seconds? _________. What is the area under the 40 to 60 second section of the flow rate? _____________. The sum of the three areas is ______________. The final water level is ______________.

* * * * *

* * * * *

2 gallons, yes, 1.5 gallons, yes, 1 gallon, 4.5 gallons, 4.5 gallons

(Sec.W2.2)

35. In the following figure the flow rate declines from

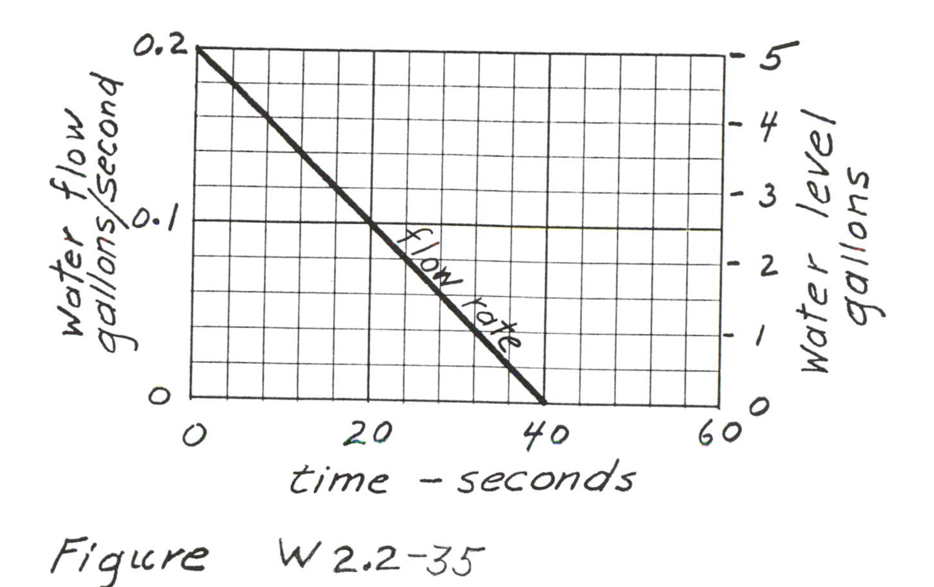

Figure W 2.2-35

0.2 gallon/second to zero in 40 seconds.

The initial slope of the water level curve will be ______________.

The final slope of the water level curve will be ______________.

The area under the flow rate curve is ______________.

If the water level starts at zero, the final height of the water level curve will be ____________ which will be reached at ________ seconds. Sketch the water level curve on the above figure.

* * * * *

* * * * *

0.2 gallons/second; zero gallons/second; 4 gallons; 4 gallons; 40

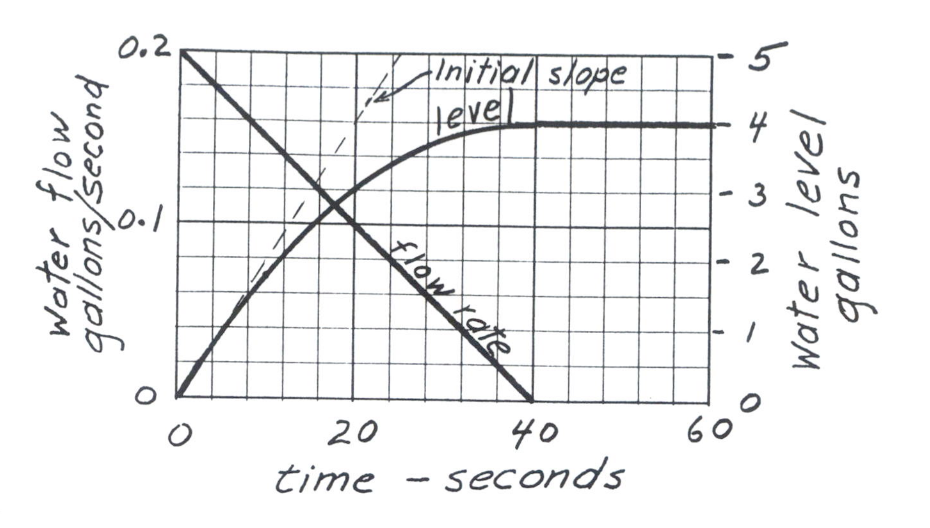

Figure W2.2-35A

36. This figure repeats Figure W2.2-7A, the flow diagram of the tank-valve feedback system, with the float replaced to operate the valve. This reconverts from an open to a feedback system.

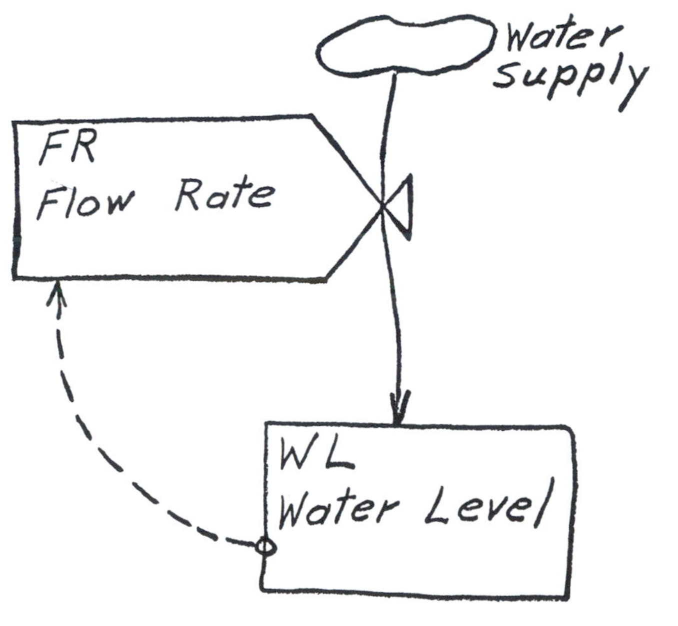

Figure W2.2-36 Flow diagram of float valve, and tank system.

The flow rate is determined by the mechanical design of the system and by the ____________ ___________.

* * * * *

* * * * *

water level

37. Suppose that the flow rate is 0.2 gallon per second when the tank is empty and declines proportionally to zero when the tank contains 4 gallons. Show the flow rate versus water level on the graph. Note that this is not flow rate versus time but instead shows how flow rate depends on water level.

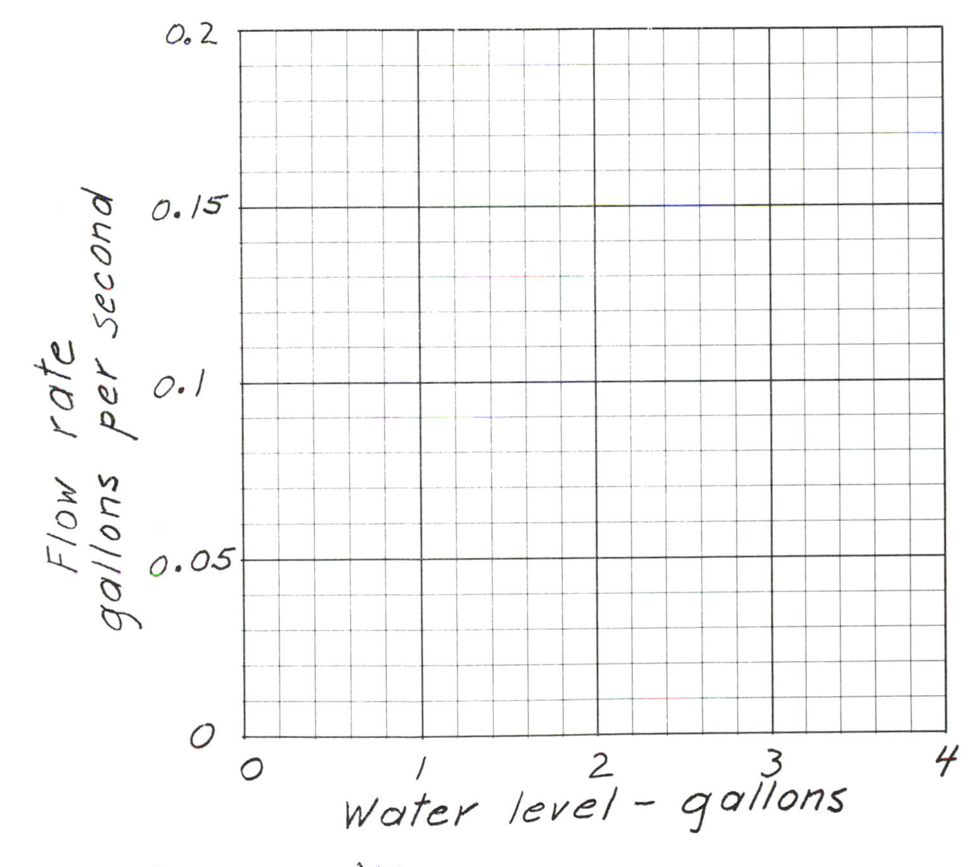

Figure W2.2-37

* * * * *

* * * * *

* * * * *

* * * * *

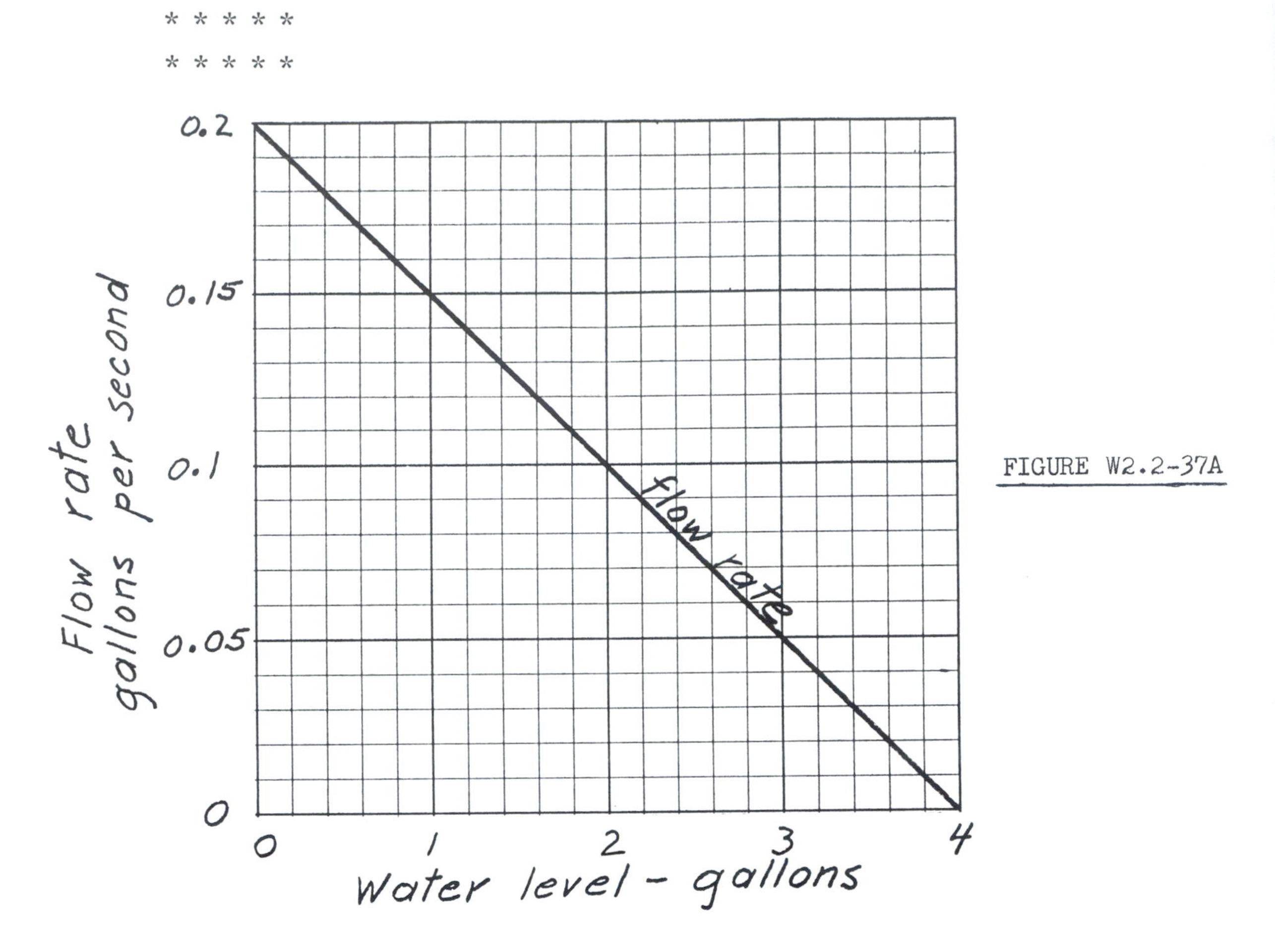

FIGURE W2.2-37A

38. Functional relationships such as shown in Figure W2.2-37A can often be expressed more conveniently as an equation from which the desired value can be computed. Here we would expect to know the water level and would want to determine the corresponding flow rate. By observing that the flow is zero when the water is 4 gallons and a maximum when water level is zero, we can see that the following form of equation is appropriate:

$$FR = \frac{1}{T}(4\text{-}WL) \qquad \text{Eq. W2.2-38}$$

To fit Fig. W2.2-37A, T=______which we can see by letting WL=0.

* * * * *

* * * * *

20 seconds (Note that 4 and WL are measured in gallons, FR is gallons/second, T must be in seconds to produce an equation with the same dimensions on each side.)

39. The equation for flow rate is then

$$FR = \frac{1}{20}(4\text{-}WL) \qquad \text{Eq. W2.2-39}$$

What is the flow rate when the water level is 2.5 gallons?

__________________________.

* * * * *

* * * * *

0.075 gallon/second

40. Does one get the same value from Figure W2.2-37A? __________.

* * * * *

* * * * *

Yes. 0.075 gallon/second

41. What is the significance of T in Equation W2.2-38 which has the value "20 seconds"? It is the time required to fill the tank to the shut off level if the flow were to continue at the ________________ rate.

* * * * *

* * * * *

initial (or beginning, or 0.2 gallon/second)

42. The flow diagram can now be modified to show all the numerical values which are used in computing the flow rate FR.

$$FR = \frac{1}{20}(4\text{-}WL) \qquad \text{Eq. W2.2-42}$$

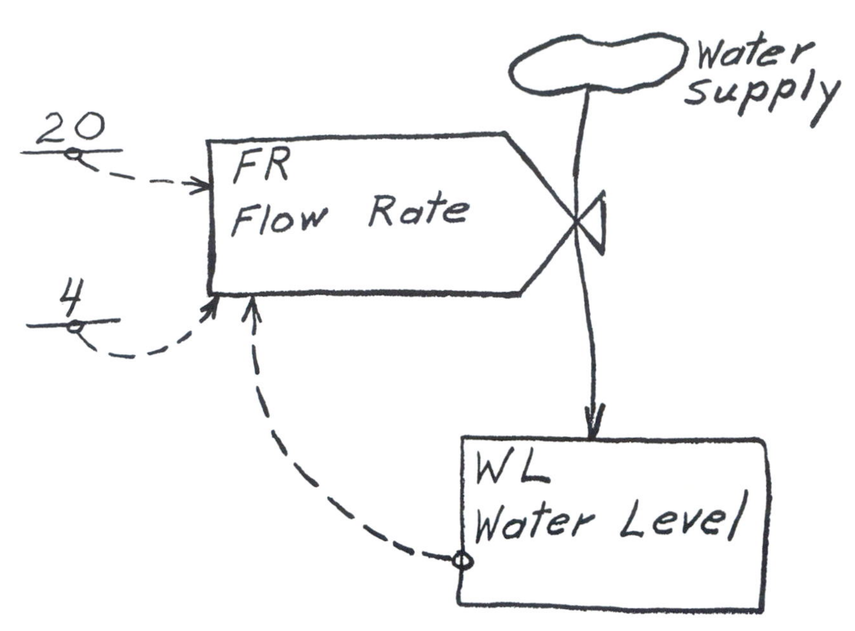

Figure W2.2-42 Flow diagram of float valve and tank system.

In the decision (here the flow rate) of a negative feedback loop, there is always a goal toward which the decision is attempting to move the system. Here the goal is to keep the water level at ________________ gallons.

* * * * *

* * * * *

4

43. We can compute, step-by-step, through time the water level by finding flow rate from Figure W2.2-37A or from Equation 2.2-42, assuming flow rate remains constant for a brief period, calculating the change in water level in the brief period, using the new water level to find a new flow rate, and continuing. If the tank is empty what is the flow rate? ______________.

 * * * * *

 * * * * *

 0.2 gallon per second (Be sure to give the dimensions of measure.)

44. How much water will be in the tank at the end of 4 seconds, assuming the flow rate is constant at 0.2 gallon per second?

 ______________.

 * * * * *

 * * * * *

 0.8 gallon

45. Using a water level of 0.8 gallon, what is the flow rate from Figure W2.2-37A? ______________.

 From Equation W2.2-42? ______________.

 How much water will be added during the next four seconds?

 ______________.

 How much water is in the tank after the first 8 seconds?

 ______________.

 * * * * *

 * * * * *

 0.16 gallon per second, 0.16 gallon per second, 0.64 gallon, 1.44 gallons

46. The computation can be organized on a work sheet as follows:

Seconds	Change in level (gallons) CL	Water level (gallons) WL	Flow rate (gallons/sec) FR
0	XXX	0	0.200
4	0.800	0.800	0.160
8	0.640	1.440	
12			
16			
20			
24			
28			
32			
36			
40			
44			
48			
52			
56			
60			

Use the water level of 1.44 gallons to obtain the flow rate at time 8 seconds from Figure W2.2-37A or Equation W2.2-42 and enter the flow rate in the last column.

At 12 seconds, the change in level is the previous flow rate multiplied by 4 seconds. The new level is the old level plus the change in level. Complete the table, plotting the curves of water level and flow rate as each new set of points is obtained. (Be sure to plot as the computation progresses, not afterward.) It is important in learning the material that the student actually do these computations and the plotting. As the computation is done, pay attention to the way in which the variables are changing and why.

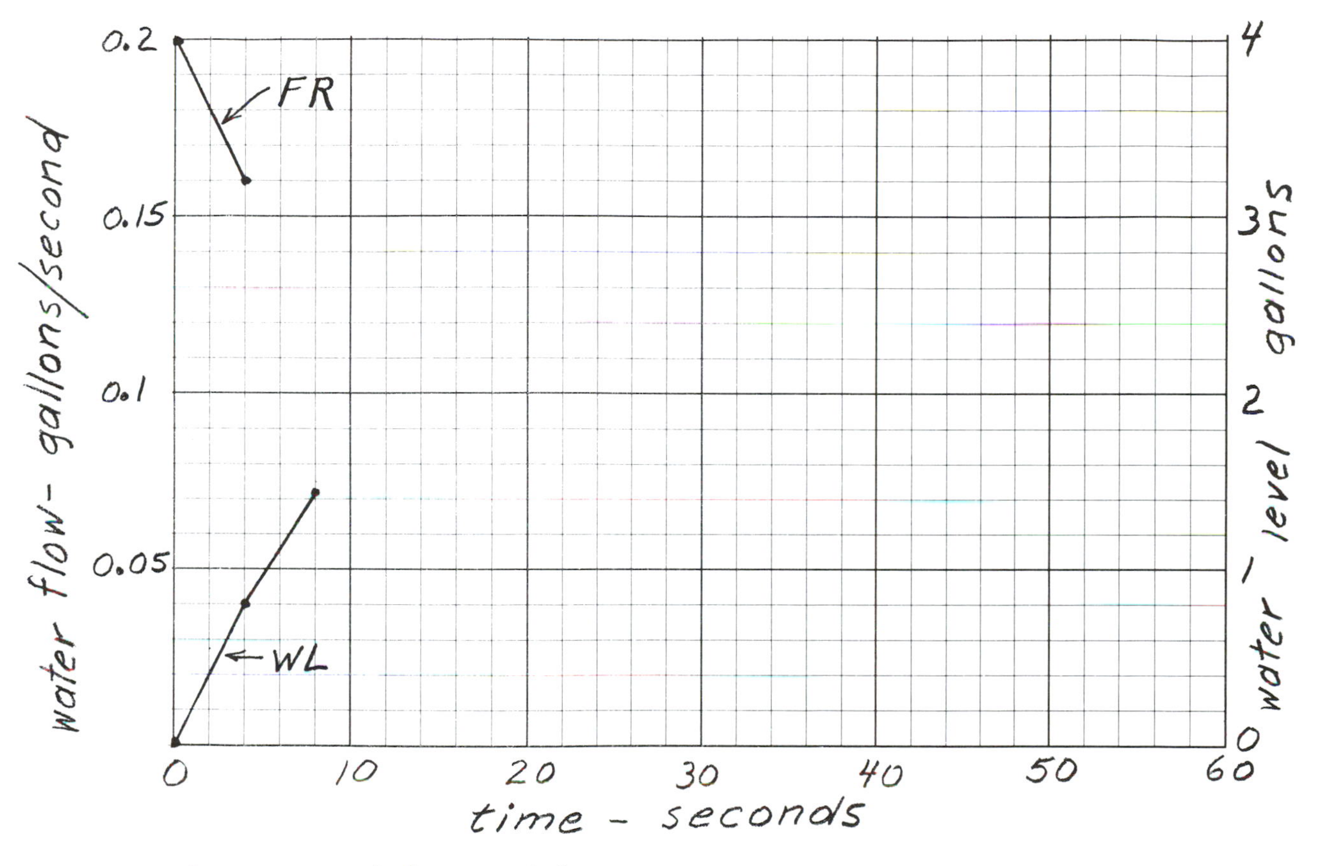

Figure W2.2-46

* * * * *

* * * * *

* * * * *

* * * * *

(Answer to Frame 46)

TIME	CL	WL	FR
0	.800	.000	.200
4	.800	.800	.160
8	.640	1.440	.128
12	.512	1.952	.102
16	.410	2.362	.082
20	.328	2.689	.066
24	.262	2.951	.052
28	.210	3.161	.042
32	.168	3.329	.034
36	.134	3.463	.027
40	.107	3.571	.021
44	.086	3.656	.017
48	.069	3.725	.014
52	.055	3.780	.011
56	.044	3.824	.009
60	.035	3.859	.007

Mod. B26

Table W2.2-46

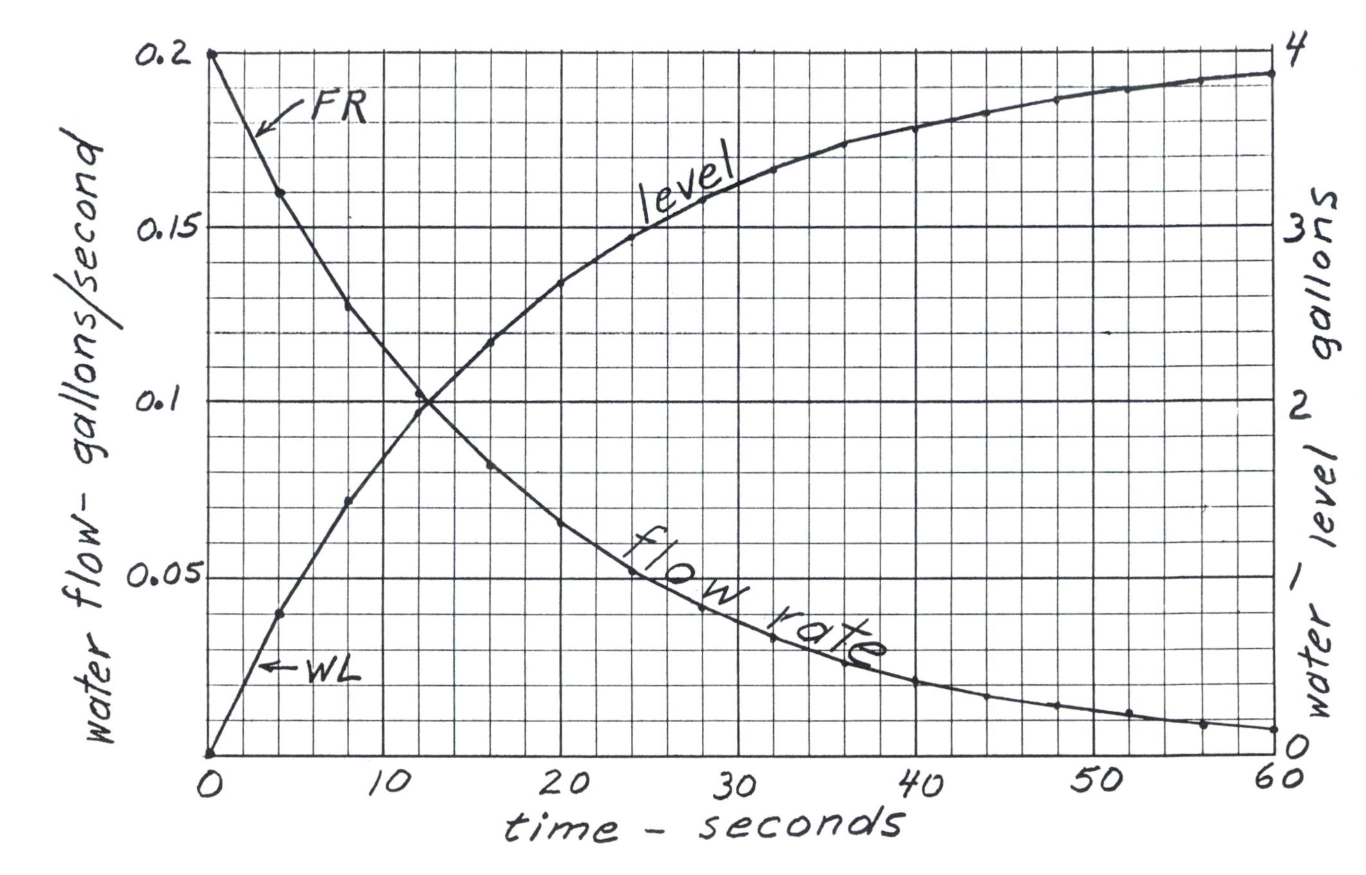

Figure W2.2-46A

47. Radio-active material spontaneously disintegrates at a rate that depends on the amount of material that remains. That is, a given fraction of the remaining material disintegrates each day. Depending on the particular kind of atom, the rate of decay can be so fast that it is difficult to observe the material before it disintegrates or so slow that it takes thousands of years to lose half of the material. Assume we have one milligram of material which decays 5% per day.

    a. Draw the flow diagram for the pertinent feedback system.

    b. Draw the graph of decay rate in milligrams per day versus the material in milligrams up to one milligram.

    c. Corresponding to the water level goal of 4 gallons in Figure W2.2-42 and Equation W2.2-42, what goal is the radio-active material moving toward?

    d. Write the equation for decay rate in terms of amount of material.

    e. Use a solution interval of five days and compute the decay rate and amount of material as time progresses until the computational procedure and the nature of the results become clear.

W2.3 Second-Order, Negative-Feedback Loop

Delays exist throughout all systems. Information is delayed because it can not be gathered, analyzed, or transmitted instantaneously. Long delays are often encountered before a person recognizes that a situation has changed even if information about a change is at hand. Action resulting from decisions often requires long periods to become effective. Delays have a major influence on the dynamic behavior of feedback systems and for this reason it is essential that we be able to represent delays in formulating the structure of a system. One of the simplest and most useful delay representations is the "first-order, exponential" delay as exemplified by the goods on order GO and the receiving rate RR in text Figure 2.3a. This ordering delay section of the system is worth examining separately because it shows behavior that will recur throughout our study of systems.

1. In this figure is reproduced only the ordering delay section from the text.

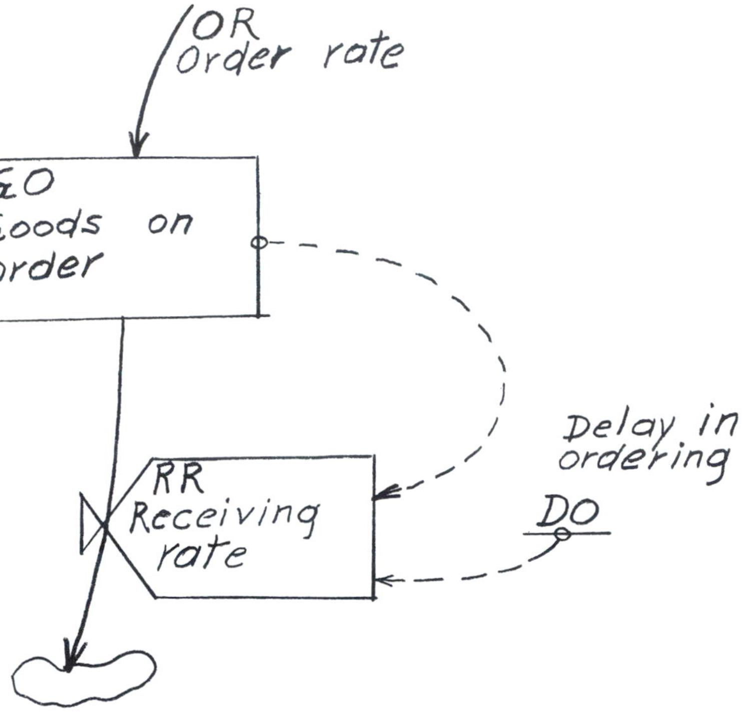

Figure W2.3-1 First-order, exponential delay.

Goods on order are increased by the order rate and depleted by the receiving rate. As given in text Equation 2.3-1 the receiving rate is determined by

$$RR = \frac{GO}{DO} \qquad \text{W2.3-1}$$

$$DO = 10$$

RR--Receiving rate (units/week)
GO--Goods on order (units)
DO--Delay in ordering (weeks)

Can the arrangement in the above figure, taken by itself, be construed as a feedback system? ____________________.

* * * * *

* * * * *

Yes. Goods on order GO determines receiving rate RR which influences goods on order.

2. Is it a negative feedback loop? ________________.

* * * * *

* * * * *

Yes. An <u>increasing</u> value of goods on order <u>increases</u> receiving rate which <u>reduces</u> goods on order. There is a reversal of sign or direction of action in traversing the loop.

3. This negative loop is acting to control what variable?

____________________.

* * * * *

* * * * *

Goods on order.

4. The feedback loop has a goal of moving the value of goods on order toward _______________.

* * * * *

* * * * *

zero. (Receiving rate will remove units from goods on order until the goods on order have vanished.)

5. Suppose that the goods on order are initially zero and that an order rate of 800 units/week suddenly begins at the 4th week and continues as shown on this graph. Supply appropriate scales and sketch the curves you would expect for goods on order GO and receiving rate RR. Do this by inspecting Figure W2.3-1 and by thinking about the process, not by computation.

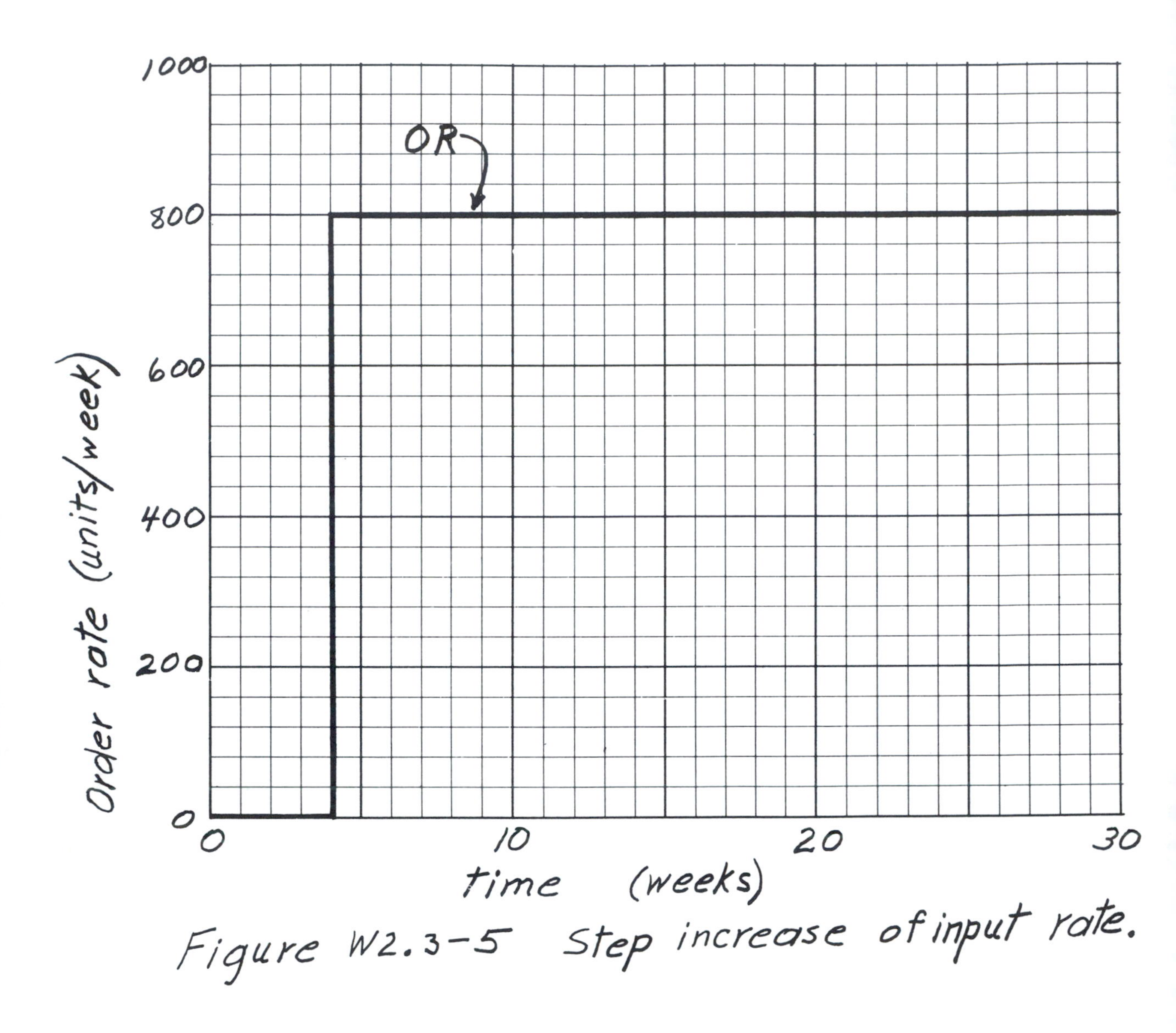

Figure W2.3-5 Step increase of input rate.

6. Now check your estimated curves on the preceding figure by computing the following table and plotting the results (as they are computed) on the above figure for comparison with the estimates.

Time (weeks)	Change in goods on order (units)	Goods on order (units)	Order rate (units/ week)	Receiving rate (units/ week)
	CG	GO	OR	RR
0		0	0	0
2	0	0	0	0
4	0	0	800	0
6	1600	1600	800	160
8	1280		800	
10			800	
12			800	
14			800	
16			800	
18			800	
20			800	

* * * * *

* * * * *

TIME	CG	GØ	ØR	RR
E+00	E+00	E+00	E+00	E+00
.000	0.	0.	0.	0.
2.000	0.	0.	0.	0.
4.000	0.	0.	800.	0.
6.000	1600.	1600.	800.	160.
8.000	1280.	2880.	800.	288.
10.000	1024.	3904.	800.	390.
12.000	819.	4723.	800.	472.
14.000	655.	5379.	800.	538.
16.000	524.	5903.	800.	590.
18.000	419.	6322.	800.	632.
20.000	336.	6658.	800.	666.
22.000	268.	6926.	800.	693.
24.000	215.	7141.	800.	714.
26.000	172.	7313.	800.	731.
28.000	137.	7450.	800.	745.
30.000	110.	7560.	800.	756.
32.000	88.	7648.	800.	765.
34.000	70.	7719.	800.	772.
36.000	56.	7775.	800.	777.
38.000	45.	7820.	800.	782.
40.000	36.	7856.	800.	786.

B821

B27-1

Table W2.3-6

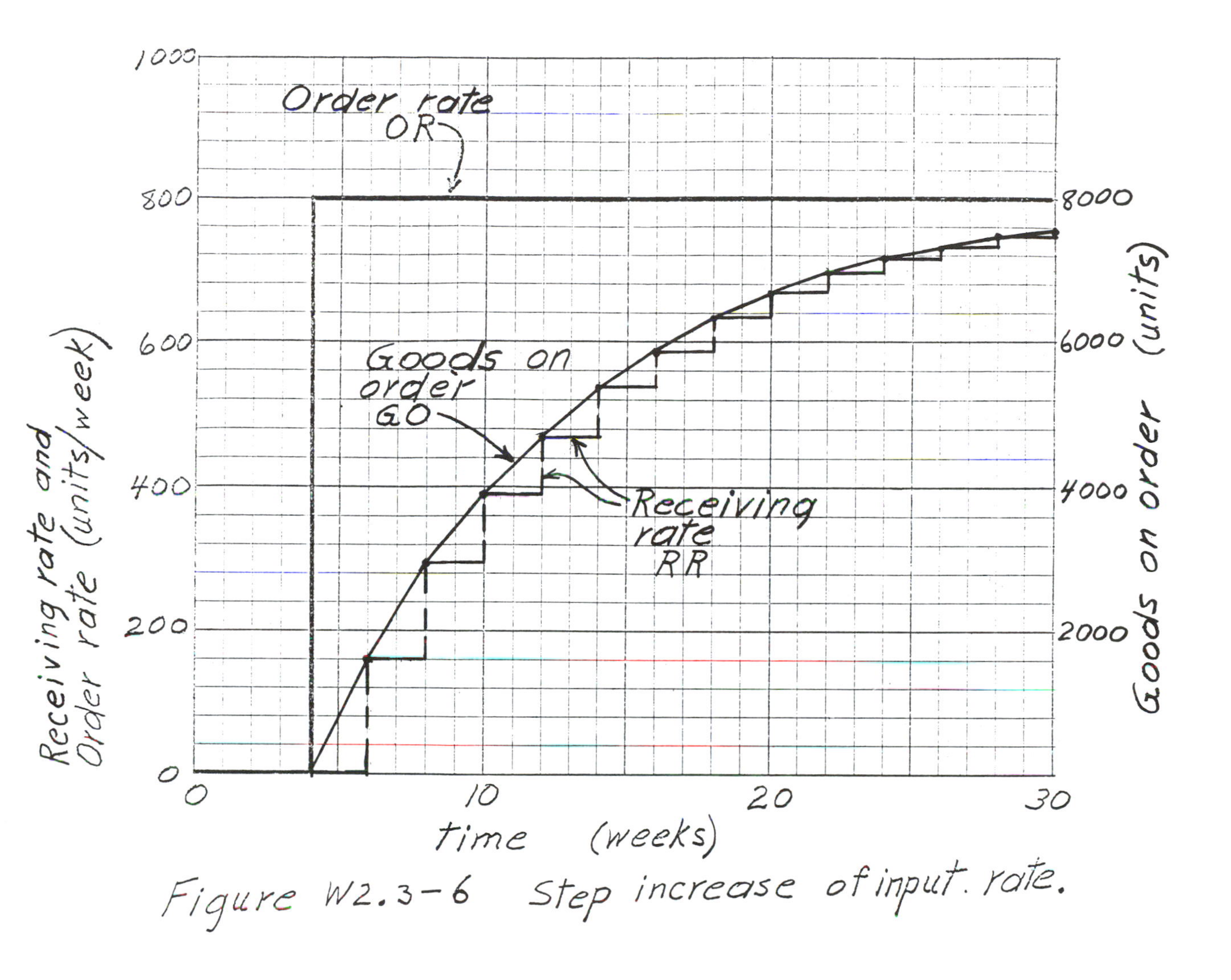

Figure W2.3-6 Step increase of input rate.

In the above figure, it is only coincidental that the steps in the RR curve fall on the GO curve. They are plotted to left and right scales respectively and a different choice of scales would have caused the curves to separate. Note how the output receiving rate RR of the delay rises slowly and represents a lag behind the time of the rise in the input OR. Note also how this particular form of delay "rounds off the corners" of the input order rate so that the output receiving rate does not respond with the same suddenness as the input. Also note how the goods on order curve has been drawn as a continuous curve of straight-line segments while the receiving rate has been drawn as a staircase sequence of receiving rates which are constant for the interval between computations and then suddenly change to a new value. This shows one of the differences between level variables and action variables as they are being computed and used in this book. We assume the rates of flow (or action) are constant between successive computations and change to new values at the time of recomputation. The level variables, on the other hand, are being acted on by the action variables throughout the interval between computations and so are changing during the interval.

This distinction between the continuous change in the level variable and the step change in the action variable is shown here only to make clearer the fundamental difference between the two types of variables. The difference is not significant in evaluating system behavior because the solution interval will always be short enough that the step variation in action variables can be ignored, as it will be in most later time graphs.

7. As before, start with zero goods on order and assume that at the end of four weeks the order rate begins a "ramp" increase at the rate of 40 units/week/week as shown. Sketch the curves you would expect for goods on order and receiving rate. Do by inspection, not computation.

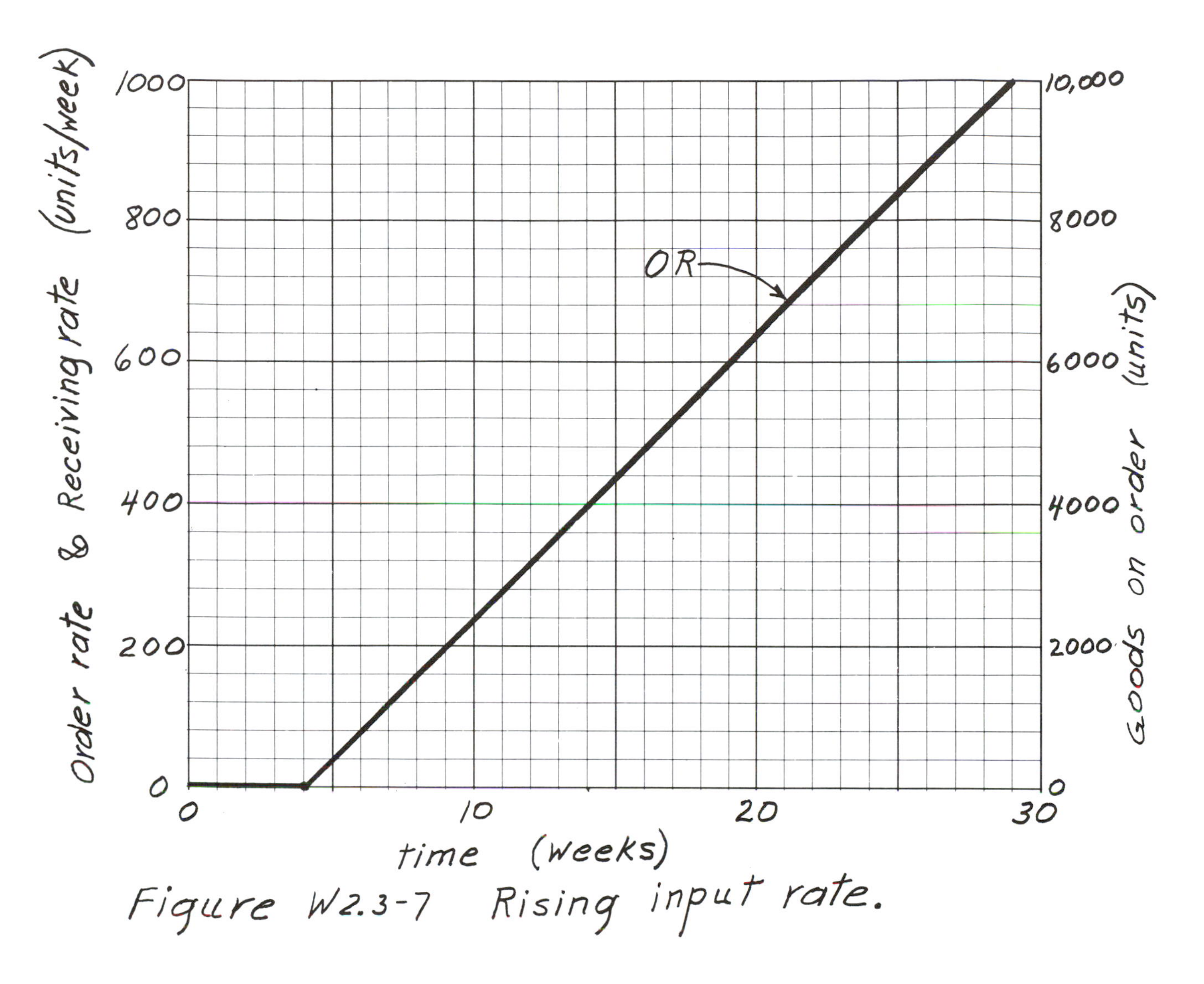

Figure W2.3-7 Rising input rate.

8. Now verify your sketched estimate of goods on order and receiving rate by the following computation. Compute only as far as necessary to be sure of the procedure and the nature of the results. Plot on the preceding figure as you compute.

Time (weeks)	Change in goods on order (units) CG	Goods on order (units) GO	Order rate (units/ week) OR	Receiving rate (units/ week) RR
0		0	0	0
2	0	0	0	0
4	0	0	0	0
6	0	0	80	0
8	160		160	
10			240	
12			320	
14			400	
16			480	
18			560	
20			640	
22			720	
24			800	
26			880	
28			960	
30			1040	

* * * * *

* * * * *

TIME E+00	CG E+00	GØ E+00	ØR E+00	RR E+00
.000	0.	0.	0.	0.
2.000	0.	0.	0.	0.
4.000	0.	0.	0.	0.
6.000	0.	0.	80.	0.
8.000	160.	160.	160.	16.
10.000	288.	448.	240.	45.
12.000	390.	838.	320.	84.
14.000	472.	1311.	400.	131.
16.000	538.	1849.	480.	185.
18.000	590.	2439.	560.	244.
20.000	632.	3071.	640.	307.
22.000	666.	3737.	720.	374.
24.000	693.	4429.	800.	443.
26.000	714.	5144.	880.	514.
28.000	731.	5875.	960.	587.
30.000	745.	6620.	1040.	662.
32.000	756.	7376.	1120.	738.
34.000	765.	8141.	1200.	814.
36.000	772.	8913.	1280.	891.
38.000	777.	9690.	1360.	969.
40.000	782.	10472.	1440.	1047.

B822 B27-2

Table W2.3-8

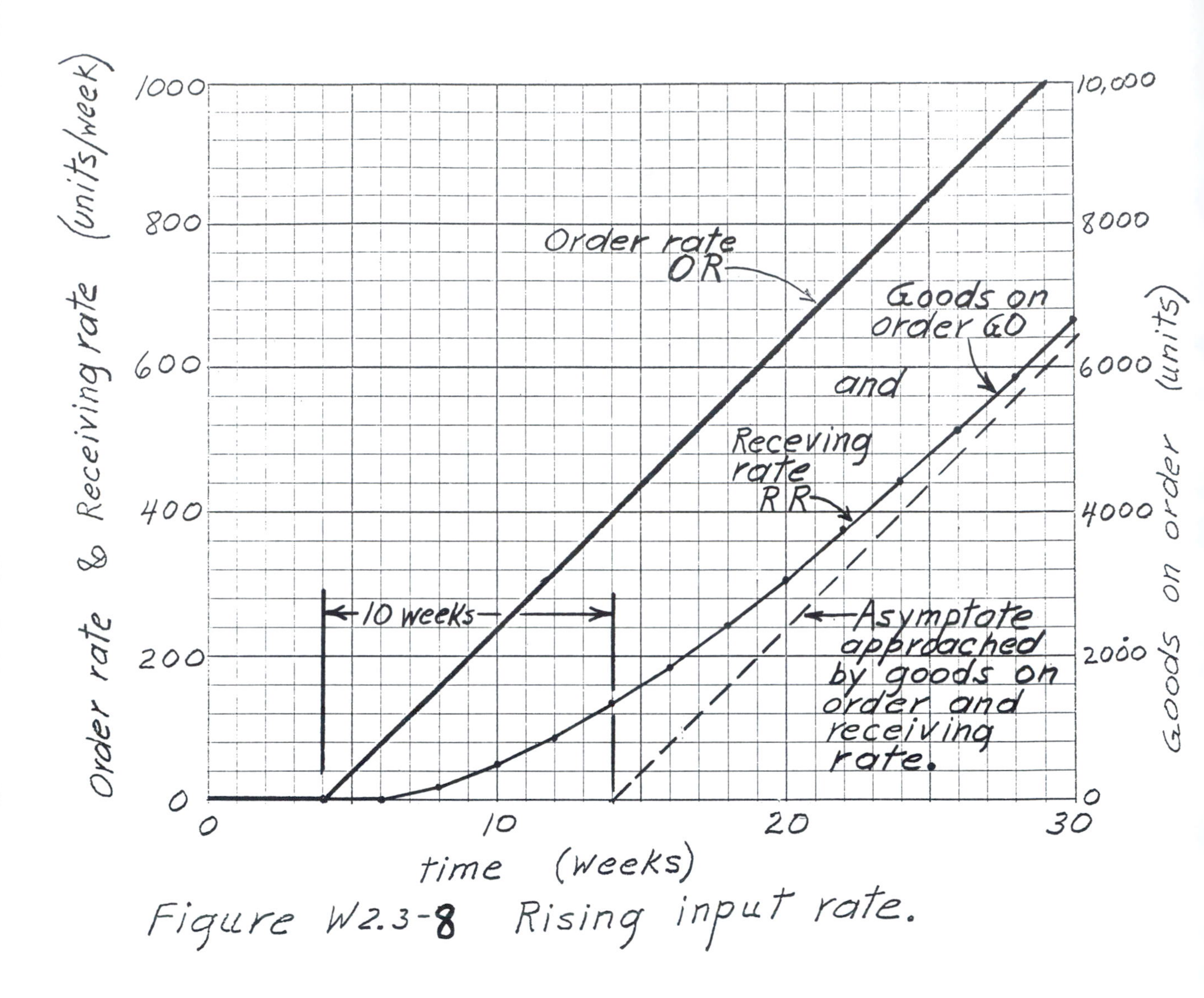

Figure W2.3-8 Rising input rate.

Note how the rising receiving rate is delayed by 10 weeks (the value of DO) after the order rate. As before, there is an initial "transient" interval while the output receiving rate is adjusting to the new conditions imposed by the input ordering rate. The receiving rate is here plotted as a smooth curve rather than the "stair steps" representing the actual computation. Receiving rate RR and goods on order GO are variables of different dimensions and are plotted to different scales; it is only coincidental, due to choice of scales that they are superimposed.

9. It is necessary to have values for only the (level/action) variables as a basis for computing successive conditions of the system.

* * * * *

* * * * *

level

10. If the ______________ variables are known the ______________ variables can be computed from them, if the governing equations and constants are available.

* * * * *

* * * * *

level, action (or rate)

11. Examine the flow diagram and the dynamic behavior in text Figures 2.3a and 2.3b. The value of the delay in ordering DO was 10 weeks. Suppose instead that DO is changed to 5 weeks. The oscillation in the system will die out (more/less) quickly.

* * * * *

* * * * *

more. This might reasonably be assumed from the fact that there is no oscillation when the delay is reduced all the way to zero as in text Figures 2.2a and 2.2c.

Compare the following figure with Figure 2.3b in the text to see how reducing the delay in ordering from 10 to 5 weeks has caused the oscillation to attenuate more quickly.

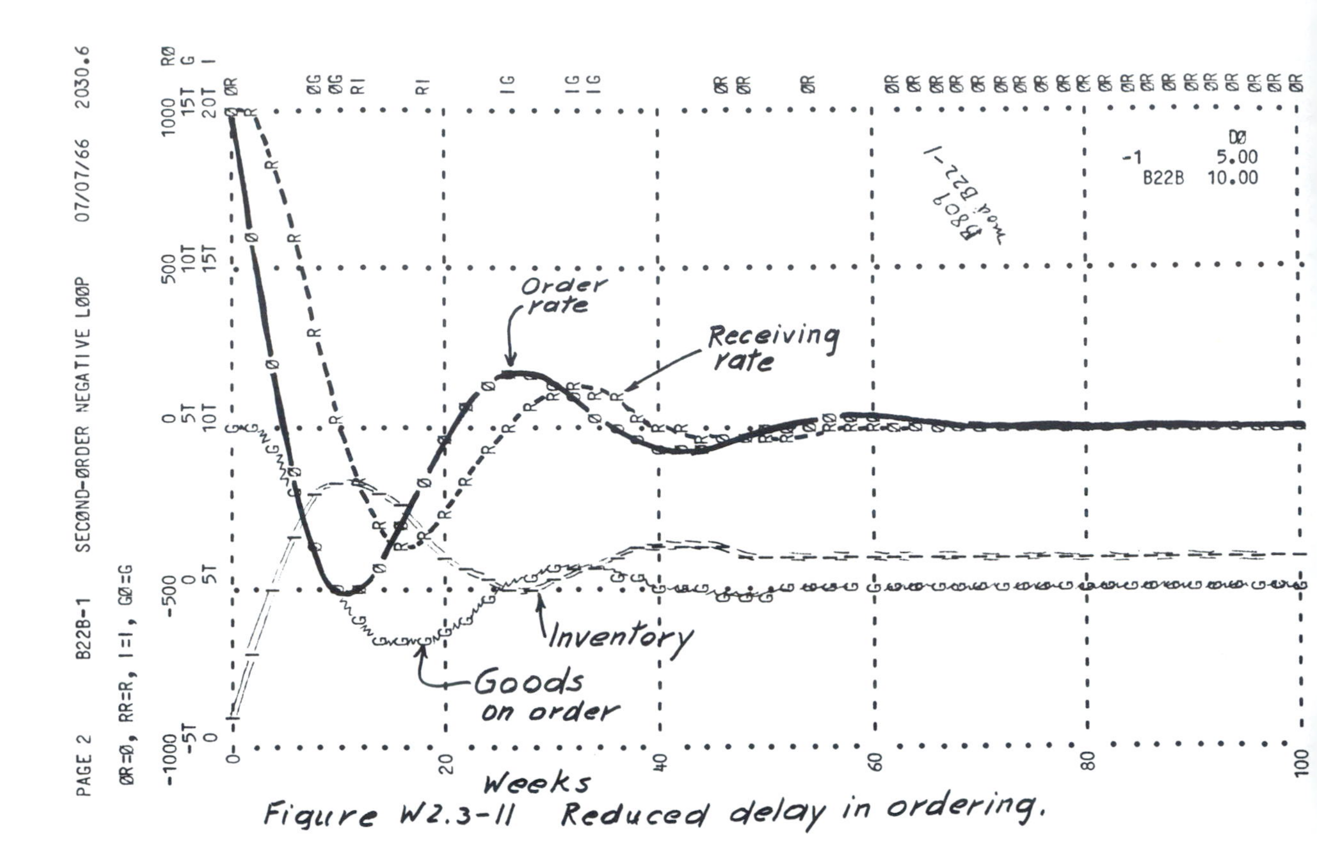

Figure W2.3-11 Reduced delay in ordering.

12. Now suppose that the delay in ordering DO is increased from the original 10 weeks to 20 weeks. The oscillation in the system will die out (more/less) quickly.

* * * * *

* * * * *

less

13. With DO equal to 20 weeks, the period, that is, the time between peaks of the oscillations, will be (longer/shorter) than for DO equal 10 weeks. In text Figure 2.3b the period is approximately ________________ weeks.

* * * * *

* * * * *

longer (as might be inferred by the fact that corrective action takes longer to traverse the feedback loop. Also, a shorter delay in Figure W2.3-11 above produced a shorter period than in text Figure 2.3b.)

44

14. The following figure shows the behavior of the system of Section 2.3 of the text except that DO = 20 weeks. Compare carefully with text Figure 2.3b. The period has increased from 44 to about __________ weeks.

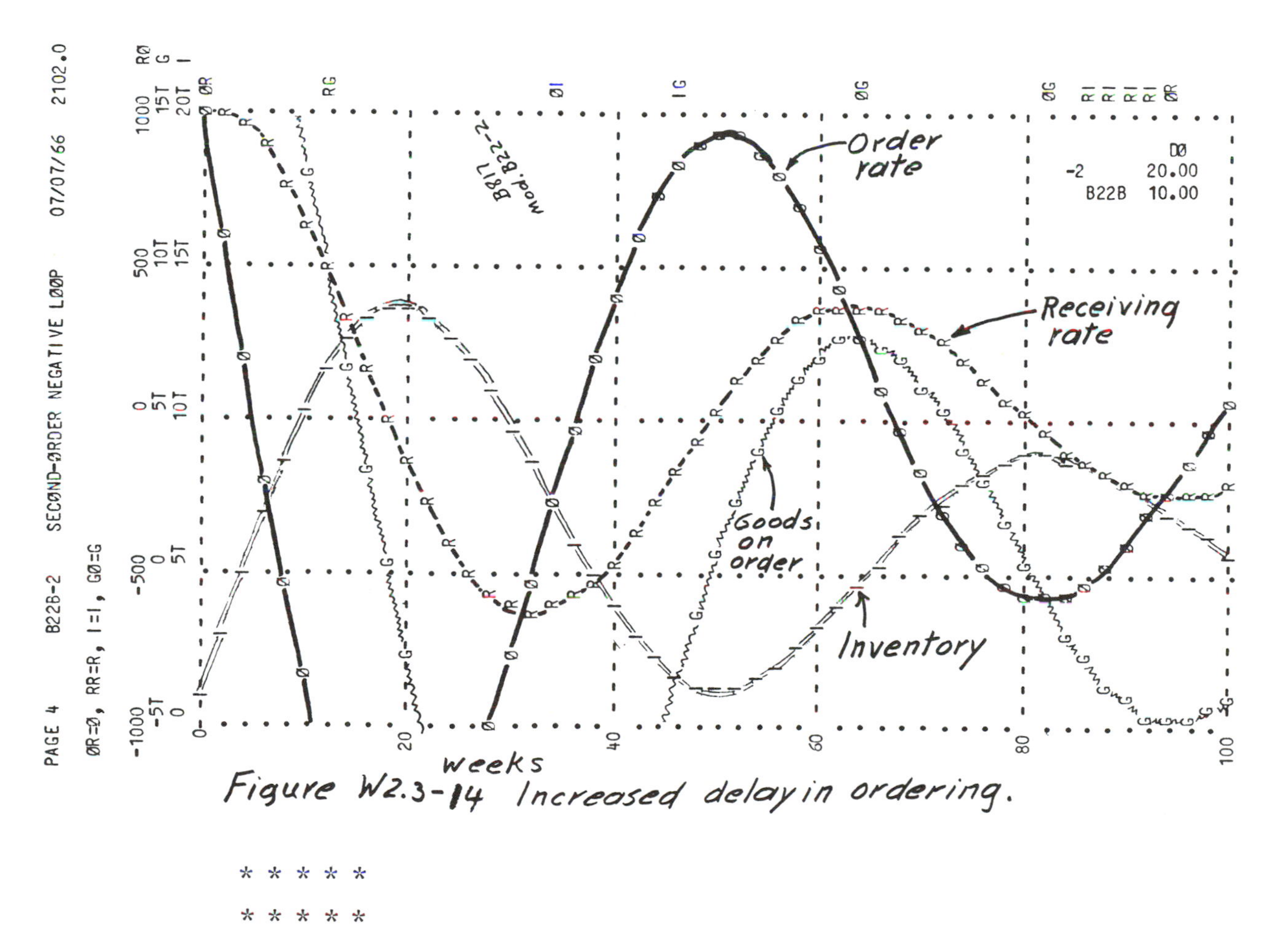

Figure W2.3-14 Increased delay in ordering.

* * * * *

* * * * *

62 as measured between the peaks of the inventory curve.

W2.4 <u>Positive-Feedback</u> <u>Loop</u>

1. In the system of text Figures 2.4a and 2.4c, the number of salesmen would grow (more/less) rapidly if SDT in Equation 2.4-1 were changed from 50 weeks to 100 weeks.

   * * * * *

   * * * * *

   less (For any number of salesmen, the hiring rate will be less. SDT can be interpreted as the time necessary, at the present hiring rate, to double the present number of salesmen.)

2. On the following figure draw the line (as in text Figure 2.4b) that represents the relationship between salesmen and salesmen hiring rate that is given by text Equation 2.4-1 when SDT equals 50 weeks.

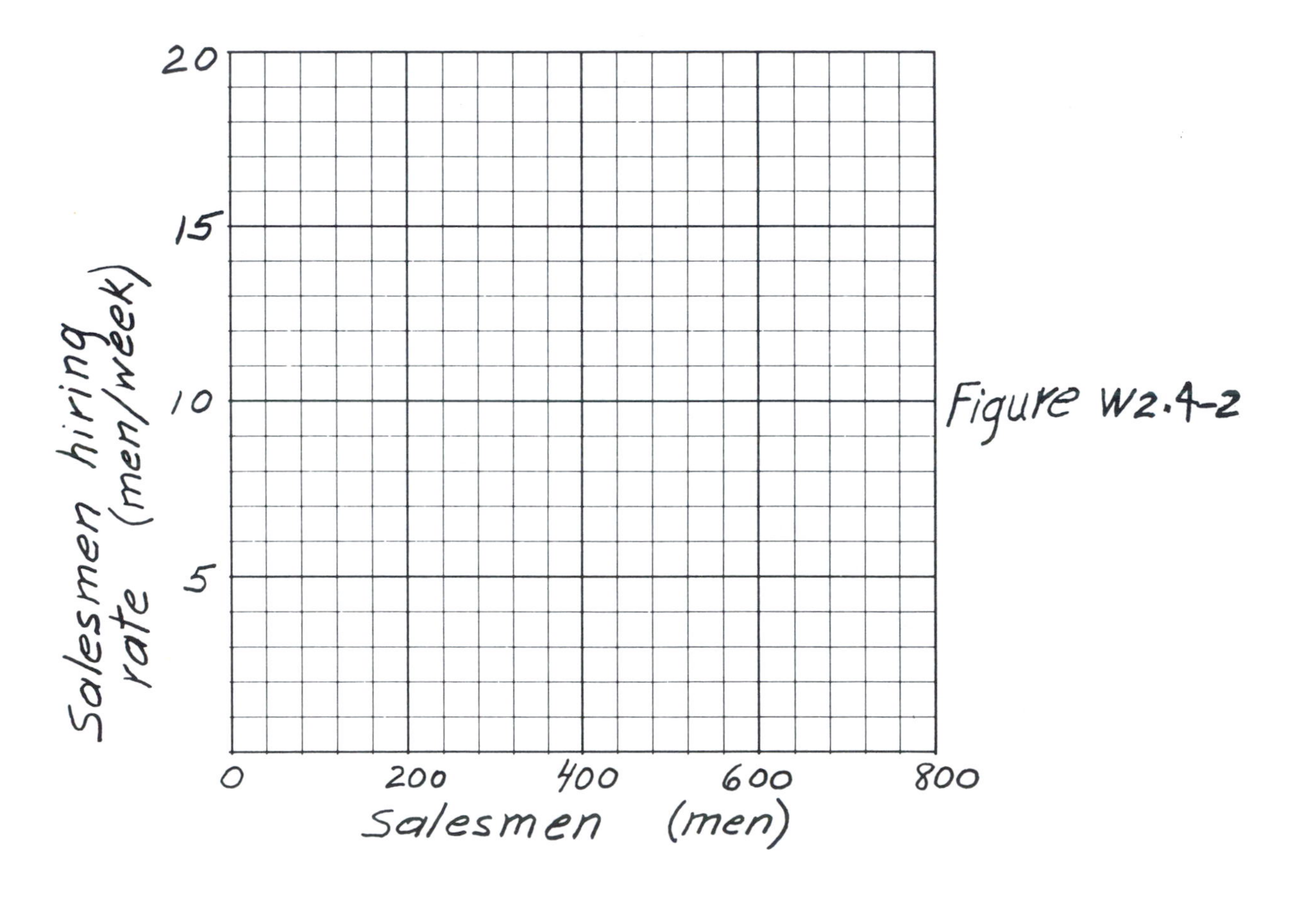

Figure W2.4-2

(Sec.W2.4)

* * * * *

* * * * *

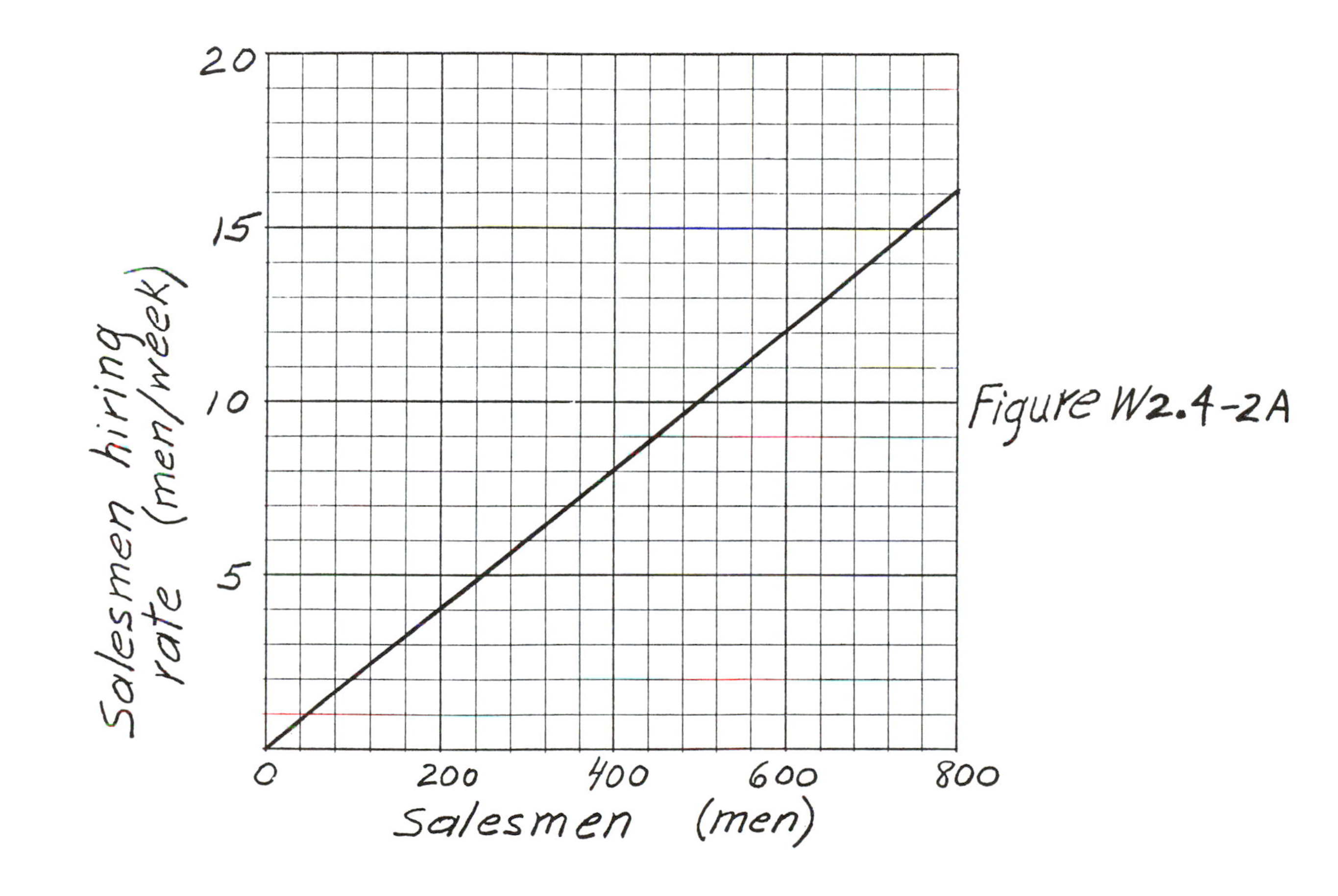

Figure W2.4-2A

3. Suppose the salesmen hiring relationship were as shown here, being a simplified approximation to Curve C of text Figure 2.4b.

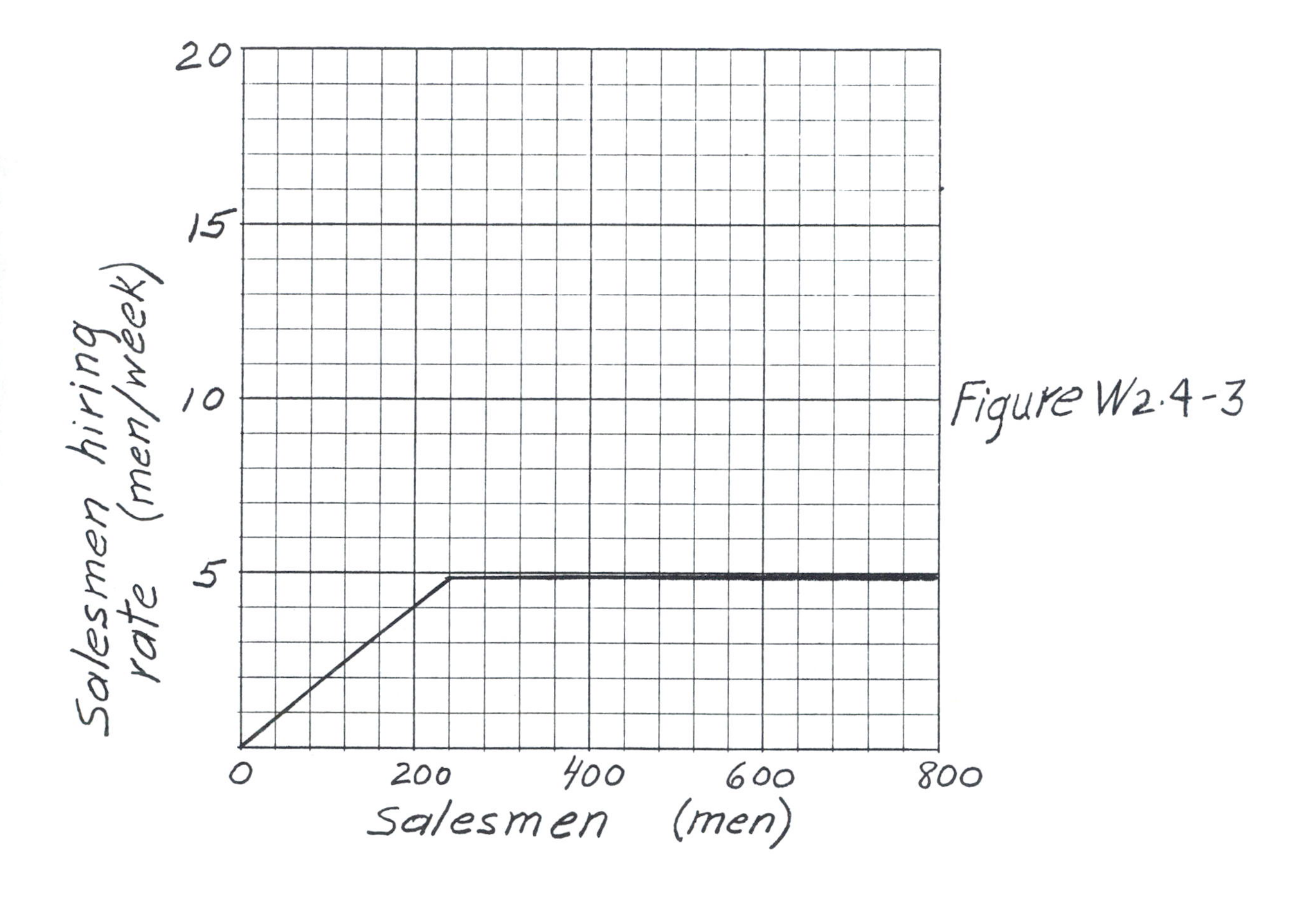

Figure W2.4-3

(Sec.W2.4)

Show on the following, copied from text Figure 2.4c, how the salesmen growth curve would differ from that previously obtained.

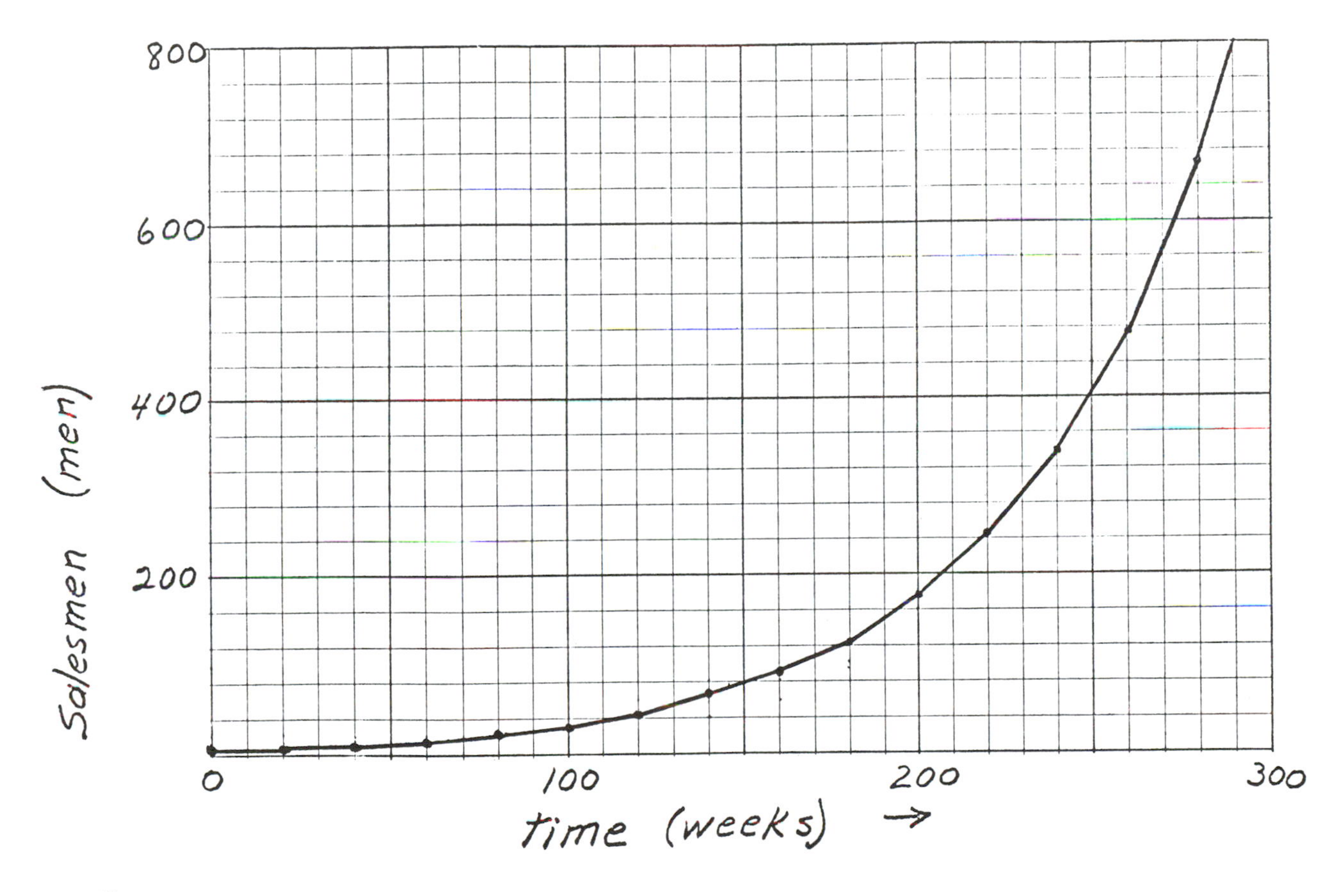

Figure W2.4-3A

* * * * *

* * * * *

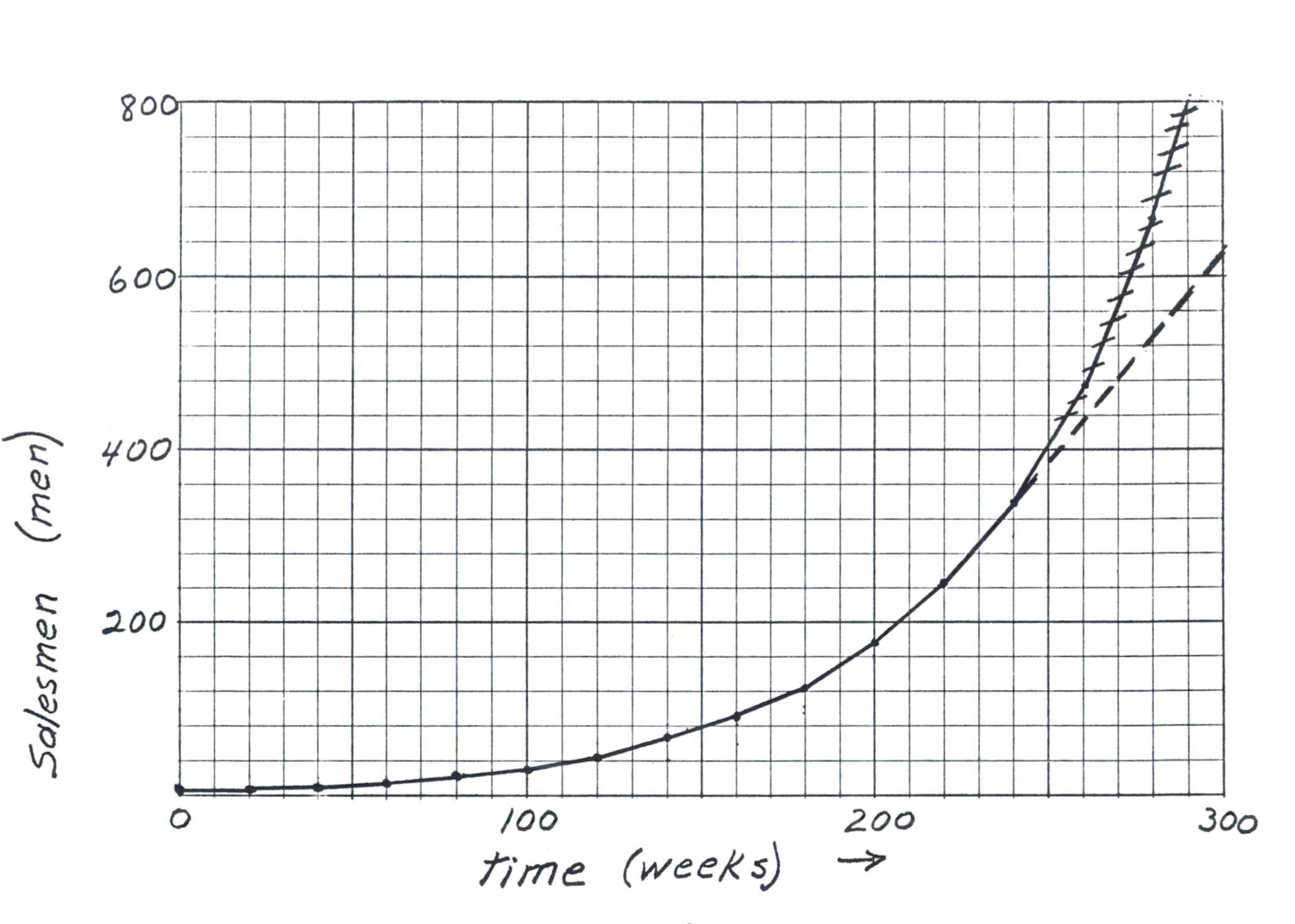

Figure W2.4-3B Positive feedback with limitation on growth rate.

When salesmen are above 240 men, the hiring rate is constant at about 5 men per week which gives a straight line increase for the number of salesmen of 5 men/week.

4. Suppose the growth in salesmen looked like this:

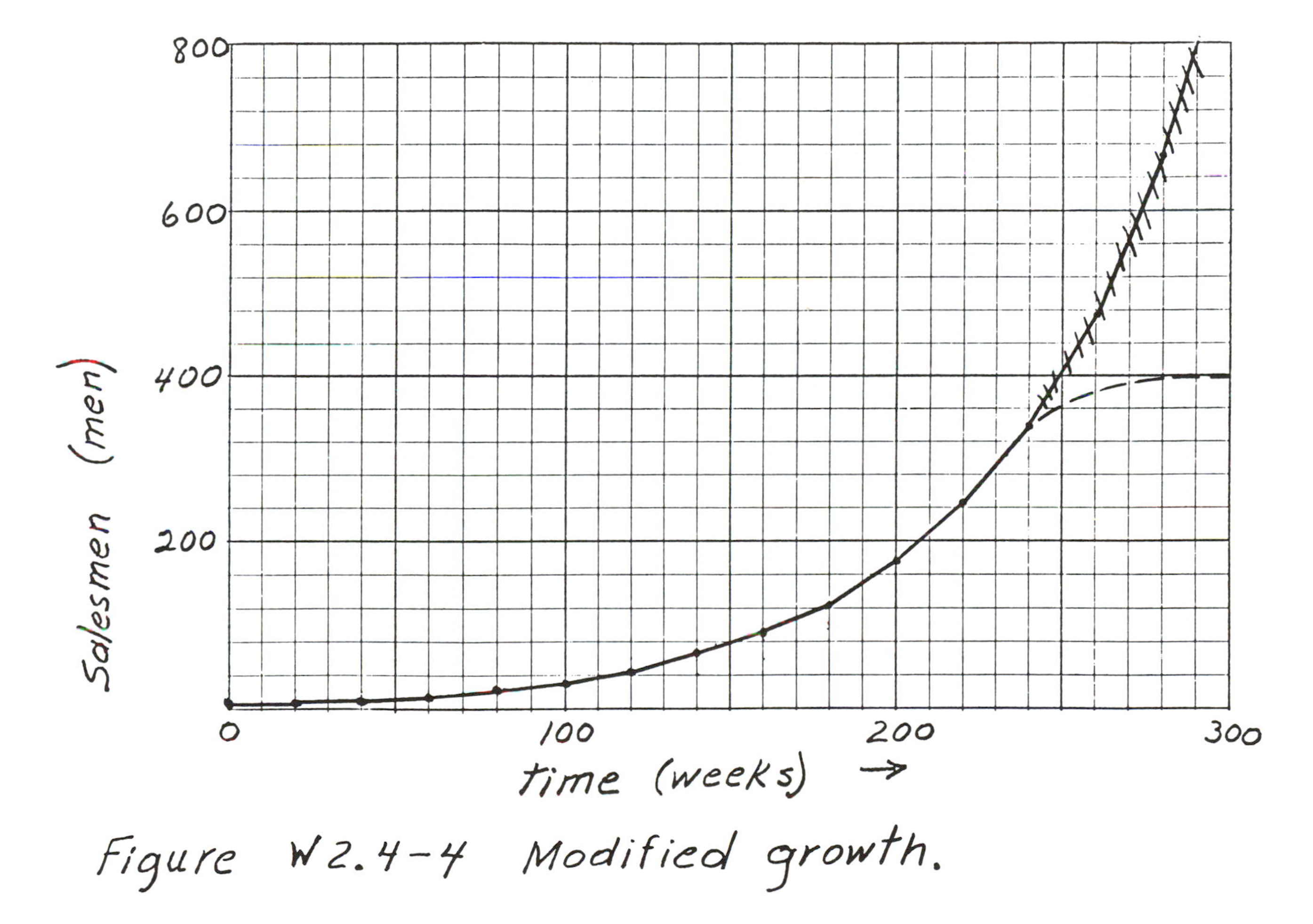

Figure W2.4-4 Modified growth.

Sketch on the following graph the salesmen hiring relationship that would give the above growth curve.

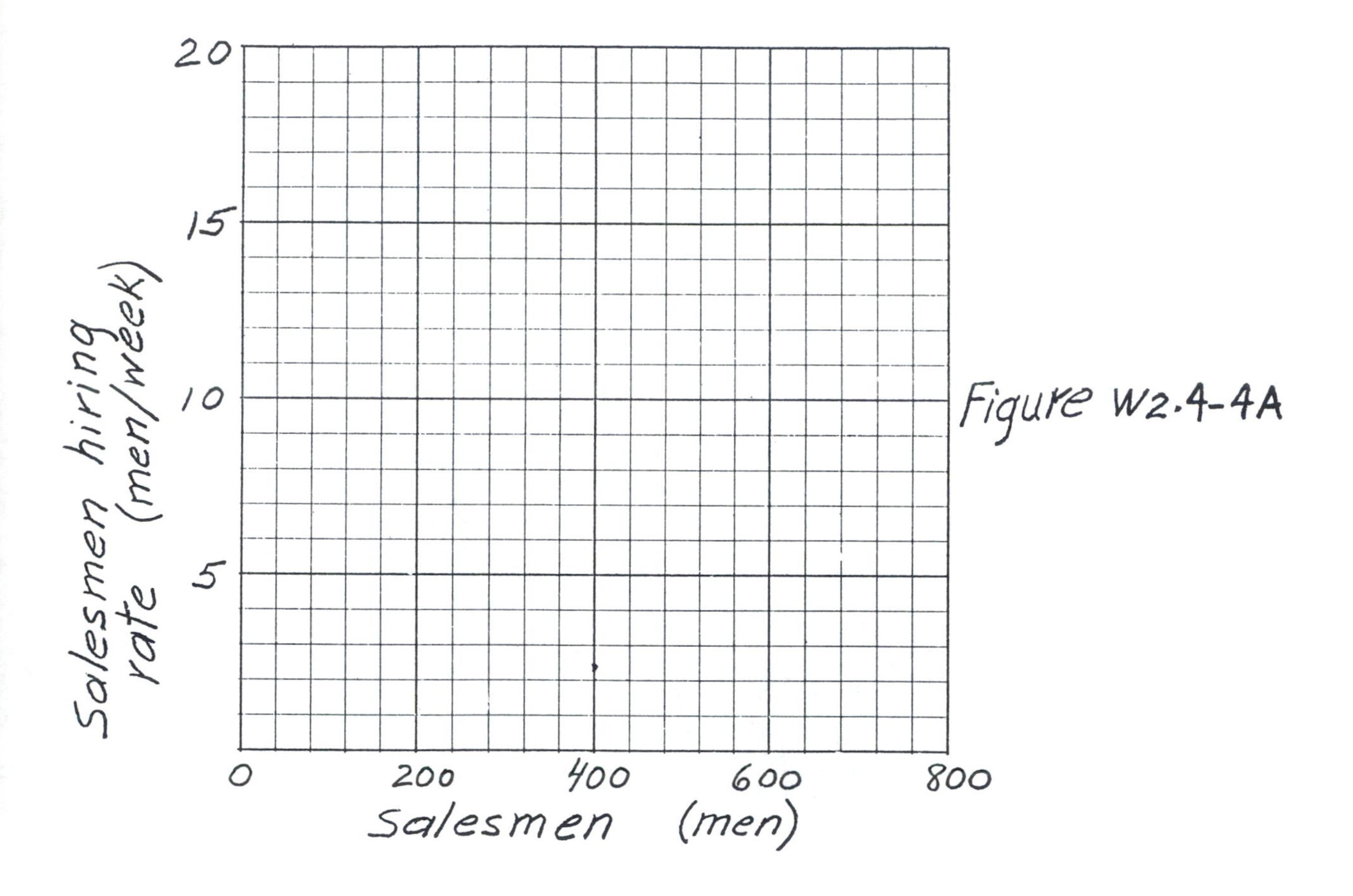

Figure W2.4-4A

(Sec.W2.4)

* * * * *

* * * * *

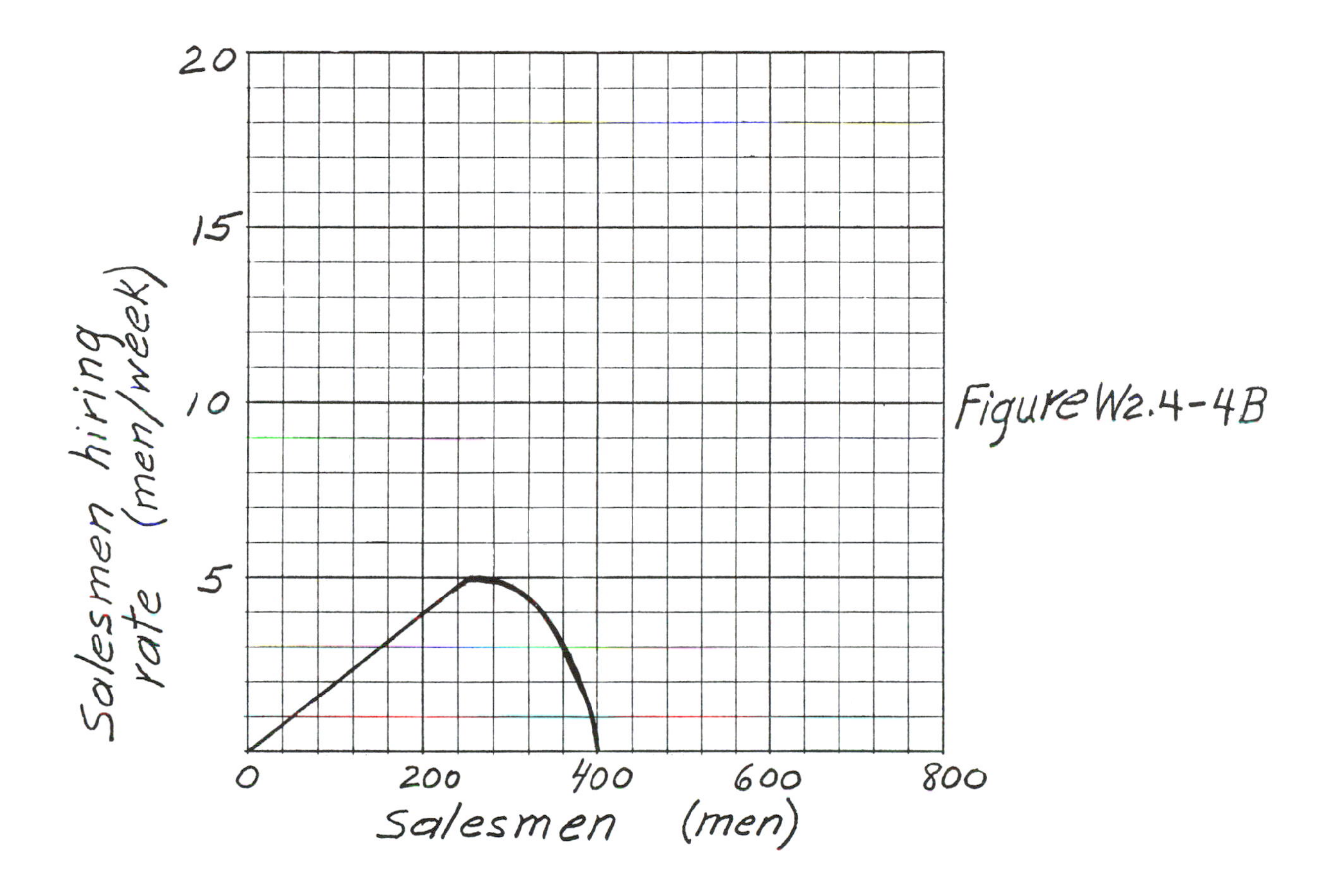

Figure W2.4-4B

Note that in Figure W2.4-4 above the original growth curve of text Figure **2.4c** persisted to a level of **240** salesmen. Then in Figure W2.4-4 above growth rate began to decline until it reached zero at 400 salesmen (as indicated by the constant level of 400 salesmen in Figure W2.4-4). Above 400 salesmen there is no growth so there can be no more than 400 salesmen.

5. The following figure is like text Figure 2.4a except that the system describes the discharge rather than the hiring of salesmen.

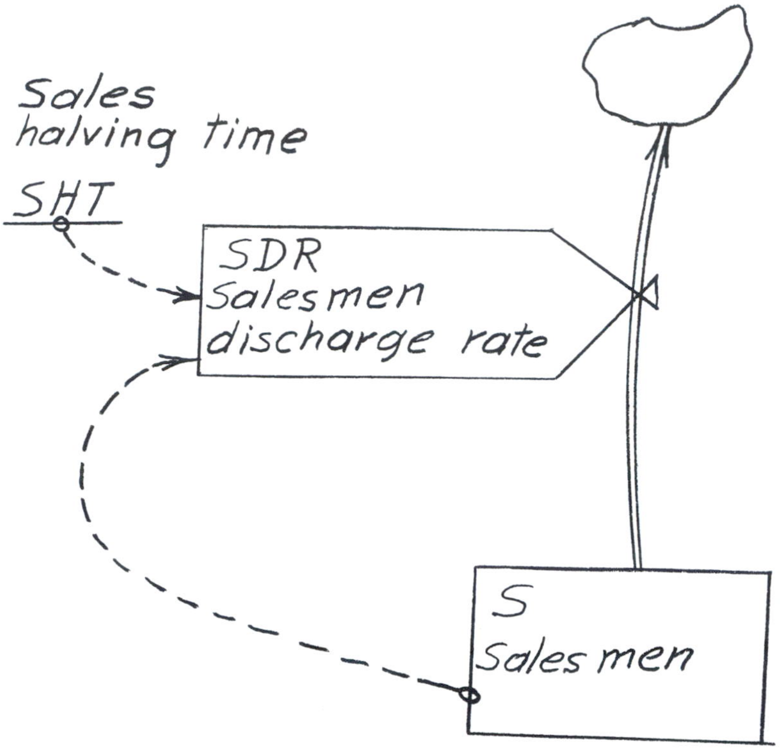

Sales halving time
SHT

SDR
Salesmen discharge rate

S
Salesmen

Figure W2.4-5 Reversal of salesman flow.

Corresponding to Equation 2.4-1

$$SDR = \frac{1}{SHT}(S) \qquad \text{Eq. W2.4-5}$$

$$SHT = 50$$

SDR--Salesmen discharge rate (men/week)
SHT--Sales halving time (weeks)
S--Salesmen (men)

The system pictured above is a (positive/negative) feedback loop.

* * * * *

* * * * *

negative (The diagram and equation are similar to those in section 2.4 for a positive loop. In text Figure 2.4a, an increase in salesmen increased the hiring rate which increased salesmen. However, here in the modified system, an increase in salesmen would increase the discharge rate which would decrease salesmen. There is now a sign reversal in the loop that arises from redefining the direction of the flow controlled by the decision.)

6. In the negative feedback loop of the preceding figure, the goal of the system is to produce a value of ________ for ________.

* * * * *

* * * * *

zero, salesmen (In Equation W2.4-5 the goal is implicit and does not appear as it did in text Equation 2.2-1 because the value here is zero. As can be seen in the preceding Equation W2.4-5, salesmen will be discharged until there are zero salesmen.)

7. The following system shows a backlog of unfilled orders which is increased by orders entered OE and decreased by orders completed OC.

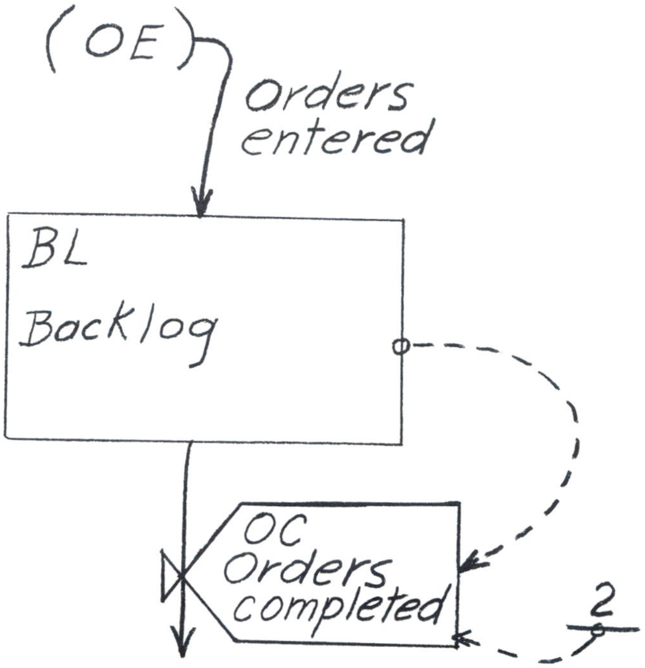

Figure W2.4-7

Suppose that the orders completed rate is given by the equation

$$OC = \frac{BL}{DD} \qquad \text{Eq. W2.4-7}$$

$$DD = 2$$

OC--Orders completed (units/month)
BL--Backlog (units)
DD--Delivery delay (months)

The preceding figure and equation represent a (positive/negative) feedback loop.

* * * * *

* * * * *

negative. [An increase in backlog increases the orders completed rate which decreases backlog. There is an odd number--one--of reversals in sign (or sense or direction of action) in traversing the loop.]

8. The goal of the system in the preceding frame is to control ________________ toward a goal of ____________.

* * * * *

* * * * *

backlog, zero

9. The following feedback loop is often used to represent a delay in the transmission of information, or to represent an information averaging process. Delivery delay impending DDI here represents the true system level which, after a delay is recognized at DDR which is the apparent level of the system. The control at CDDR represents the process by which DDR is altered toward DDI. CDDR may be either positive or negative.

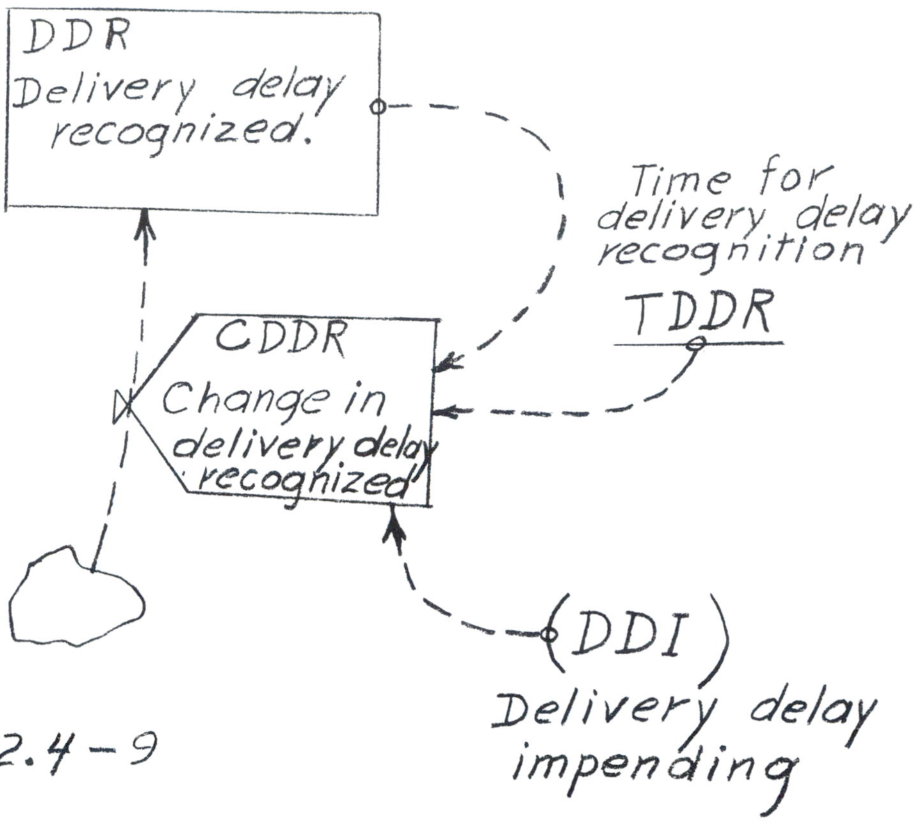

Figure W2.4-9

$$CDDR = \frac{1}{TDDR}(DDI-DDR) \qquad \text{Eq. W2.4-9}$$

CDDR--Change in delivery delay recognized (months/month)
TDDR--Time for delivery delay recognition (months)
DDI--Delivery delay impending (months)
DDR--Delivery delay recognized (months)

In the above equation we see that the change rate CDDR depends on the difference between the true level DDI and the recognized level DDR. The recognized (or "apparent" to use the terminology of text Section 1.4) level is always being gradually adjusted toward the true level. The loop in the above figure represents (positive/negative) feedback. Why?

* * * * *

* * * * *

negative. An increase in DDI would cause an increase in CDDR which would increase DDR which would <u>decrease</u> CDDR.

10. The negative feedback loop of Figure W2.4-9 has the purpose of adjusting ______________ toward ___________ as a goal.

* * * * *

* * * * *

DDR, DDI

11. Consider the following feedback loop.

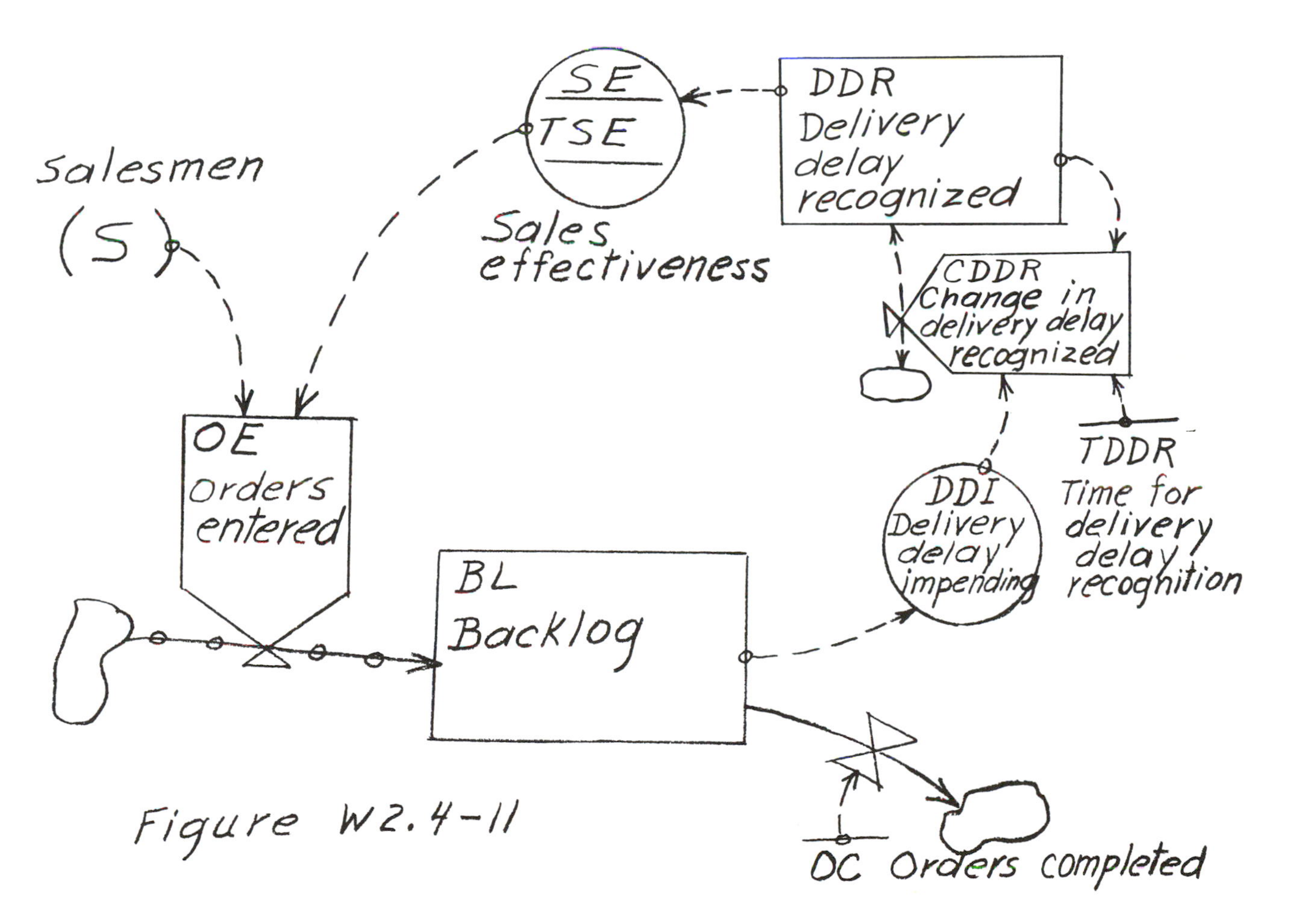

Figure W2.4-11

Orders entered OE increase the backlog BL which is decreased by the constant orders completed rate OC. The delivery delay impending DDI increases when backlog increases. The delivery delay recognized DDR follows changes in DDI. Sales effectiveness SE is related to DDR as shown in Figure W2.4-12.

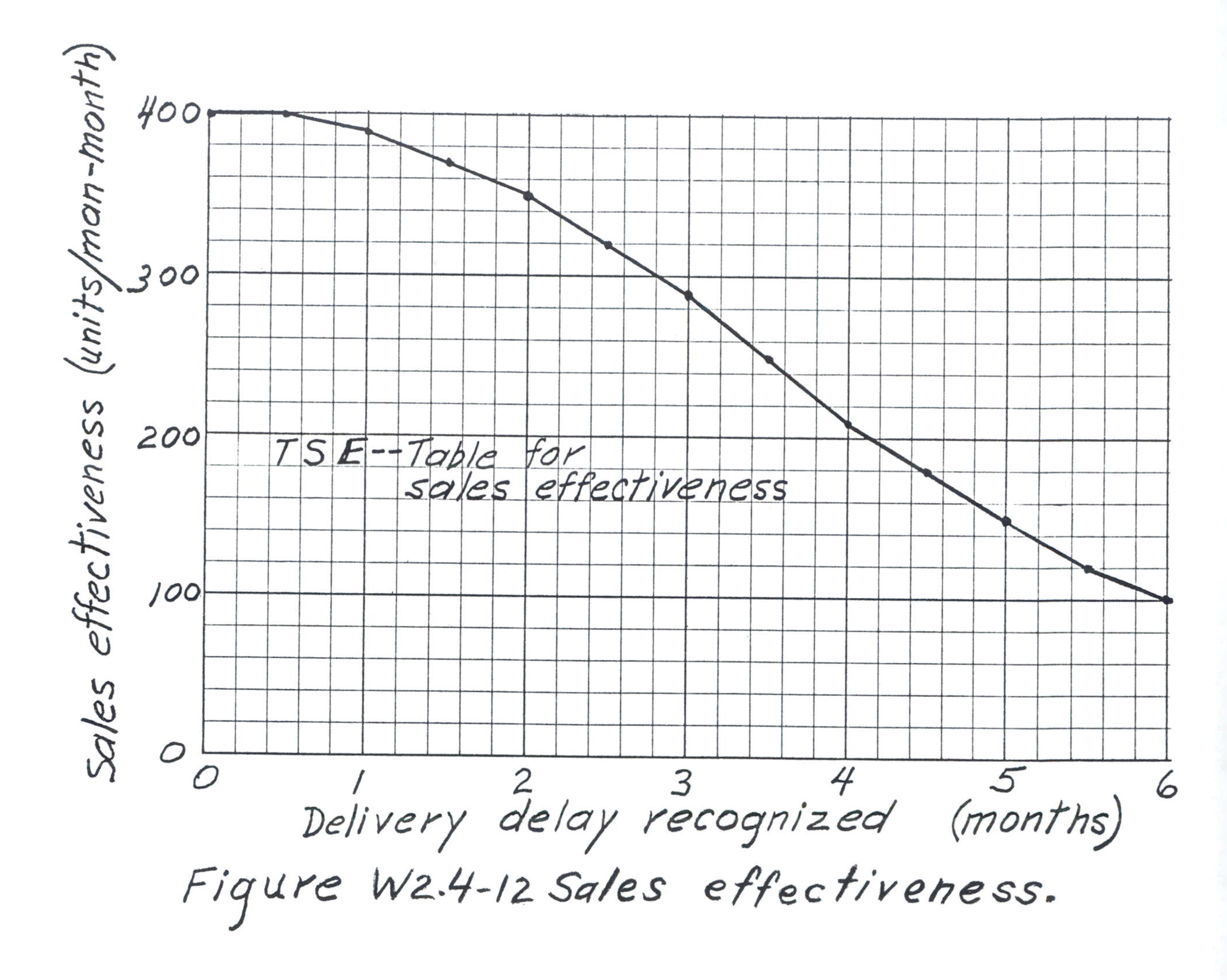

Figure W2.4-12 Sales effectiveness.

Orders entered OE are proportional to sales effectiveness. Suppose that conditions are such that the system operates with a sales effectiveness between 200 and 300 units per man-month. This is a (positive/negative) feedback loop.

* * * * *

* * * * *

negative. (Rising sales effectiveness increases orders entered which increases backlog which increases delivery delay impending which increases delivery delay recognized which decreases sales effectiveness (See Figure W2.4-12). There is an odd-number--one--reversal in the direction of action around the loop.)

12. The system in the preceding frame is controlling ____________ in an attempt to make it equal to ____________.

* * * * *

* * * * *

orders entered rate, orders completed rate

(If orders entered are below the rate of orders completed, backlog will fall until sales effectiveness rises enough to make orders entered equal to orders completed and vice versa. But as in any negative feedback loop having two or more level variables--here backlog and delivery delay recognized--the system may fluctuate around the value it is seeking.)

13. If the payment rate PR for the total salaries of all men in the company is measured in "dollars per month," what is the proper unit of measure for the average individual salary of one man? ________________.

If

$$PR = (M)(AIS) \qquad \text{Eq. W2.4-13}$$

PR--Payment rate (dollars/month)
M--Men (men)
AIS--Average individual salary (   )

does your answer give the same units on each side of the equation?

* * * * *

* * * * *

dollars per man-month (or dollars/man/month).

(Note that the salary is for one man working one month. The usual abbreviated answer of "dollars/month" assumes a "man-month" but will not give the correct units because

dollars/month ≠ (men)(dollars/month)
(does not equal)

= (men)(dollars/man-month)

W2.5 Coupled Nonlinear Feedback Loops

Whether a feedback loop acts as a positive feedback with growing and explosive results, or acts as a negative, goal-seeking loop depends on the magnitudes and algebraic signs of the numerical values that describe the loop. Many of these numerical values change with the condition of the system. They often change in such a way that they convert a positive loop to a negative loop and vice versa. Consider the positive feedback loop on the left side of text Figure 2.5a.

1. Text Equations 2.5-1, 2.5-2, 2.5-3, 2.5-4, and 2.5-5 give the values of salesmen, orders booked, budget, indicated salesmen and salesmen hiring. Text Equation 2.5-2 can be substituted into text Equation 2.5-3 to give

$$B = (S)(SE)(10) \qquad \text{Eq. W2.5-1}$$

B--Budget	(dollars/month)
S--Salesmen	(men)
SE--Sales effectiveness	(units/man-month)
10--the value of RS	(dollars/unit)

Here we are interested in budget as it depends on the number of salesmen and on the value of sales effectiveness which can change and can alter the character of the positive feedback loop. Equation W2.5-1 above can be substituted into text Equation 2.5-4 to give

$$IS = \frac{(S)(SE)(10)}{2000} = \frac{(S)(SE)}{200} \qquad \text{Eq. W2.5-1A}$$

IS--Indicated salesmen (men)
S--Salesmen (men)
SE--Sales effectiveness (units/man-month)
10--the value of RS (dollars/unit)
2000--the value of SS (dollars/man-month)
200--coefficient (units/man-month)

In turn, the preceding equation can be substituted into text Equation 2.5-5 to give

$$SH = \frac{1}{SAT}\left[\frac{(S)(SE)}{200} - S\right] \qquad \text{Eq. W2.5-1B}$$

$$= \frac{1}{20}\ (S)\left(\frac{SE}{200} - 1\right)$$

SH--Salesmen hiring (men/month)
SAT--Salesmen adjustment time (months)
S--Salesmen (men)
SE--Sales effectiveness (units/man-month)
20--value of SAT (months)
200--coefficient (units/man-month)

This substitution maneuver has, in effect, combined the major positive loop of text Figure 2.5a with its smaller internal negative loop that involved only salesmen S and salesmen hiring SH. Ignoring the need to have an "orders booked" output to go to the backlog in text Figure 2.5a, the entire left half of the figure is represented by the pool of salesmen and the above Equation W2.5-1B and is diagrammed in the following figure.

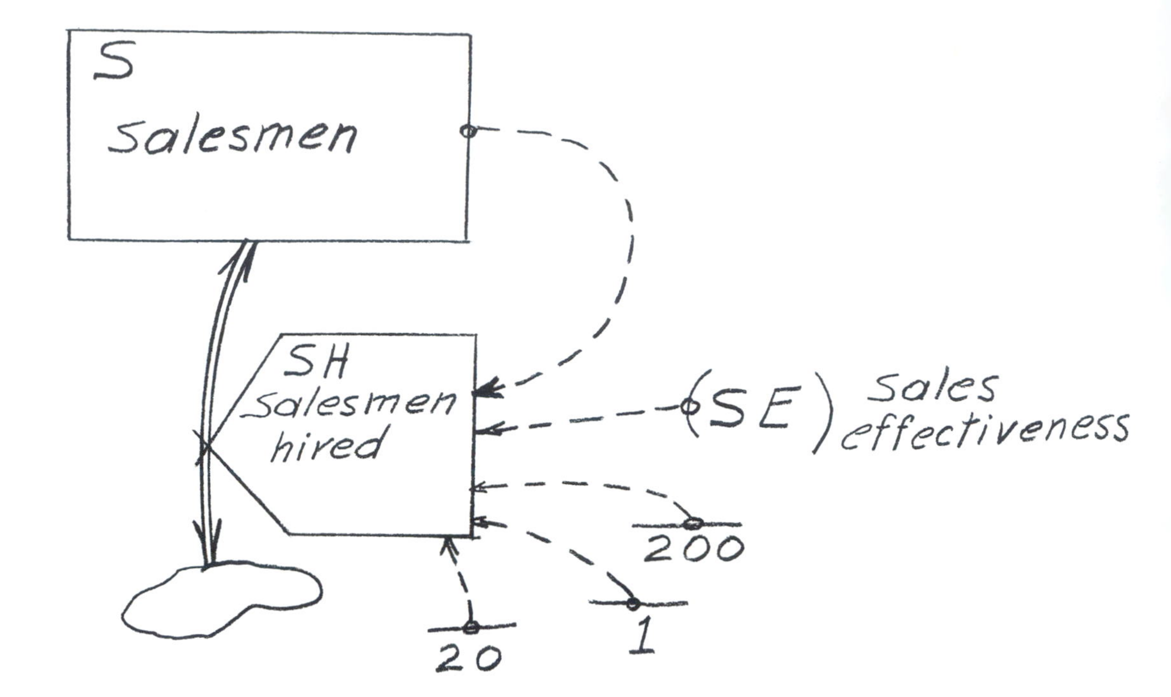

Figure W2.5-1 Combined positive and negative feedback loops.

Suppose that sales effectiveness SE is greater than 200 in the above Equation W2.5-1B. Then salesmen will (increase/decrease). An increase in salesmen will cause an (increase/decrease) in salesmen hiring SH. An increase in salesmen hiring causes an (increase/decrease) in the rate of growth of salesmen. This is a (positive/negative) feedback loop.

* * * * *

* * * * *

increase, increase, increase, positive

2. In the above Equation W2.5-1B, if SE = 300, and if the organization has not been started and there are no salesmen, the equation suggests what growth rate for salesmen? ___________.

   * * * * *

   * * * * *

   none (Growth is proportional to present salesmen and none exist.)

3. Suppose that SE = 100 in above Equation W2.5-1B. If there are 100 salesmen, the number will be (increasing/decreasing). The salesmen hiring rate SH will be (positive/negative). If salesmen are increased in a positive direction, sales hiring rate moves in a (positive/negative) direction. This is a (positive/negative) feedback loop.

   * * * * *

   * * * * *

   decreasing, negative, negative, negative

4. The negative loop of the preceding frame is moving the system toward ______________ salesmen as a goal.

   * * * * *

   * * * * *

   zero

5. The shift from a positive to a negative feedback loop occurs when sales effectiveness SE equals ____________ units per man-month.

   * * * * *

   * * * * *

   200

6. In the operation of the system of text Section 2.5, at what time does the sales effectiveness reach 200 units per man-month?

   * * * * *

   * * * * *

   At 100 months.

7. In above Equation W2.5-1B, as sales effectiveness SE is increased above 200 units per man-month, the rate of growth of salesmen is (increased/decreased).

* * * * *

* * * * *

increased

8. The solid line on the following figure is the sales effectiveness curve from text Figure 2.5c which produced the system behavior in text Figure 2.5d. Suppose the sales effectiveness were different as shown by the dotted line. The initial growth rate in salesmen would be (faster/slower). The saturation point where growth stops would occur (later/earlier).

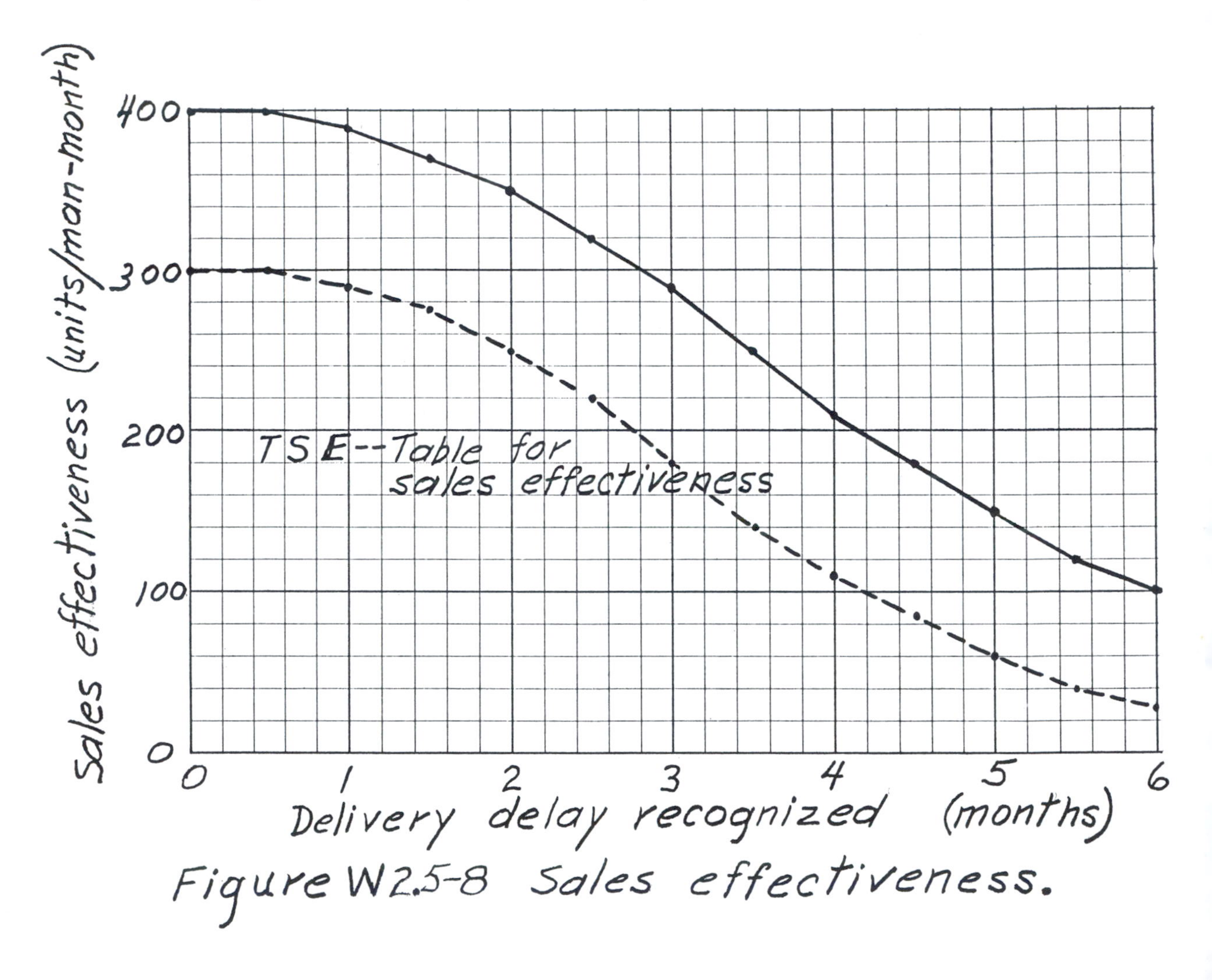

Figure W2.5-8 Sales effectiveness.

* * * * *

* * * * *

slower, later

If the change to the lower sales effectiveness is made, the plotted results of system operation are shown in the following figure which should be compared with text Figure 2.5d.

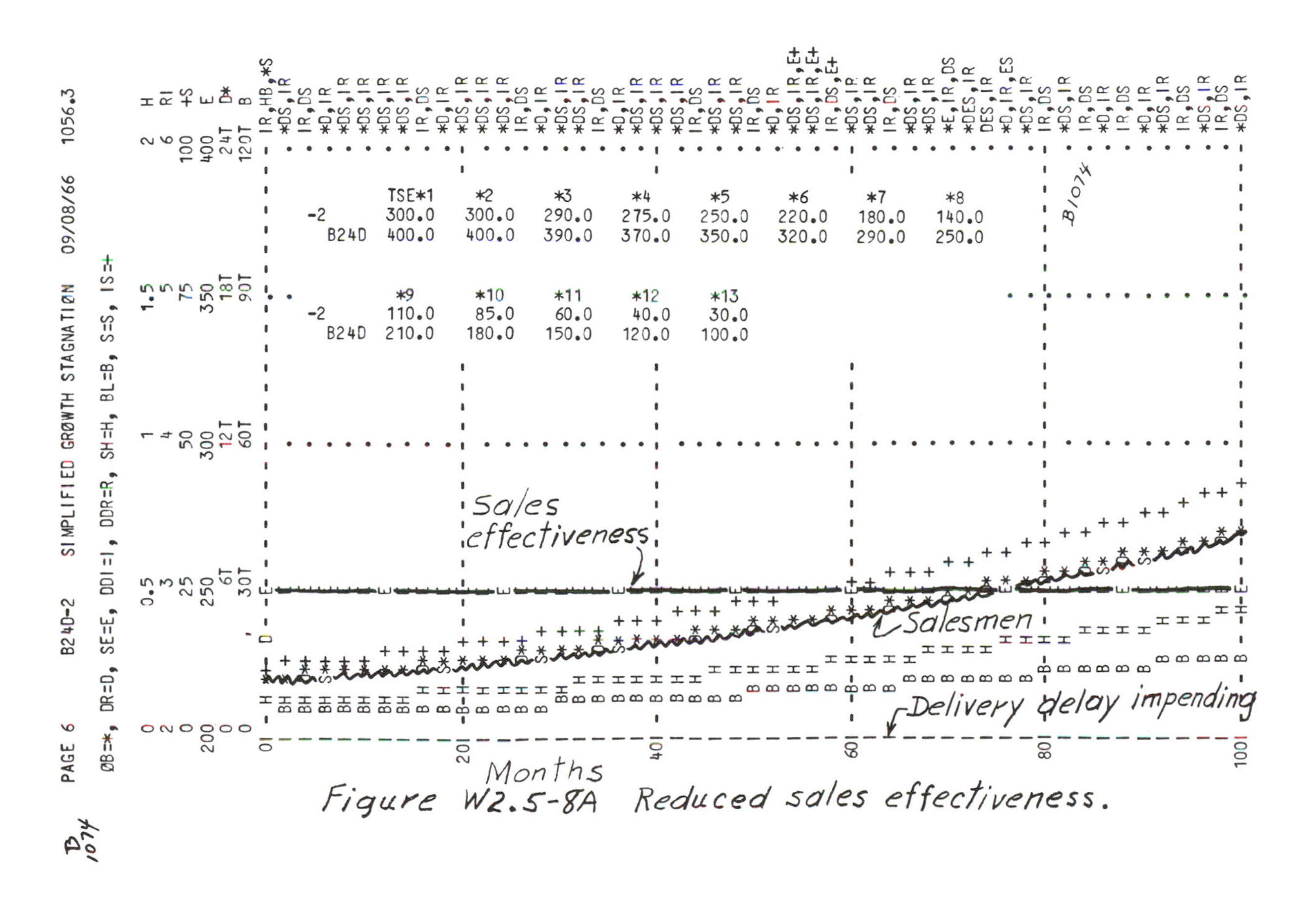

Figure W2.5-8A Reduced sales effectiveness.

Because, for any given delivery delay, the sales effectiveness is lower, the positive feedback loop generates expansion of salesmen more slowly. It takes longer for the sales rate to reach the available production capacity.

9. Instead of changing the sales effectiveness relationship that determines how the customers respond to delivery delay, suppose that the characteristics of the manufacturing process are different as shown in the following figure.

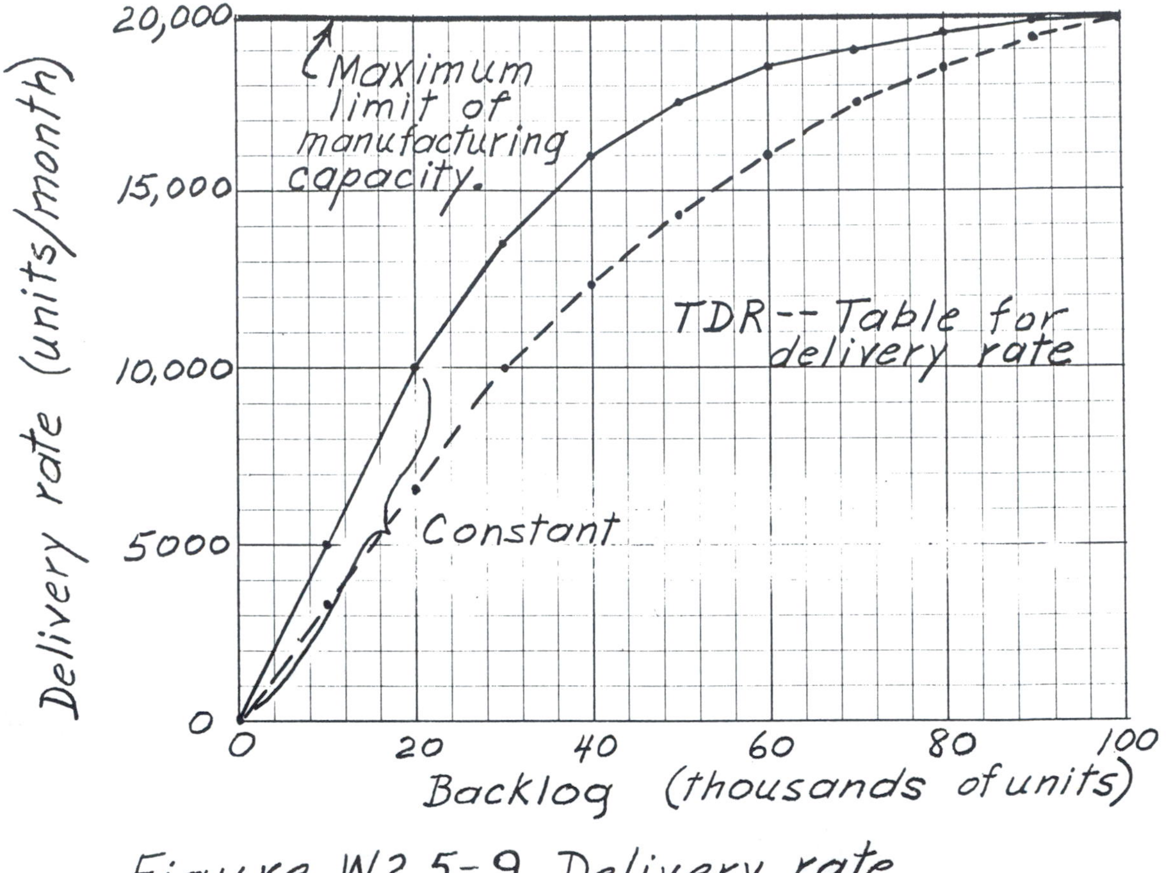

Figure W2.5-9 Delivery rate

The solid line shows the factory delivery rate to backlog relationship used for text Figure 2.5d. The dotted curve shows a factory with a (longer/shorter) delivery delay.

* * * * *

* * * * *

longer (for any backlog the delivery rate is lower and more time will be required for the backlog to be delivered)

(Sec.W2.5)

10. With the above longer delivery delay and the original sales effectiveness of text Figure 2.5c, the sales growth will be (faster/slower) than in text Figure 2.5d. The system will be operating at a (higher/lower) sales effectiveness.

* * * * *

* * * * *

slower, lower

The change in growth is seen by comparing the following figure with text Figure 2.5d.

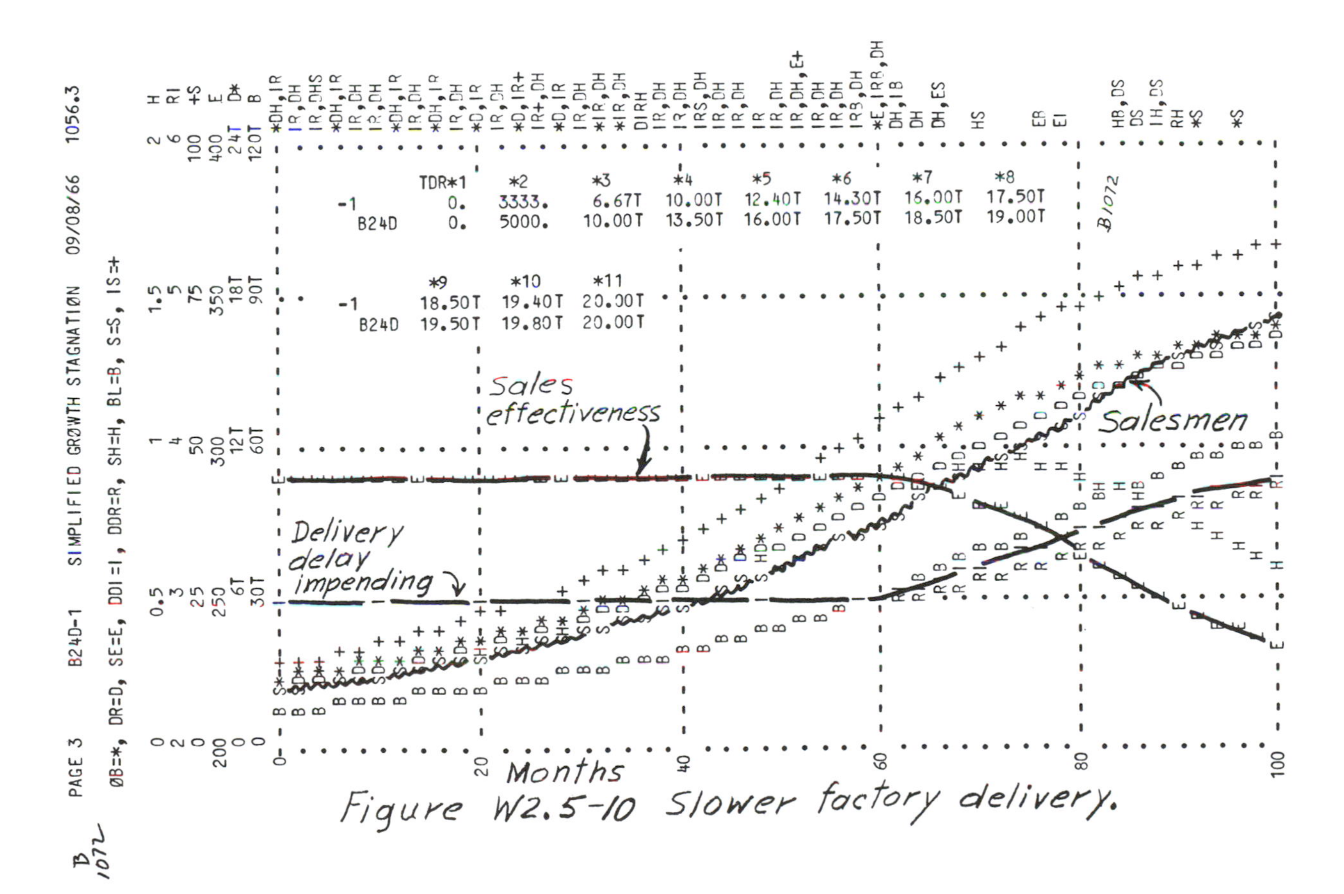

Figure W2.5-10 Slower factory delivery.

Because the factory delivery delay is longer (less delivery rate for any given backlog, see text Equation 2.5-10) the product is less attractive, the sales effectiveness is lower, sales are lower, and the budget to expand sales force is lower.

11. Consider again the sales effectiveness relationship. The solid line in the following figure repeats text Figure 2.5c.

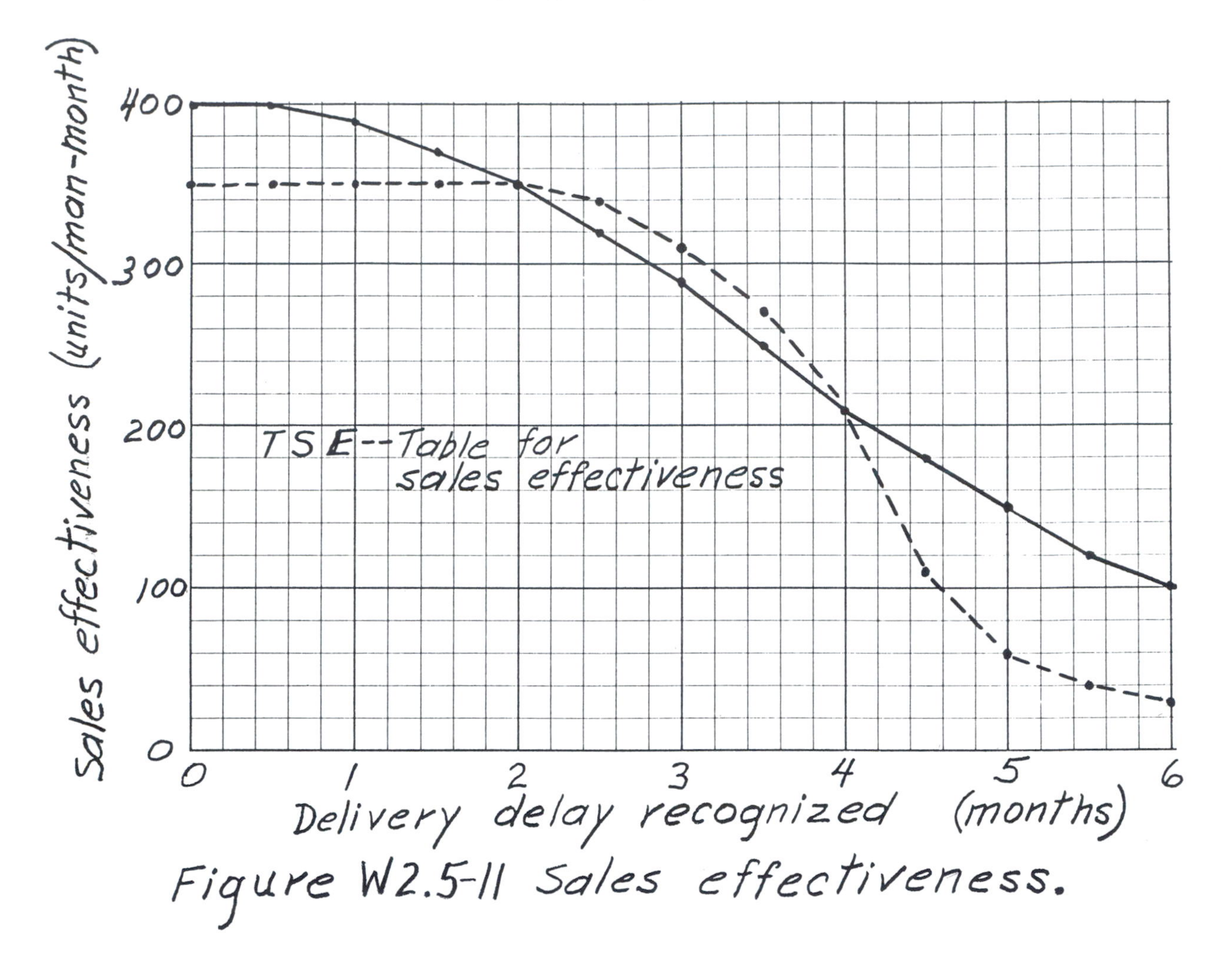

Figure W2.5-11 Sales effectiveness.

The dotted curve shows an assumed change in the reaction of the market. Compared with text Figure 2.5d, there (will/will not) be a substantial change in the sales growth rate during the first 60 months. The steeper section of the above sales effectiveness curve has its principal effect (before/after) the factory has reached its maximum production rate. The steeper section of the sales effectiveness shows a market that is (more/less) sensitive to changes in delivery delay. The fluctuation in demand will be (more/less) violent after the factory reaches full capacity.

* * * * *
* * * * *

will not (the sales effectiveness curve is still at 350 units per man-month in the vicinity of the initial delivery delay of 2 months)

after (because the delivery delay does not rise to the region of the steeper part of the curve until the factory is near full load)

more (in the vicinity of 4 months delivery delay the market as shown by sales effectiveness changes more with changes in delivery delay than before)

more (in general as the sensitivity of a loop increases, that is, as the extent of a correction of an imbalance is greater, the tendency to oscillate increases)

The behavior of the system with the new sales effectiveness relationship is shown below. Compare with text Figure 2.5d and workbook Figure W2.5-8A.

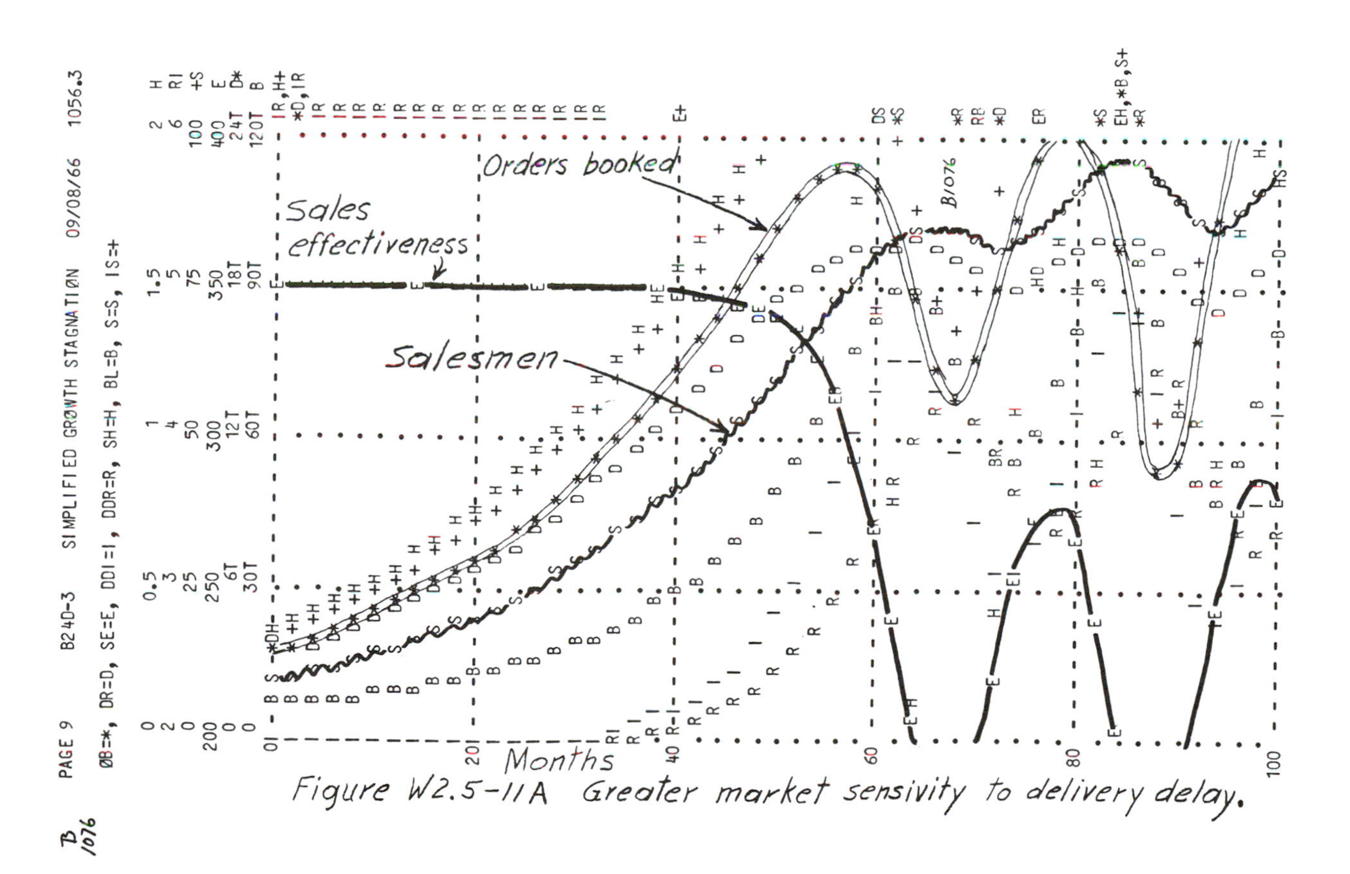

Figure W2.5-11A Greater market sensivity to delivery delay.

In this system response, the growth section is almost identical to text Figure 2.5d. However, following month 60, the oscillation is wider and the peaks are closer together because of the greater sensitivity of the market to delivery delay. Such a greater sensitivity would represent a product which is not unique, where other sources exist, where the customer can more easily shift between suppliers, where delivery delay is more easy to determine, and where delivery delay is more important to the customer. Customers then tend to move quickly toward suppliers with good delivery and away from suppliers with poor delivery.

12. In workbook Section W2.3, Frame 13, the delay within a second-order, negative-feedback loop was increased. This caused the oscillation to become (smaller/larger) and the period between peaks to become (longer/shorter).

    * * * * *

    * * * * *

    larger, longer

13. Suppose in the right hand negative loop of text Figure 2.5a that the time for delivery delay recognition TDDR is increased from 6 months to 12 months. The market then lags further behind in recognizing the actual condition of delivery delay. This will have a (substantial/slight) effect on the sales growth in the first 50 months. The principal effect will be (before/after) the factory approaches full capacity. The sales fluctuations after full production is reached will be (larger/smaller) and the time between peaks will be (longer/shorter).

* * * * *

* * * * *

slight (since the delivery delay is not changing, it does not matter how long the market takes to recognize changes. The initial condition of recognized delay is here taken. at the true initial delay)

after (because then the factory delivery delay is changing)

larger (as in Section W2.3)

longer (In general, increasing the delay in a simple negative feedback loop makes the operation more unstable and with a longer period between peaks of the disturbance.)

The effect of the longer time for delivery delay recognition is shown in the following figure. Compare with text Figure 2.5d.

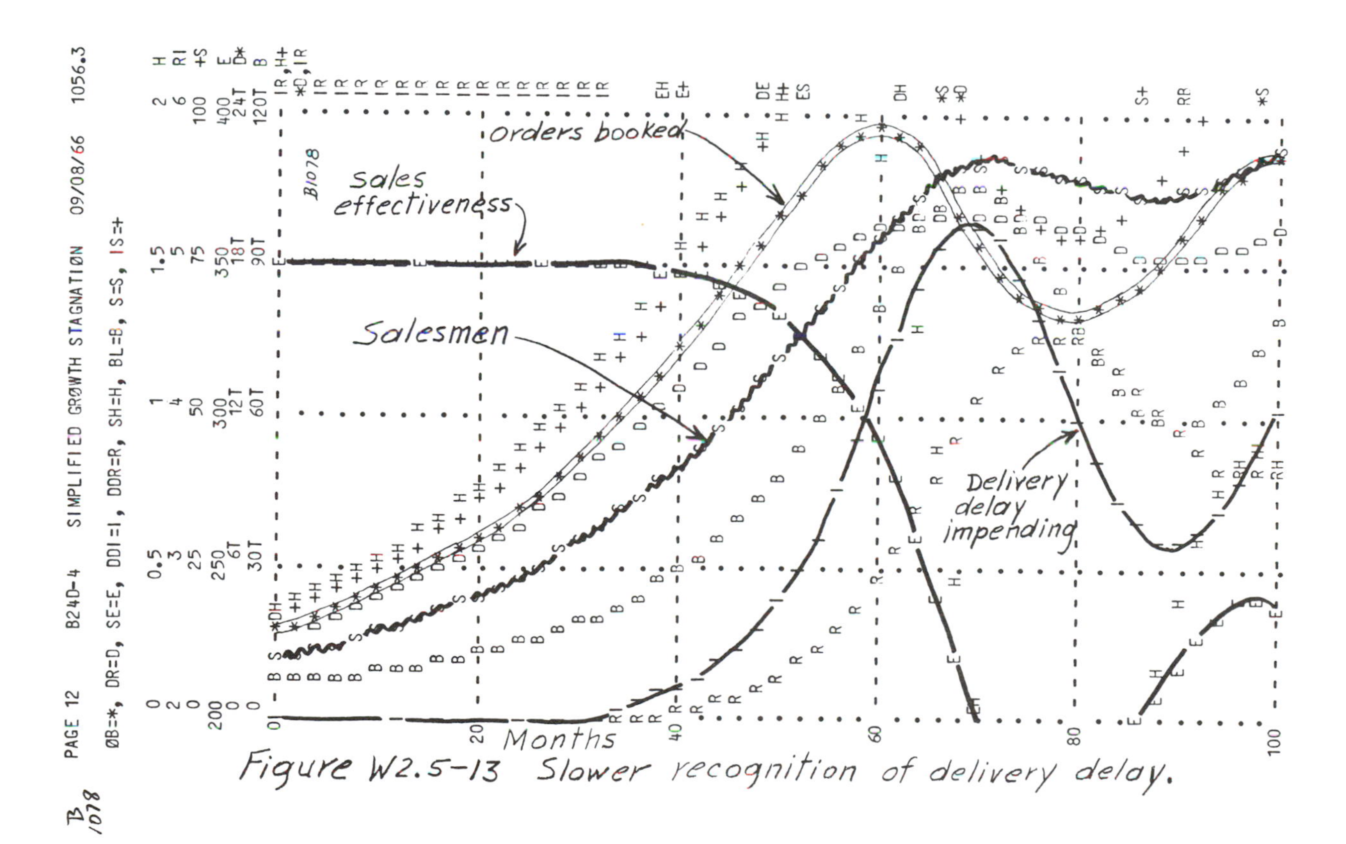

Figure W2.5-13 Slower recognition of delivery delay.

CHAPTER W3

MODELS AND SIMULATION

W3.1 Models

1. Models can be classified as "static" or "dynamic" models. A floor plan layout of machines in a factory is a ____________ model. The replica of an airplane for studying performance and stability in a wind tunnel is a _________ model.

   * * * * *

   * * * * *

   static, dynamic

2. Dynamic models can be classified as "physical" models or "abstract" models. A miniature ship in a towing basin is a ___________, ___________ model. The equations for a positive-feedback growth process in text Section 2.5 represent a __________,__________ model.

   * * * * *

   * * * * *

   dynamic, physical

   dynamic, abstract

3. Abstract models can be classified as "mental," "verbal," "pictorial," "diagrammatic," or "mathematical" depending on whether the true system is represented by mental images, word descriptions, pictures, flow diagrams, or mathematical equations. Our memory of a familiar street corner is a _________,_________ model.

   * * * * *

   * * * * *

   static, mental

4. Our impressions of the plays in a football game are stored in memory as a ____________, ___________ model.

* * * * *

* * * * *

dynamic, mental

5. The description in a service manual of operation of an automobile air conditioner is a ____________, ___________ model.

* * * * *

* * * * *

dynamic, verbal

6. A photograph of an automobile in a studio is a __________,_______ model.

* * * * *

* * * * *

static, pictorial

7. The sketch with arrows and other artist's symbolism for motion showing the circulation of ocean currents is a ____________, _______________ model.

* * * * *

* * * * *

dynamic, diagrammatic

8. A flow diagram as in text Figure 2.5a showing the two major loops and three minor loops of a sales growth and saturation process is a ____________, ___________ model.

* * * * *

* * * * *

dynamic, diagrammatic

9. Text Equations 2.5-1 through 2.5-13 describing the same sales growth and saturation system form a ______________, __________ model.

* * * * *

* * * * *

dynamic, mathematical

W3.2 The Basis of Model Usefulness

1. "Managers quickly adjust their mental image of the amount of inventory in the warehouse to bring the image into agreement with the latest reported financial figures, but they only slowly adjust their image of how well their customers like the quality of the product the company sells." The preceding is a general and vague statement of the difference between two processes of adjusting the apparent information levels on which decisions might be based. A similar adjustment process was described in text Equation 2.5-11. Write here the corresponding equation for the change of mental impression of inventory, choosing your own estimate of a suitable parameter value.

   CMII = ______________________________

   TCMII = ______________

   CMII--Change in mental image of inventory (units/month)
   TCMII--Time to change mental image of inventory (months)
   RI--Reported inventory (units)
   MII--Mental image of inventory (units)

   * * * * *

   * * * * *

   $$CMII = \frac{1}{TCMII}(RI-MII)$$

   TCMII = 2 months. (The figure might be longer than 2 months. People do not read the reports immediately. They may not change their decision-making basis until they realize that others are responding to the newly reported information. Plans may already be committed that will take several months to change.)

2. In the preceding frame, it is suggested that managers "only slowly adjust" impressions of what customers think about the company. Corresponding to the previous frame, write the equation for the rate of change of management perception of customer attitude toward product quality.

CICS--Change in image of customer satisfaction (quality units/month)
TCICS--Time to change image of customer satisfaction (months)
RCS--Reported customer satisfaction (quality units)
ICS--Image of customer satisfaction (quality units)

* * * * *

* * * * *

$$CICS = \frac{1}{TCICS} (RCS-ICS)$$

TCICS = 30 months (Here the suggested difference "only slowly adjust" is the difference between 2 months and 30 months. The model statement forces a precise and quantitative meaning where otherwise only general qualitative description would usually be used. Of course, the verbal description could be made precise by specifying "first-order, exponential adjustment processes with time constants of 2 and 30 months, respectively." The 30 month adjustment time may seem long but the self-image of how well one is doing is hard to displace. Furthermore, information about what the customers think is more tenuous and less persuasive than is the numerical inventory information from the accounting statements. The less certain the information, and the more controversial the process of acquisition and interpretation, the longer it takes for the new information to be accepted.)

3. An indefinite statement of ordering policy might be "orders depend on sales and inventory." A more precise verbal statement would be, "the weekly order rate equals average sales plus

one-eighth of the excess of desired inventory over actual inventory." Write the mathematical model statement of the preceding verbal model for ordering policy.

OR = ______________________

OR--Order rate (units/week)
AS--Average sales (units/week)
DI--Desired inventory (units)
AI--Actual inventory (units)

* * * * *

* * * * *

$OR = AS + \frac{1}{8}(DI-AI)$

W3.3 Simulation vs. Analytical Solutions

1. A model is used to ______________ the behavior of the real system represented by the model.

   * * * * *

   * * * * *

   simulate

2. A simulation solution (can/can not) be used to evaluate directly the condition of the system at a specified time.

   * * * * *

   * * * * *

   can not

3. An analytical solution, when obtainable, does yield the condition of the system at any point in ____________.

   * * * * *

   * * * * *

   time

4. Consider a model stated in the following terms:

   a. A tank receives and accumulates the flow of water from a pipe.

   b. The pipe delivers 5 gallons per minute to the tank.

   c. The tank initially contains 120 gallons.

The amount of water could be computed for each succeeding minute by adding 5 gallons to the preceding value. Complete the following table.

Time (minutes)	Gallons in tank	gallons/minute from pipe
0	120	5
1	125	5
2	130	
3		
4		
5		

The above is a ________________ solution to the water level in the tank.

* * * * *

* * * * *

Time (minutes)	Gallons in tank	gallons/minute from pipe
0	120	5
1	125	5
2	130	5
3	135	5
4	140	5
5	145	5

simulation

5. The water level in the tank could also be expressed as:

$$WL = 120 + 5t$$

where t is the time in minutes. This is an __________ solution corresponding to the ____________ solution in Frame 4. Unlike the model statements in Frame 4, this equation contains __________ explicitly as an independent variable. If time, t, equals 5 minutes in this equation, the water level is ____________. Does this agree with the simulation result in Frame 4? ____________.

* * * * *
* * * * *

analytical, simulation, time, 145 gallons, yes

6. With rare exceptions, an analytical solution is not possible for a nonlinear system. One indication that a system is nonlinear is when it contains a product or a ratio of two variables. By the test of whether or not two variables are multiplied, is text Equation 2.4-1 nonlinear? ______________.

   * * * * *

   * * * * *

   no. (SDT is a constant. Only S is a variable in determining the salesmen hiring rate SHR.)

7. By the multiplication or division of variables test, which equations in text Section 2.5 are nonlinear? _______________.

   * * * * *

   * * * * *

   Equation 2.5-2 and 2.5-10

   (In Equation 2.5-2, both S and SE are variables. In Equation 2.5-10 both BL and DR are variables. In the other equations, variables enter as sums or as multiplications by constants.)

8. It is often more convenient to show on a graph how two variables are related rather than write an equation for how one depends on the other. Suppose that y depends on x as shown in Figure W3.3-8.

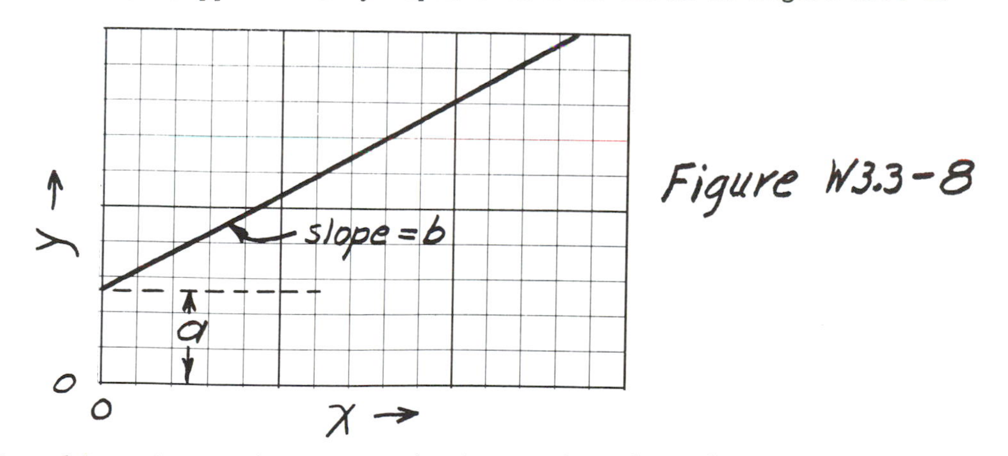

Figure W3.3-8

For this relationship, write by inspection the relationship of y to x:

y = ________ + ________.

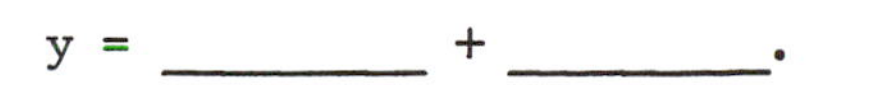

Is this a linear equation? ____________________.

* * * * *

* * * * *

$y = a + bx$

Yes. (The variable x is multiplied only by a constant.)

9. In Figure W3.3-9, the relation between y and x is

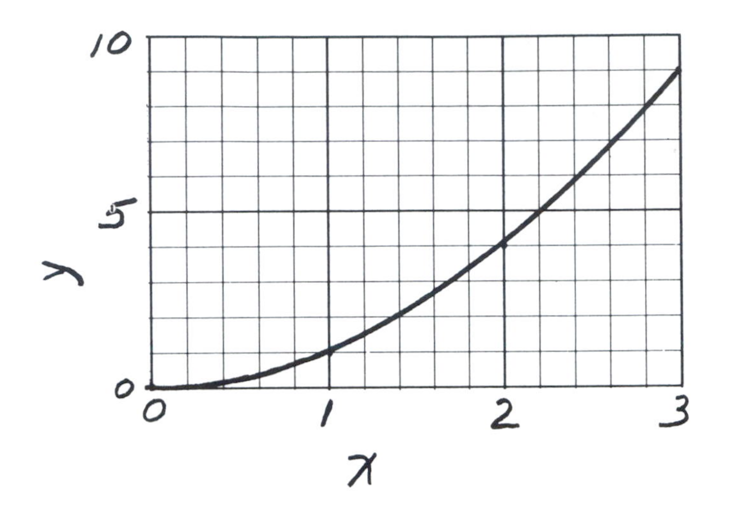

Figure W3.3-9

$$y = x^2$$

Is this a linear equation? ______________.

* * * * *

* * * * *

No. (It is x times x, the product of two variables, here the same variable repeated.)

10. Any graphical relationship between two variables that is not a straight line is nonlinear. Such nonlinearities occur in text Section 2.5 for generating which variables? ______________

____

* * * * *

* * * * *

delivery rate in Figure 2.5b

sales effectiveness in Figure 2.5c

(With these, and the two nonlinear equations identified in Frame 7, there are four sources of nonlinearity in the system of text Section 2.5, any one of which would prevent our obtaining an analytical solution. Simulation by step-by-step computation remains as the only attack.)

CHAPTER W4

STRUCTURE OF SYSTEMS

W4.1 Closed Boundary

1. Suppose we wish to construct a model of a commodity market to show how supply and demand reactions to changes in price can cause fluctuating production and price. Is it necessary in such a model to include government policies for price support of the commodity?

   ________________.

   * * * * *

   * * * * *

   No. (Government price support policies are not required to cause fluctuating production and price. Instability has occurred in commodity systems long before government price support has been attempted.)

2. If we wish to show why international efforts to stabilize the price of tin have failed, must our commodity model include the policies established by treaties and the stabilization authority? ______________.

   * * * * *

   * * * * *

   Yes. (Here the objective of the model deals with the interaction between the regulatory function and the basic supply and demand situation.)

3. Suppose we wish to construct a model of a distribution system for consumer durable goods (refrigerators, automobiles, furniture) to show how such a system can amplify any changes in retail purchase rate. Which of the following must be included in the model?

   a. National military budget

   b. Retail inventories

   c. Consumer confidence

   d. Consumer disposable income

e. Purchase policy at distributor

f. Factory new product research

g. Delay in filling orders at the factory

* * * * *

* * * * *

b, e, g (The others may have an influence on the retail purchase rate, but that is not the question. The model is to show how the distribution system amplifies changes in retail sales. It does not matter what causes the changes.)

4. Suppose we wish to model the processes of assigning engineers to a research project to study how project cost and work efficiency depend on the timing of putting men on the job and reassigning them as the job nears completion. Are the information channels from market to research department involved? These are the information channels that indicate which products are needed by the customers. ________________.

* * * * *

* * * * *

No. (The efficiency in executing a specific project can be independent of the suitability of that product for the market.)

5. But suppose the research department model is addressed to how well the product stream from research matches the demands of the market. Must the customers and information channels to and from the market be included? __________________.

* * * * *

* * * * *

Yes. (Now the system involves the selection of research department tasks in accordance with the mismatch between the customer's needs and the company's present products. The customers are part of the system.)

W4-2 Feedback Loop--Structural Element of Systems

1. Water runs from a hole in a tank. What action is being governed by the decision process? ________ ________.
   What information input is controlling the decision? ________
   ________.

   * * * * *

   * * * * *

   The rate of water flow.

   The water level. (The feedback loop is from water level to flow rate to water level.)

2. We switch on the motor of a vacuum cleaner. The motor quickly comes up to the normal operating speed. The torque to accelerate the motor at any instant depends on how far the motor speed is below the normal speed determined by the design and by the electric power frequency. Is there a feedback loop in this situation? ________.
   Identify the loop. ________
   ________.

   * * * * *

   * * * * *

   Yes.

   Motor speed to torque to speed. (The relationship between actual speed and design speed determines the torque which implies the acceleration (action) measured in revolutions per minute per minute, which changes the speed (level) measured in revolutions per minute.)

3. A student wishes to maintain a B grade in a course. He repeatedly decides how much to study. What is the feedback loop surrounding the decision? ________
   ____
   ____
   ________.

* * * * *

* * * * *

The decision controlling his study time (action) changes his knowledge (level) about the subject. His perception of his knowledge causes further modification of study time.

4. A teacher wishes to set the pace of the subject matter for maximum learning rate. There is an optimum amount by which the material being presented should be ahead of the comprehension of the students. If the material is too advanced, the students fail to understand and are discouraged. If the material is presented too slowly it lacks challenge and the students become bored. What feedback loop governs the rate at which new material is introduced? ____

____

____.

* * * * *

* * * * *

The teacher compares the level of the material being taught with the level of accomplishment of the students to control the rate of introduction of new material.

5. Management must decide how much to expand the field service department which maintains the company's products that are in use by the customers. What feedback loop surrounds this decision?

____

____

____

____

____.

* * * * *

* * * * *

The decision to add servicemen (action) increases the size of the service department (level) which causes a change (action) in the customer satisfaction (level) which is observed by management to further change the number of servicemen.

W4.3 Levels and Rates--the Substructure Within Feedback Loops

1. A house owner devotes a part of his time to repairing his property to maintain its physical condition. What is the level variable in the feedback loop that surrounds his decision? ______________.
The rate variable? ____________________________________.

* * * * *

* * * * *

Condition of the house

Fraction of his time establishing the repair rate.

2. One sleeps a part of each day to control his degree of weariness. What is the decision (rate) in this feedback loop that controls one's condition? ______________________________________.
What is the associated level? ______________________________.

* * * * *

* * * * *

hours per day of sleep. (Determining the negative flow of "weariness per day.")

degree of weariness

3. An executive buys more manufacturing equipment when the backlog of unfilled orders rises too high. What level variables exist in this feedback loop? ____________________________________
____.

* * * * *

* * * * *

Amount of manufacturing equipment in use, and the backlog of unfilled orders.

4. In the system of the preceding Frame 3, what decisions (or rates) are involved? ____
____.

* * * * *

* * * * *

The rate of purchase of equipment, the rate of production of orders.

5. As stated in Frame 3, on what level does the equipment purchase rate depend? ____

____.

* * * * *

* * * * *

The backlog of unfilled orders

6. In the system of Frame 3, on what level does the production rate depend? ____.

* * * * *

* * * * *

The amount of equipment. (The production rate will also depend on the order backlog when the backlog falls so low that the equipment can not be fully utilized.)

7. Suppose the expenditure rate for advertising depends on the order backlog and on the financial condition (say for simplicity, the bank balance) of the company. The advertising rate changes the customer attitude toward the product. The customer attitude influences orders. Order backlog controls shipments. Shipments add to accounts receivable. Accounts receivable control the flow of cash to the bank balance. The bank balance is depleted by payments to satisfy accounts payable which arise from the manufacturing expense rate and the advertising expense rate. In the system as described, what levels are mentioned? ______________

____

____

* * * * *

* * * * *

order backlog, bank balance, customer attitude, accounts receivable, accounts payable.

8. In Frame 7, what rates cause changes in the accounts payable?

____

____.

* * * * *

* * * * *

Manufacturing expense rate, advertising expense rate, and the payments from the bank balance. (The first two increase accounts payable, the last decreases.)

9. In Frame 7, on what level does the order rate depend? ____.

* * * * *

* * * * *

customer attitude

10. In Frame 7, what rates cause changes in the level of bank balance? ____.

* * * * *

* * * * *

outward payment rate to satisfy accounts payable, and inward cash flow to satisfy accounts receivable.

11. As stated in Frame 7, does any flow rate depend on another flow rate? __________. Does the present value of any level depend directly on the present value of another level? _______.

* * * * *

* * * * *

no, no

12. In Frame 7, consider the path from advertising rate to order rate to backlog to accounts receivable to advertising rate. Restate the path along this feedback loop identifying all of the alternating rates and levels. Underline the levels.

____

____

____

____

____.

* * * * *

* * * * *

advertising rate, <u>customer attitude</u>, order rate, <u>backlog</u>, shipment rate, <u>accounts receivable</u>, cash flow, <u>bank balance</u>, advertising rate. (Note that no two rates appear sequentially, nor do two levels follow one another.)

13. In the system of Frame 7, how many variables must have initial values to completely specify the condition of the system? ________. In which frame have these been listed? ______________________.

* * * * *

* * * * *

5, Frame 7 (the system levels)

W4.4 <u>Goal, Observation, Discrepancy, Action--Sub-substructure Within a Rate</u>

1. The policy governing the number of janitors in a building is stated in terms of a degree of cleanliness to be maintained. What are the components of this policy? ________________________

____

____

____

____

____

______________________________________.

* * * * *

* * * * *

goal = the cleanliness standard

observed condition = the cleanliness achieved

discrepancy = the difference between goal and condition

action = hiring rate of janitors that depends on the discrepancy between the cleanliness standard and that achieved.

2. The policy has been to hire each month one half the workers needed to bring the work force up to the authorized number.

(Sec. W4.4)

What are the policy components? ________________________________

____

____

____

____

* * * * *

* * * * *

goal = authorized number

condition = actual work force

discrepancy = authorized minus actual

action = half the discrepancy per month

3. The "policy" governing the flow of current from a constant-voltage battery VB into a series resistance-capacitance circuit depends on constants describing the resistance R and the capacitance C and can be stated as:

$$I = \frac{VB - \frac{Q}{C}}{R}$$

I--current into capacitor (coulombs/second)
VB--Voltage from battery (volts)
Q--charge on the capacitor (coulombs)
C--Capacitance of the capacitor (coulombs/volt)
R--resistance (volts/coulomb/second)

On what system level does the current I depend? ________________

* * * * *

* * * * *

the charge Q. (Q is the charge resulting from the accumulation of the current flow.)

4. In Frame 3, what is the goal of the control process? ____________

____

* * * * *

* * * * *

VB (To bring the voltage across the capacitor, Q/C, up to the battery voltage VB.)

5. The "policy" describing the outcome of a serious disease states that, of those afflicted, 4% recover each day and 1% die. On what flow rates does the number afflicted depend? ____________

____

____

* * * * *

* * * * *

1. Those catching the disease (men/day)
2. Those recovering (men/day)
3. Those dying (men/day)

6. In Frame 5, what are the components of the recovery rate policy?

____

____

____

____

______________________________________.

* * * * *

* * * * *

goal = zero afflicted persons

condition = number of afflicted persons

discrepancy = goal minus actual = 0 minus actual

action = 4% of discrepancy (negative number indicating flow out of afflicted group.)

CHAPTER W5

EQUATIONS AND COMPUTATION

W5.1 Computing Sequence

1. When starting computations at a new time step, the equations of what type are evaluated first? ____________________.

   * * * * *

   * * * * *

   level equations

2. The arbitrary convention used here represents the "present" time for which the equations are being evaluated by the time designator ____________.

   * * * * *

   * * * * *

   K

3. The computations involve the values of levels at time designators __________ and __________ and rates in the intervals __________ and __________.

   * * * * *

   * * * * *

   J, K, JK, KL

4. Levels at time designator __________ do not enter into the computations.

   * * * * *

   * * * * *

   L

5. Which principle in text Chapter 4 might suggest the computing sequence wherein levels and rates are assigned to separate groups for alternate processing? Principle No. __________.

* * * * *

* * * * *

Principle 4.3-1 (levels and rates represent different types of variables in the substructure of a feedback loop.)

6. The symbol _________ is used to indicate the length of time between successive evaluations of the system equations.

* * * * *

* * * * *

DT (for "difference in time" or "Δ time" or "delta time" corresponding to the terminology used in several professional fields.)

7. Which principle in text Chapter 4 suggests that the rate equations can be evaluated independently of one another? _______________________.

* * * * *

* * * * *

Principle 4.3-6 (Rates depend only on levels and constants, not on other rates.)

8. The solution interval DT (does/does not) need to equal the unit of time used as a measure of system variables.

* * * * *

* * * * *

does not (as will be discussed in a later section, DT should be a fraction--less than half--of the shortest delay existing in the system.)

9. Which principle in text Chapter 4 suggests that the level equations can be evaluated in any sequence? _______________.

* * * * *

* * * * *

Principle 4.3-3 (Levels changed by the rates. Levels do not depend on other levels.)

(Sec.W5.1)

10. On the following figure put on the time designators to show at what point the computation is being done.

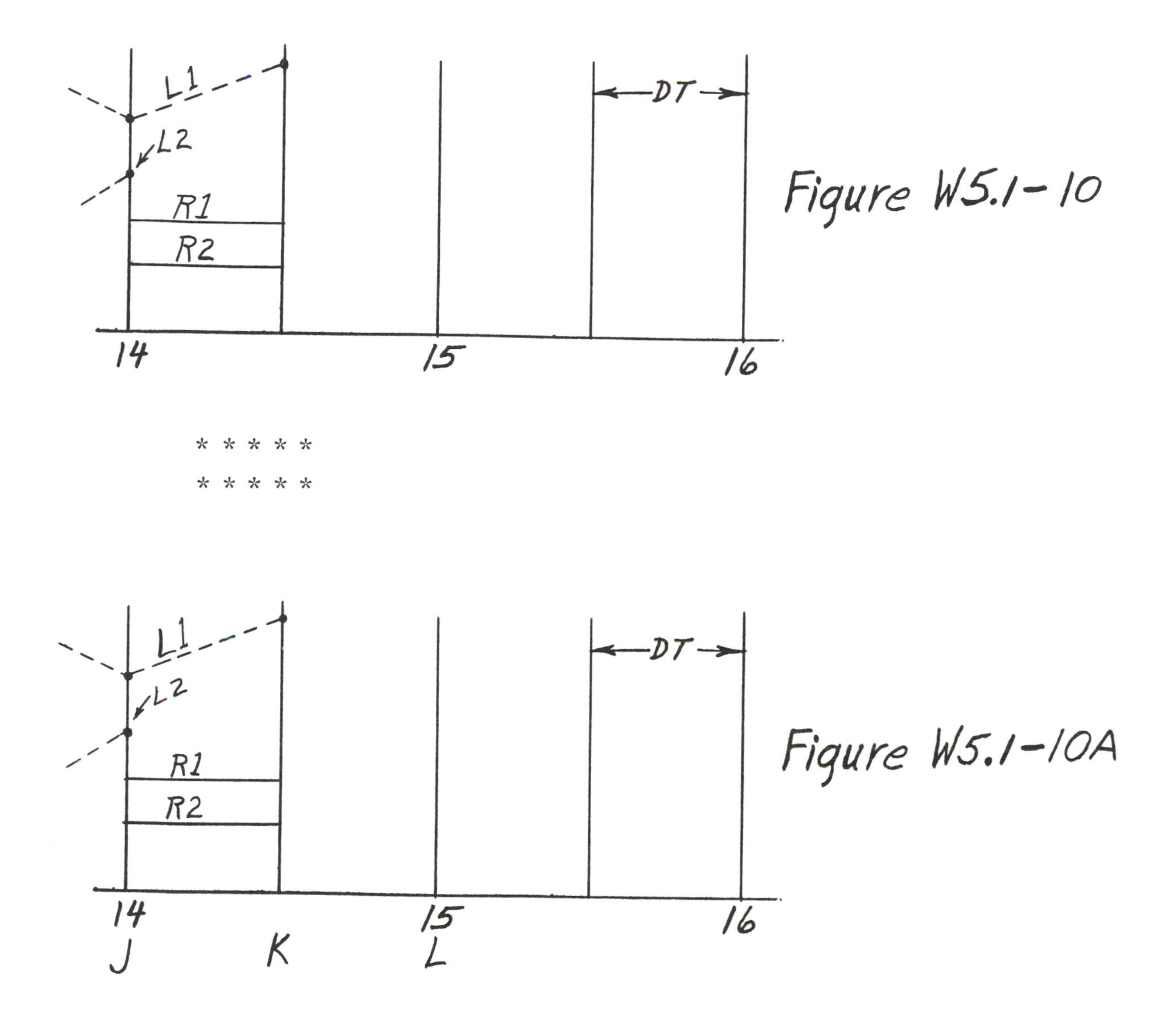

11. On the following figure there are two possible locations for the time designators. Put on both sets.

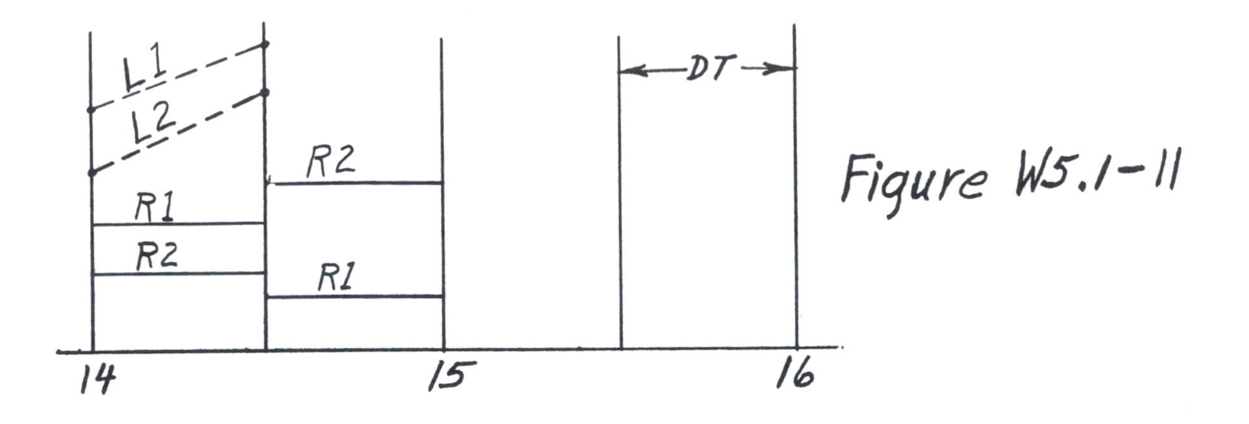

Figure W5.1-11

* * * * *
* * * * *

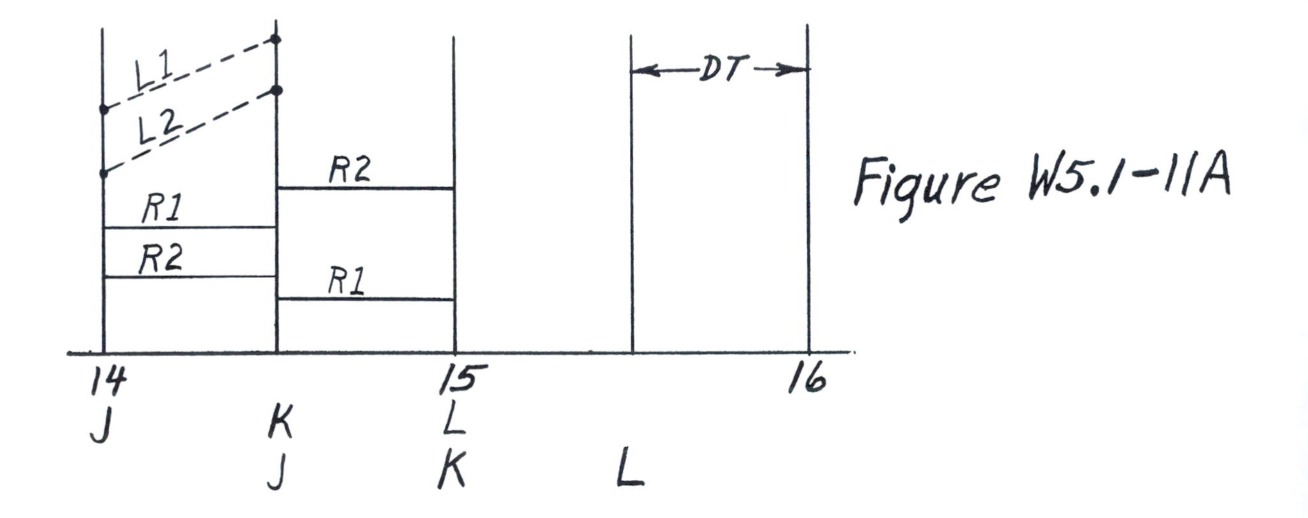

Figure W5.1-11A

(The computation at time 14+DT has been completed. Time K can be at the completed 14+DT or at the ready-to-begin time 15.

(Sec.W5.1)

12. If level L is increased by rate R, if R = 80 men per month, and if DT = 0.1 month, compute the change in L. _________.

    * * * * *

    * * * * *

    8 men (Be sure the measure "men" is given.)

13. In the following figure, where the variables are plotted to scale, rate R1 decreases level L and rate R2 increases L. Plot the level L from time J to time K.

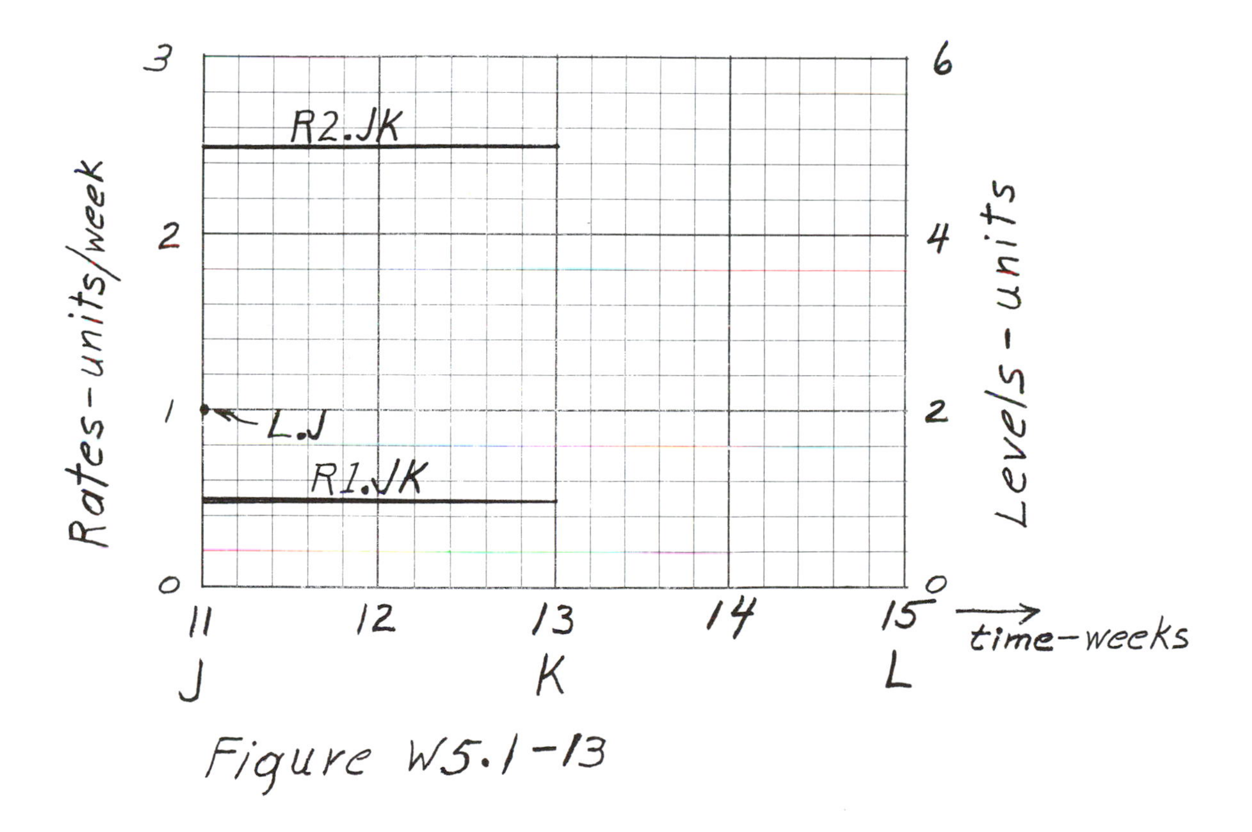

Figure W5.1-13

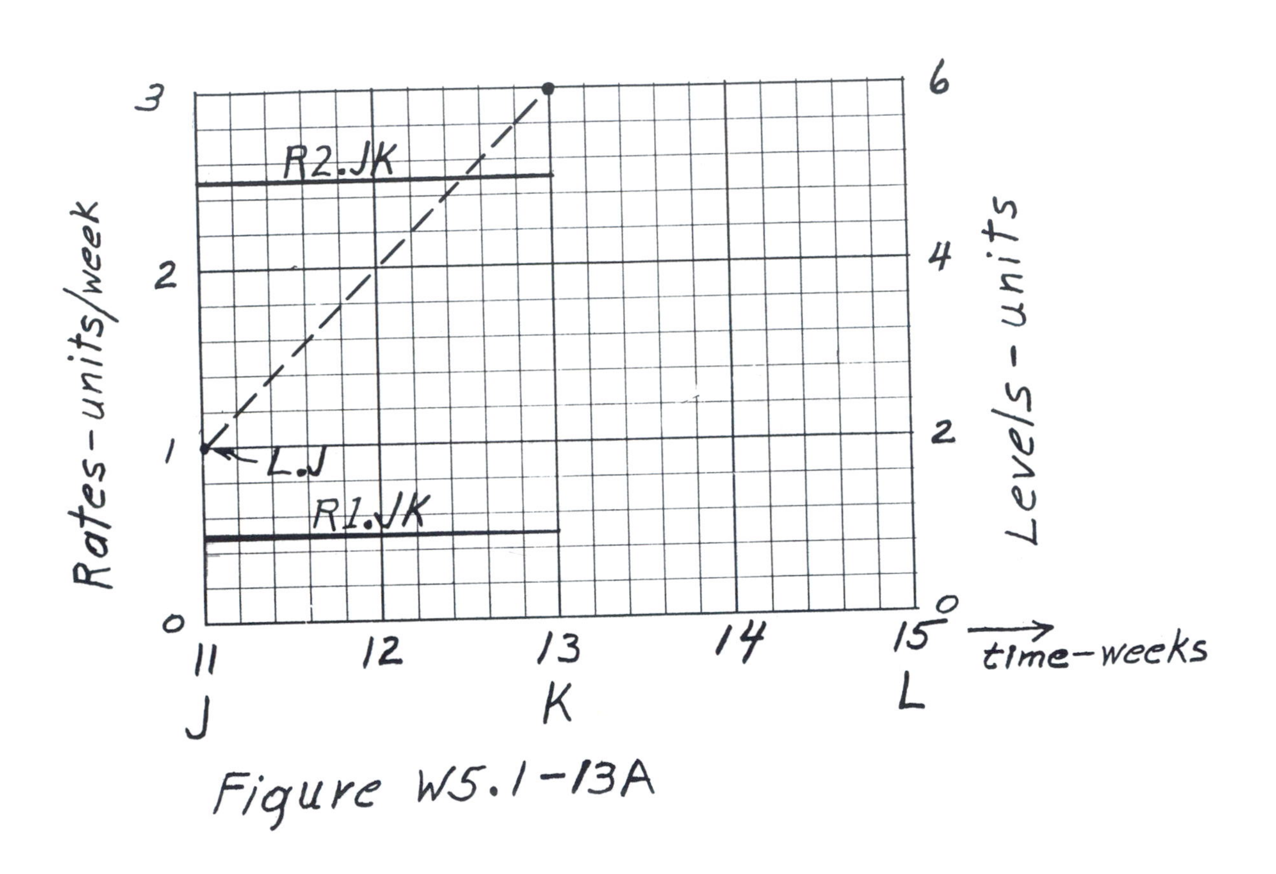

Figure W5.1-13A

(Note that here the solution interval DT is twice the basic time unit of the system. The change in L is

$$2(R2.JK-R1.JK) = 2\ (2.5-.5) = 4$$

W5.2 Symbols in Equations

1. Which of the following symbols are not permitted by the conventions adopted in the text?

1.	L7.K	5.	AL.J
2.	RTE.L	6.	5M8.JK
3.	4BC.K	7.	AM67ABK
4.	B47.LM	8.	BC.K

* * * * *

* * * * *

2. If this is a rate, K has been omitted from the time postscript; if a level, the postscript L is not permitted

3. First character must be alphabetic

4. Time interval LM not permitted for rates

6. First character must be alphabetic

7. If a constant, too many characters, not more than six. If a level, the period omitted before K.

2. Which of the following symbol groups are not permitted?

1.	ABC.KL	6.	XYL7
2.	XY.K	7.	BACKLOG
3.	XY.L	8.	CASH273.K
4.	XYL	9.	CAS27.K
5.	7XYL	10.	CASH27J

* * * * *

* * * * *

3, 5, 7, 8, 10

W5.3 Level Equations

1. A backlog of unfilled orders is fed by orders booked and reduced by orders filled. Write the appropriate level equation.

____

* * * * *

* * * * *

BL.K = BL.J + (DT)(OB.JK - OF.JK)

2. A tank of fluid contains an amount of heat. There is a source of heat supplied and two paths of heat loss, heat withdrawn in the exit fluid and heat radiated. Write the level equation for heat.

   ___

   * * * * *

   * * * * *

   H.K = H.J + (DT)(HS.JK - HW.JK - HR.JK)

3. Machine tools MT are expanded by the rate of tools purchased TP and reduced by tools discarded TD and tools sold TS. Write the level equation.

   ___

   * * * * *

   * * * * *

   MT.K = MT.J + (DT)(TP.JK - TD.JK - TS.JK)

4. The labor force LF in a factory is changed by a bi-directional flow, the worker change rate WC. Write the level equation for labor force.

   ___

   * * * * *

   * * * * *

   LF.K = LF.J + (DT)(WC.JK)

5. Circle the errors in the following level equations.

   HRF.K = HRR.J + (DT)(HR.JK + HF.JK)

   RM.K = RM.J + (STR)(MF.JK + LF.JK)

(Sec.W5.3)

* * * * *

* * * * *

HRF.K = (HRR).J + (DT)(HR.JK + HF.JK)

(The new value of the level is based on the value of the same level at the previous time.)

RM.K = RM.J + (STR)(MF.JK + LF.JK)

(The rates must be multiplied by the solution interval DT in a level equation.)

6. Circle the errors in the following level equations.

LB.J = LB.J + (DT)(LS.JK - RR.KL)

PRL.K = PRL.J + (DT)(PRL.JK - PR.JK)

* * * * *

* * * * *

LB.(J) = LB.J + (DT)(LS.JK - RR.(KL))

(The equation should be computing the level at the present time K. It must be based on the past rates in the JK interval.)

PRL.K = PRL.J + (DT)((PRL).JK - PR.JK)

(On the right PRL multiplied by the time DT could not have the same units as PRL on the left side of the equation.)

7. In dealing with our main interest--the dynamic behavior of systems--we need not be concerned with the "fine structure" of the computation process, that is, the step-wise changes in flow rates and the broken-line nature of levels as they change through time. The solution interval DT is always

taken small enough that it can be ignored and the variables can be thought of, in the broad view, as changing continuously. So, consider the accompanying figure without regard to the small solution interval. A level equation accumulates (integrates) its related flows. The figure shows the only rate affecting a level. Draw the value of the level through time starting at zero.

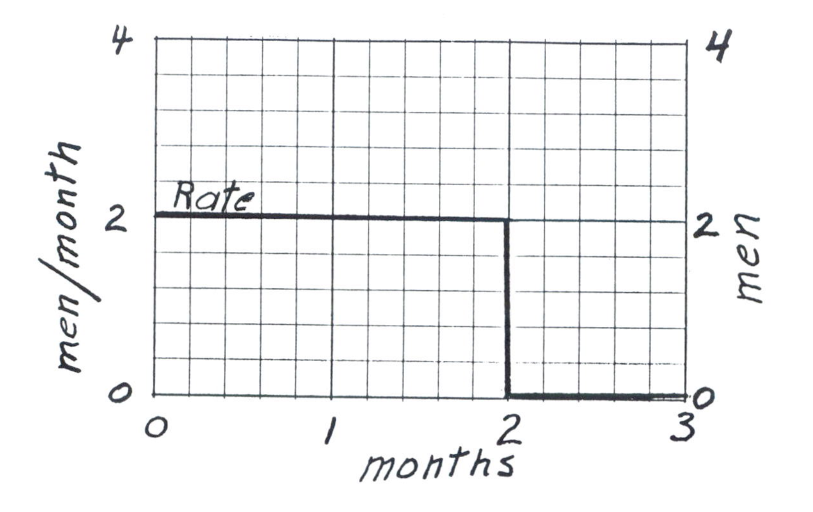

Figure W5.3-7

* * * * *
* * * * *

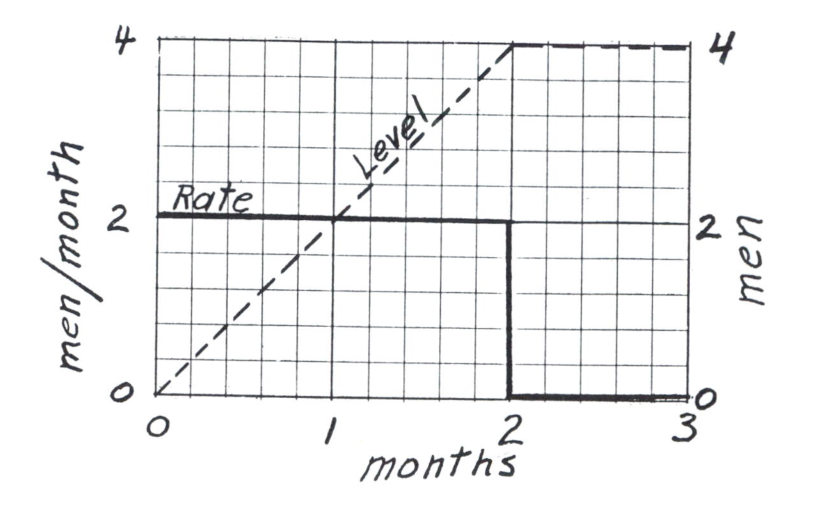

Figure W5.3-7A

Note that the constant rate (the 2 men/month from 0 until 2 months) produces a constantly rising level of men. The reader may object that men are not subdividable into less than single units, but effective men can well be a continuous quantity. For example, men undergoing training do not suddenly jump from entirely useless to fully effective but instead acquire a gradually increasing skill. Many of the "discontinuous" responses in social systems are more apparent than real.

8. Plot the level for the rate in the figure, starting at zero.

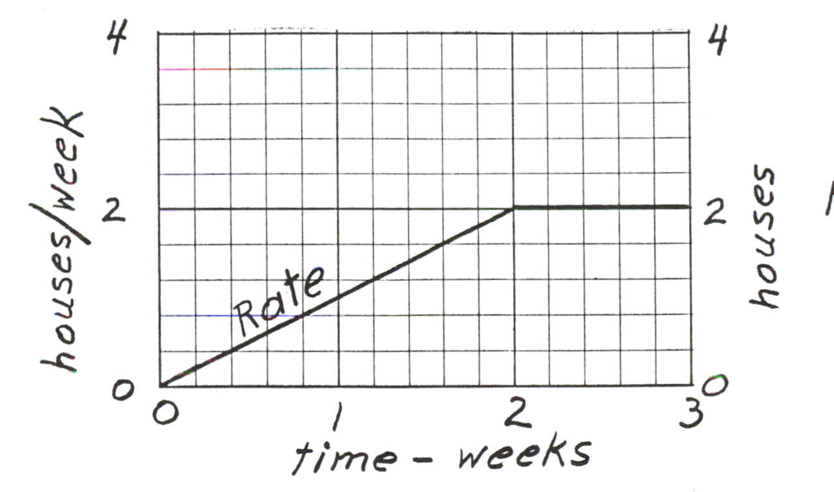

Figure W5.3-8

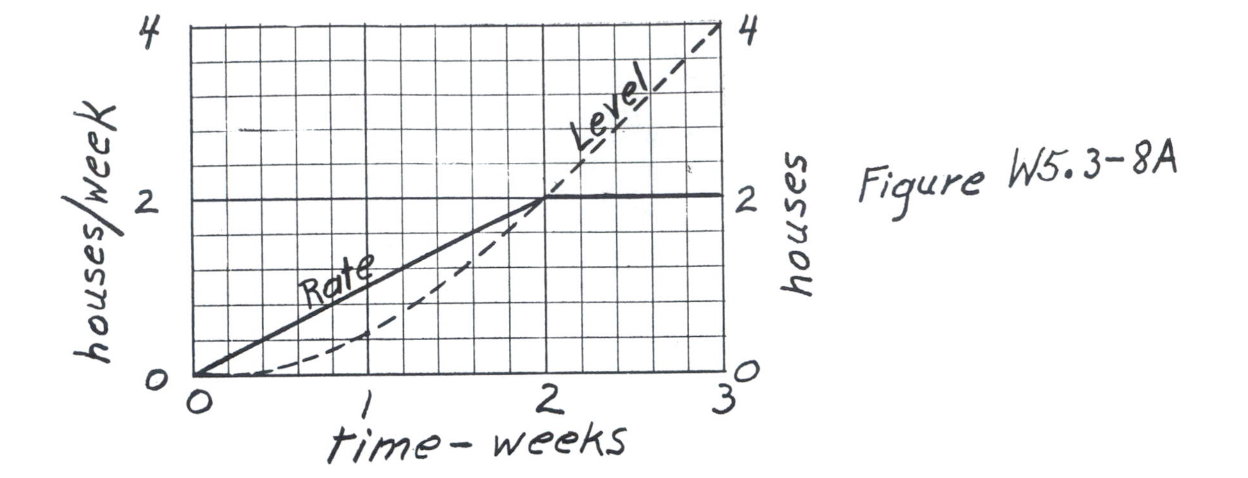

Figure W5.3-8A

Observe that, as is true for all levels, the <u>slope</u> of the level is proportional to the <u>magnitude</u> of the net rate. The higher the rate of flow, the steeper the slope of the line representing the accumulated level. Where the rate is constant, the slope of the level is constant.

9. In the previous figure, the area under the rate curve between 0 and 2 weeks is ______________. Is this the amount of change in the level curve in the same interval? ______________.

* * * * *

* * * * *

2 houses (Half the height times the width of the triangle is $\frac{1}{2}\left(2\ \frac{\text{houses}}{\text{week}}\right)\left(2\ \text{weeks}\right)$.)

yes

(Sec.W5.3)

10. The figure shows a rate for machines produced MP which rises, holds steady, and then falls. Draw to scale the corresponding level for machines M added to inventory starting from zero.

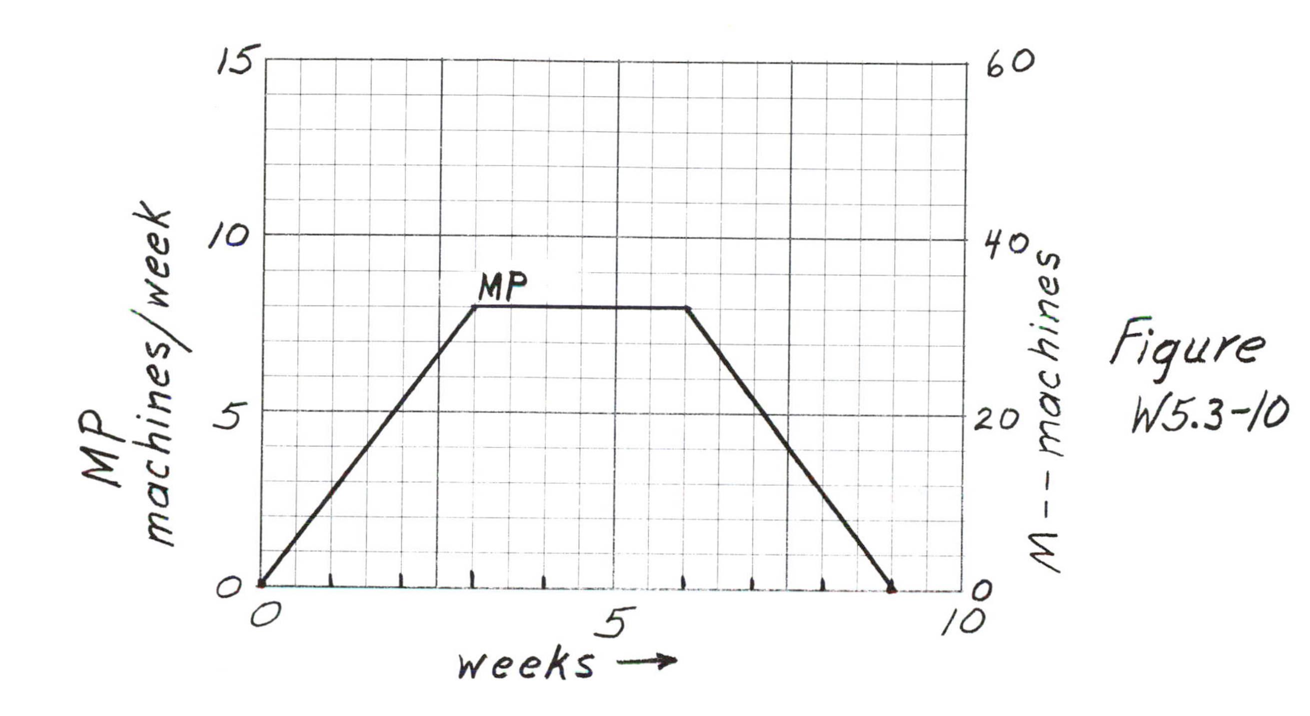

Figure W5.3-10

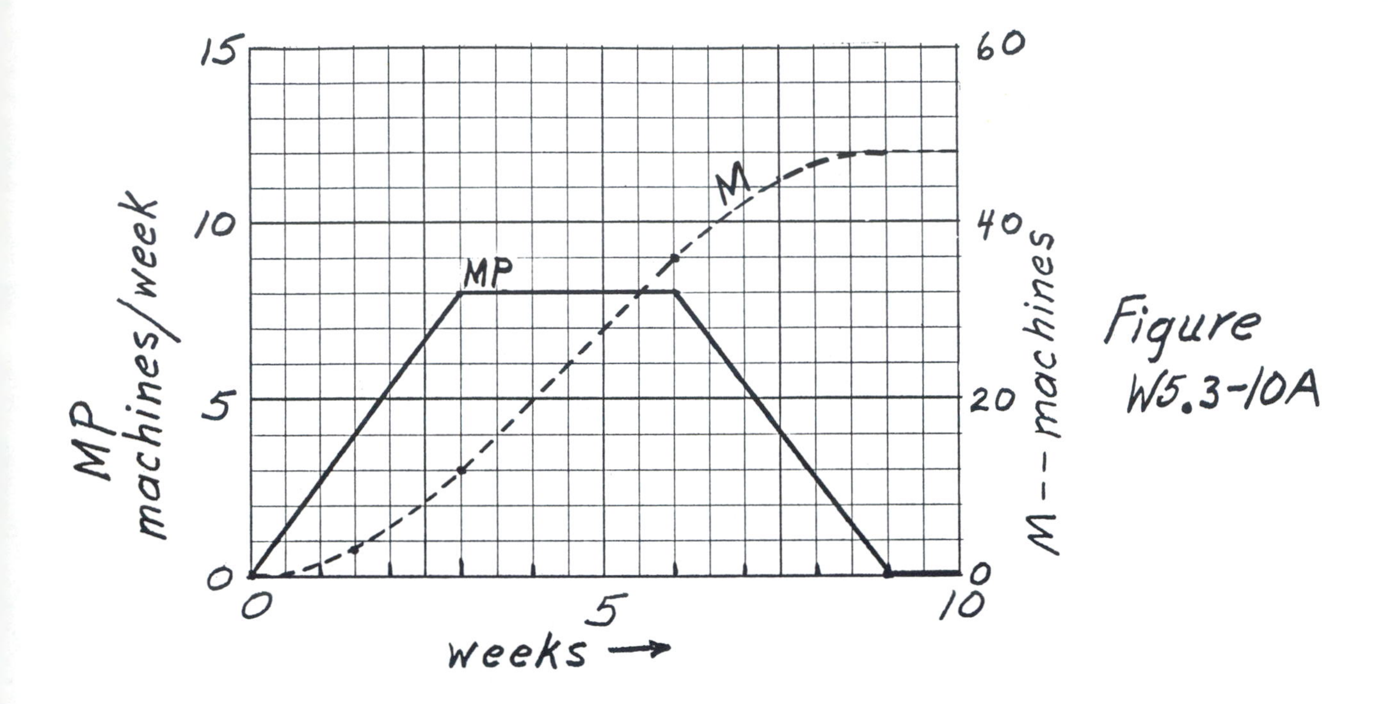

Figure W5.3-10A

Note that at the start the rate is zero, so the slope of the curve for the level is zero (horizontal). As the rate continues to rise, the curve for the level steepens. When the rate is constant at its peak value, the level continues to rise at constant slope. As the rate falls, the level becomes less steep until it is again horizontal at the time the rate returns to zero. At any point in time, the change in <u>height</u> of the level curve is equal to the <u>area</u> under the rate curve up to that time.

(Sec.W5.3)

11. The following figure has the same scales and same shape of rate variation as in the previous frame except that the scales are compressed in each direction and the variation in rate repeated in both the negative and positive directions. Suppose that the rate represents the net flow of machines delivered MD to inventory. A negative rate means more machines were shipped than manufactured. Plot the level of inventory starting from 16 machines.

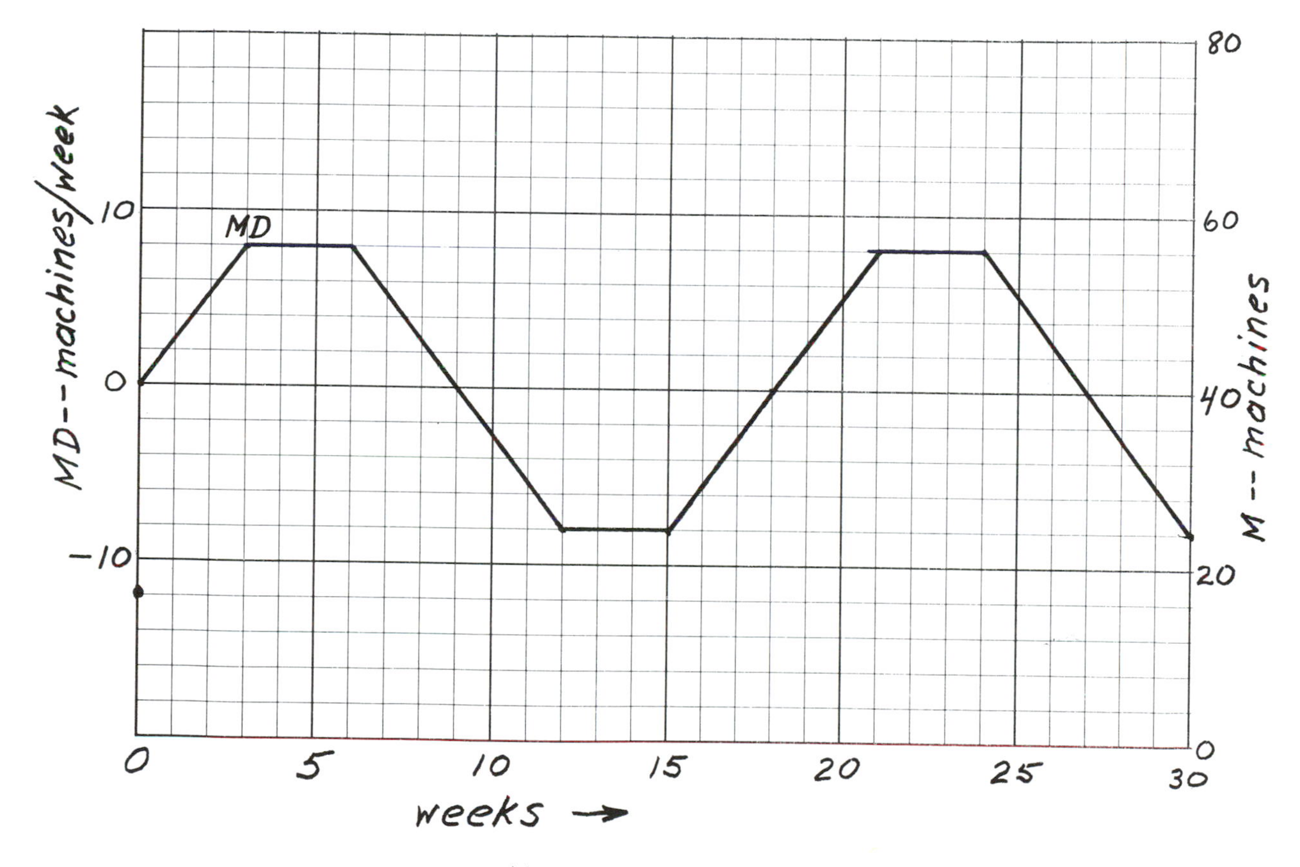

Figure W5.3-11

* * * * *

* * * * *

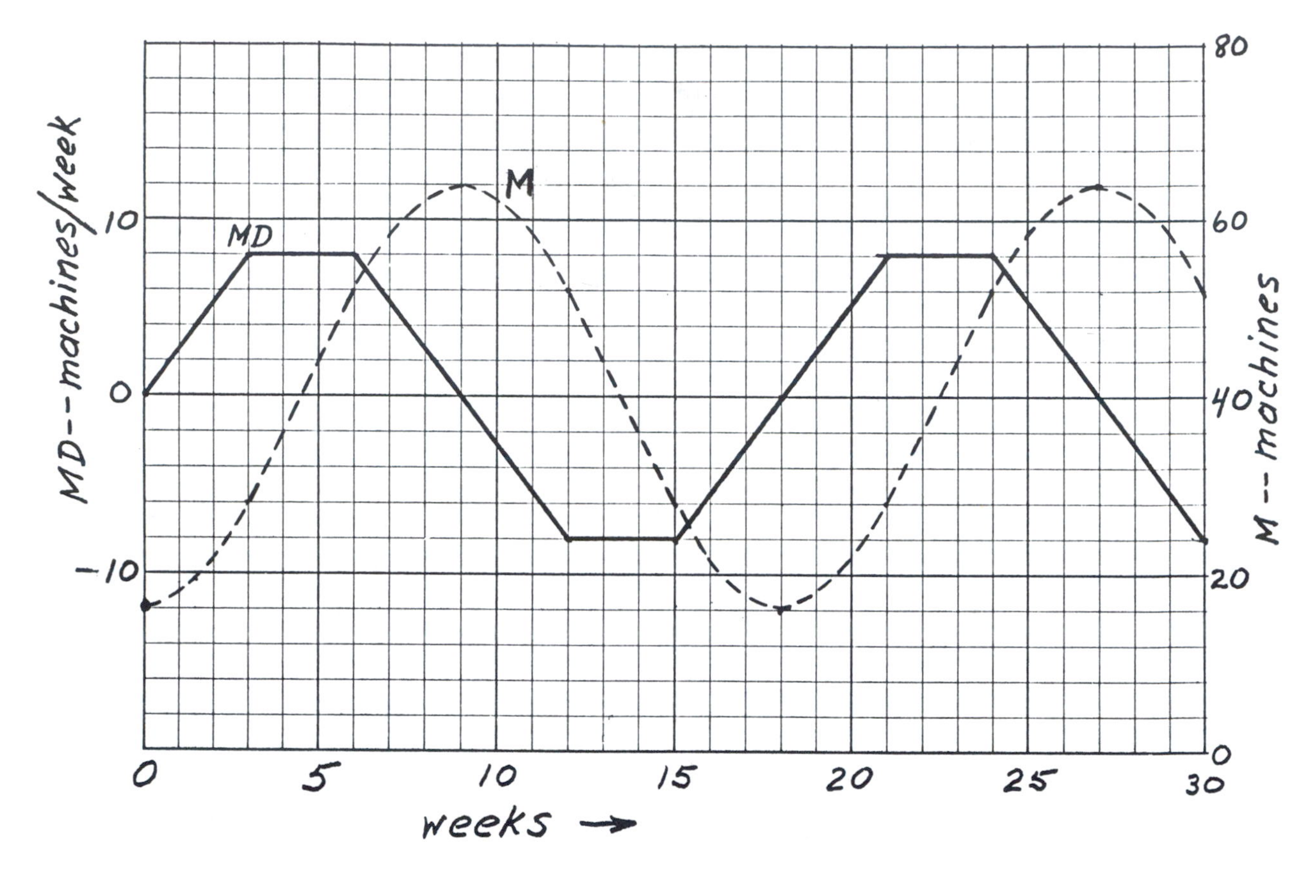

Figure W5.3-11A

Observe how the process of accumulation (integration) that occurs when a rate flows into a reservoir changes the shape and timing of the input. The inventory fluctuates as does the rate of machines delivered. However, the peak in the inventory comes later than the peak in the flow rate curve. Also note that the sharp corners on the flow rate have become rounded corners on the curve for level of inventory.

12. In Figure W5.3-11A, how many weeks does the peak in inventory M lag behind the peak of machines delivered MD? ________________. The "period" of a fluctuating curve is the length of time for a full cycle (here one cycle consists of a positive plus a negative loop in MD). What is the period of the MD curve? ________________. Is the period of the M curve the same? _____. What fraction of the period is the time lag of inventory M behind machines delivered MD? ________________.

    * * * * *

    * * * * *

    4-1/2 weeks

    18 weeks

    Yes

    one-fourth

    A level equation, acting on a symmetrically fluctuating rate, always produces a one-quarter period time delay. In general, integration (accumulation) produces a quarter period delay in any sinusoidal fluctuation.

W5.4 Rate Equations

1. Suppose that the men hired per month is proportional to the number of requisitions for men RM that are unfilled in the personnel office. Write the rate equation for men hired.

    ____

    * * * * *

    * * * * *

    MHM.KL = (P)(RM.K)

2. In Frame 1, RM is a ________________ variable.

    * * * * *

    * * * * *

    level

3. In Frame 1, the units of measure of the quantities are

MHM-- ______________________

P-- ______________________

RM-- ______________________

* * * * *

* * * * *

MHM--men/month, or men per month

P--1/months

RM--men

4. What is the physical meaning of P in Frame 1?

____

____

* * * * *

* * * * *

P is the fraction of the requisitions that are filled per month.

5. The rate equation in Frame 1 could be written as

$$MHM.KL = \frac{RM.K}{T}$$

What are the units of measure of T? ______________.

What is the physical meaning of T when written in this form as the reciprocal of P?

____

____

* * * * *

* * * * *

months

T is the average time necessary to find a man to fill a requisition.

6. Are all the following statements correct? ______________.

NPRF.KL = (1-FT)(APBTF.K)

NPRF--Net profit rate at factory (dollars/month)
FT--Fraction of gross profit to taxes (dimensionless)
APBTF--Average gross profit before tax at factory (dollars)

* * * * *
* * * * *

No. The left side is measured in dollars per month, the right side in dollars. The average profit APBTF is not properly defined as measured in dollars but instead should be in dollars per month.

7. Circle the errors in the following rate equations.

PO.KL = EAM.K/SEPO.K

$$\text{OP.KL} = \frac{\text{(POUN.K)(FO.K)(NFO.J)}}{\text{DOP}}$$

EMBE.JK = EMT.J/DEMT.K

EMH.KL = EM.JK/FEMH.K

PMR.KL = (AP.K)(DT)(FMPM)

* * * * *
* * * * *

$$\text{OP.KL} = \frac{\text{(POUN.K)(FO.K)(NFO.}\boxed{\text{J}}\text{)}}{\text{DOP}}$$

The rate depends only on levels at time K, not time J.

EMBE.(JK) = EMT.(J)/DEMT.K

The equation defines the rate for the KL period. Only levels from the K period should appear on the right.

EMH.KL = (EM.JK)/FEMH.K

Rates (assuming JK indicates a rate) should not appear on the input side (right-hand side) of a rate equation. If EM is a level, the time postscript should be K.

PMR.KL = (AP.K)((DT))(FMPM)

The solution interval DT should not appear in a rate equation, only in level equations.

8. Circle the errors in the following rate equations.

RNDM.KL = (NDM.K)(FRDMRTM.K)

RWDM.KL = (WDM.J)(FRDM.JK)

HDM.JK = (W2DM.KL)(4FIDM.K)

DCT.K = DCT.J+(DT)(CMD.JK-DMD.JK)

* * * * *

* * * * *

RNDM.KL = (NDM.K)(FRDMRTM.K)

Too many characters in the symbol.

RWDM.KL = (WDM.J)(FRDM.JK)

The level should be from time K. The last term incorrectly states a level or is a forbidden rate.

The rate is defined for the KL interval. W2DM is an incorrectly stated level or a forbidden rate. In the last term, our arbitrary convention forbids an initial numeral in a symbol.

DCT.K = DCT.J+(DT)(CMD.JK-DMD.JK)

This is a level equation, not a rate equation.

(Sec.W5.4)

9. The following figure shows how a level has changed during 15 months. Sketch on the figure the net flow rate which must have been the input to the level.

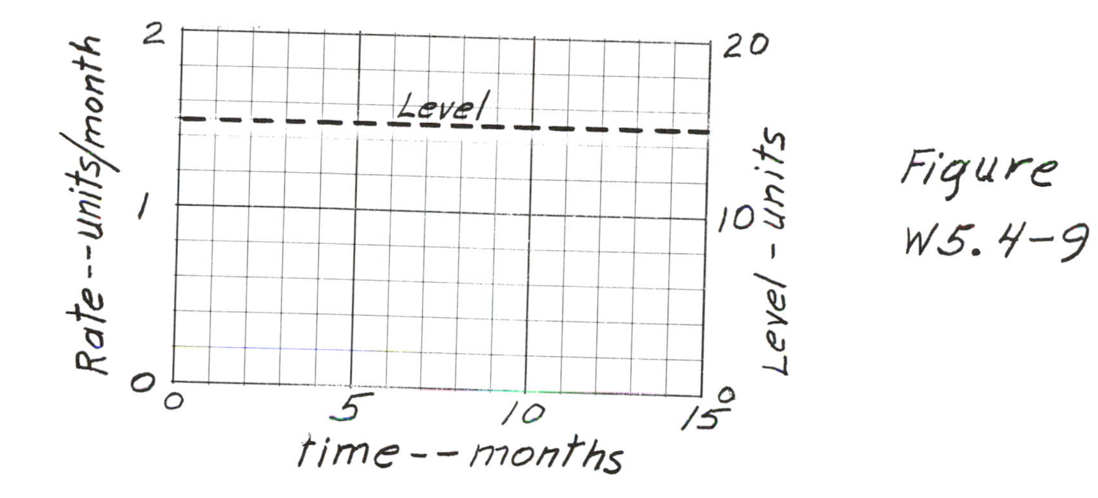

* * * * *
* * * * *

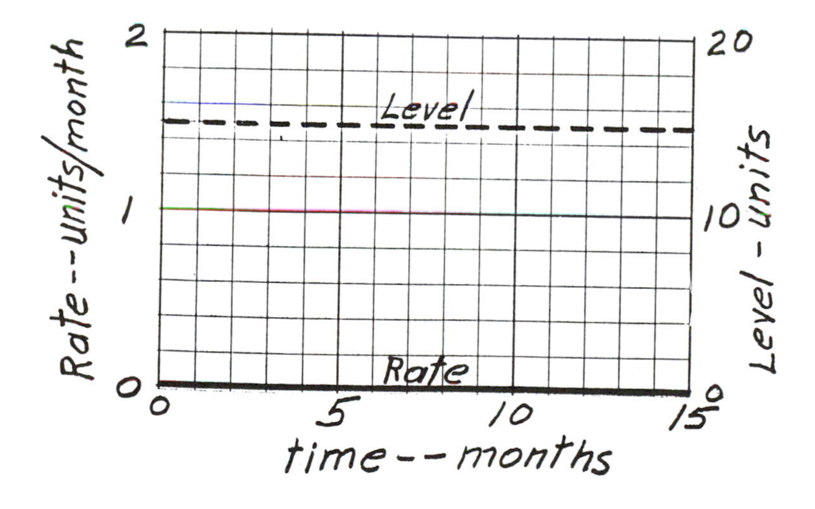

Figure W5.4-9A

The level was not changing so the net rate must have been zero.

10. The following figure shows the history of a level variable in a system. Show on the figure the net flow rate which must have existed to cause the change in the level.

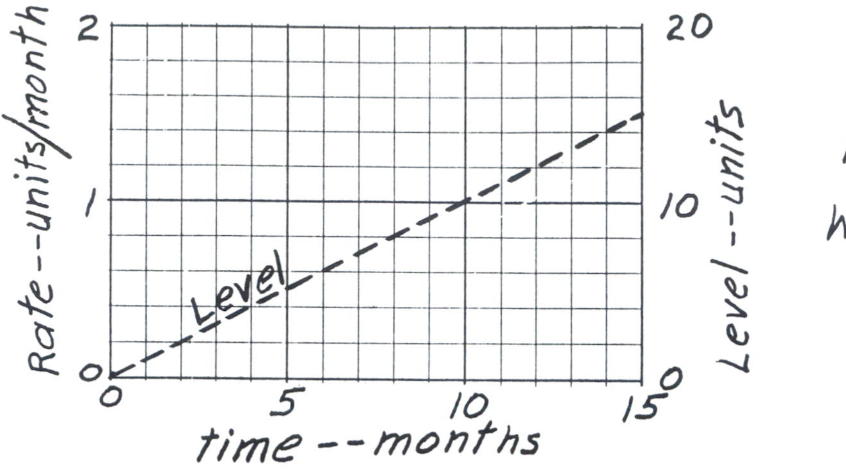

Figure W5.4-10

* * * * *

* * * * *

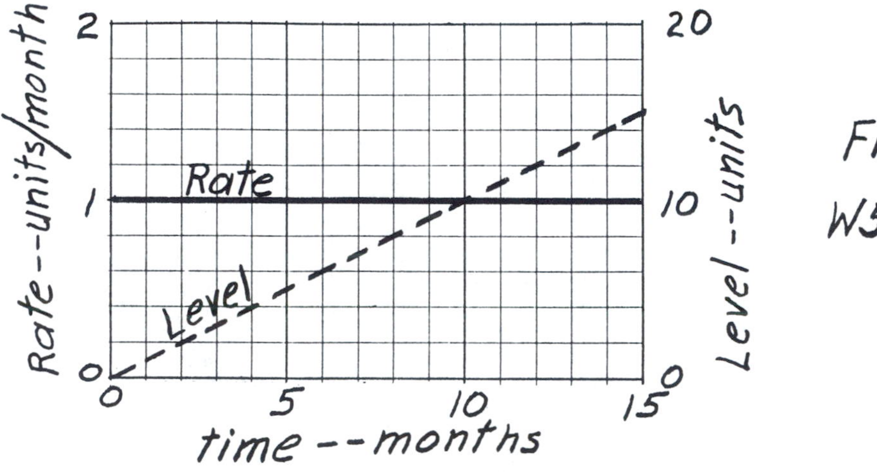

Figure W5.4-10A

The rate must have been constant to produce the uniform slope in the level curve. The rate equals the slope of the level. The area under the rate curve must equal the change in the height of the level.

(Sec.W5.4)

11. Sketch the net rate that must have acted to produce the change in level in the accompanying figure.

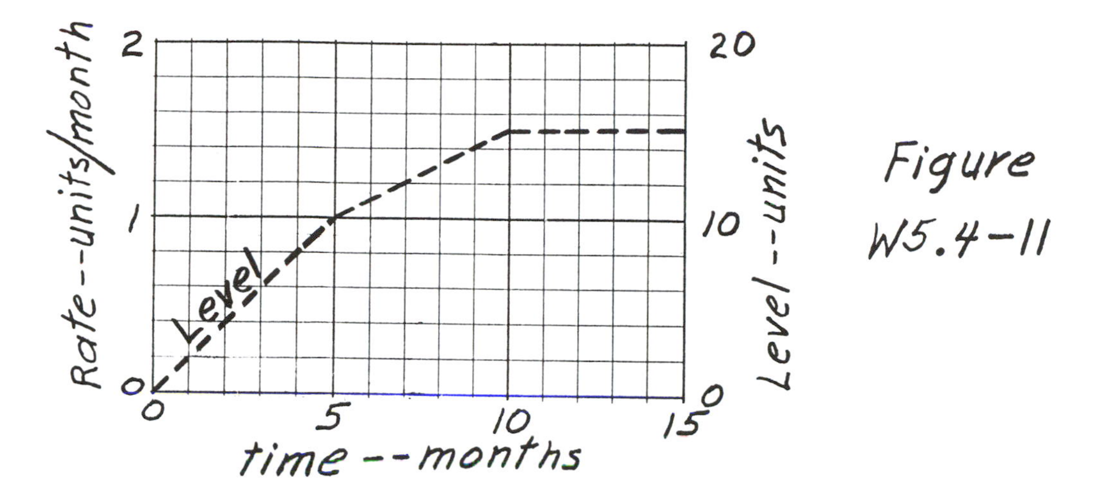

* * * * *
* * * * *

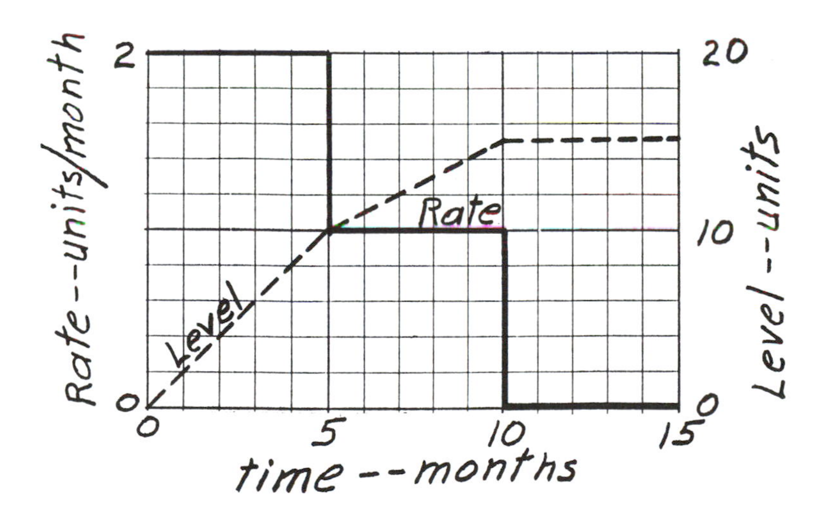

Figure W5.4-11A

12. The following fluctuation has occurred in a system level. Sketch the rate which could be the cause.

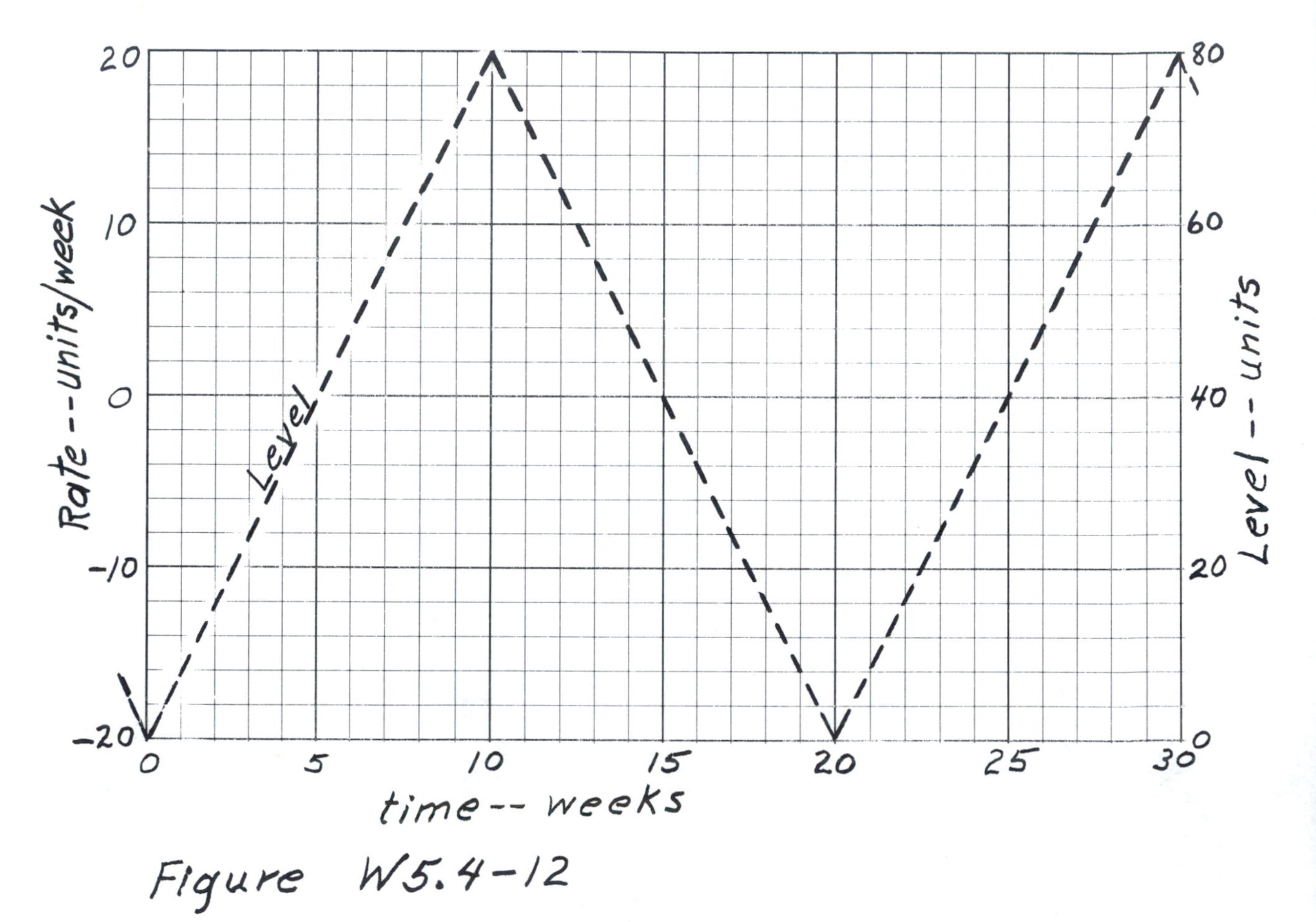

Figure W5.4-12

(Sec.W5.4)

* * * * *

* * * * *

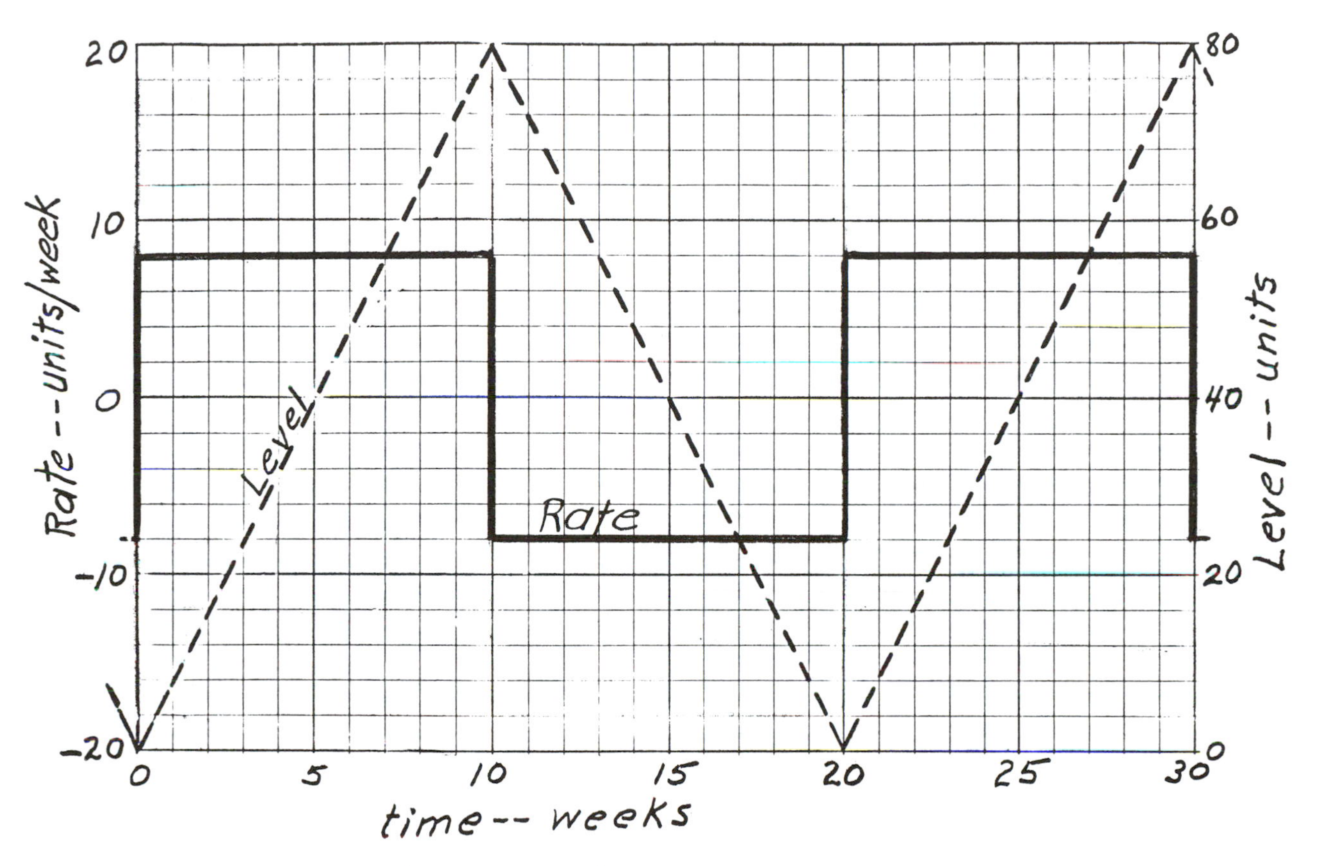

Figure W5.4-12A

13. In Figure W5.4-12A, by how many weeks does the level lag behind the rate? Measure from the center of the square-wave rate to the peak of the triangular-wave level. ________________. What is the period of the fluctuation? _____________. What fraction of a period does the level lag behind the rate? ______________.

* * * * *

* * * * *

5 weeks

20 weeks

one-quarter

Note here again, as in Frame 12 of Section W5.3, that the process of integration in a level equation makes a quarter-period phase shift with the level lagging behind the net rate. This quarter-period shift is also called a 90-degree phase shift (the full 360-degree circle is trigonometrically related to one full period.)

W5.5 Auxiliary Equations

1. Substitute the following auxiliary equation into the rate equation.

OR.KL = (ASE.K)(SEC)(DDM.K)

DDM.K = (TDDM)(MDDR.K)

____

* * * * *

* * * * *

OR.KL = (ASE.K)(SEC)(TDDM)(MDDR.K)

(Sec.W5.5)

2. Eliminate the following auxiliary equations. Write the full set of equations without auxiliaries.

MDDR.K = ADD3.K/NMDD
ASE.K = ASE.J+(DT)(SM.JK-A.JK)
OR.KL = (ASE.K)(SEC)(DDM.K)
DDM.K = (TDDM)(MDDR.K)
ADD3.K = ADD3.J+(DT)(DD2.JK-DD3.JK)

____

____

____

____

* * * * *
* * * * *

OR.KL = (ASE.K)(SEC)(TDDM)(ADD3.K/NMDD)
ASE.K = ASE.J+(DT)(SM.JK-A.JK)
ADD3.K = ADD3.J+(DT)(DD2.JK-DD3.JK)

The equations for ASE and ADD3 are level equations. The auxiliary equation for MDDR has been substituted into the auxiliary equation for DDM and that into the rate equation for OR.

3. What error exists in the following set of auxiliary equations that all are from the same system formulation:

BR.K = MB.K+S.K+DC.K
A.K = BR.K/T
S.K = (A.K+F.K)(L)

____

____

____

* * * * *

* * * * *

The equations represent a closed loop of auxiliary equations without intervening levels and rates. BR depends on S which depends on A which depends on BR from the first equation. There is no sequence that leads only from values of levels to the computation of rates.

4. Mark the errors in the following set of equations.

BL.K =	BL.J+(DT)(OR.JK-PR.JK)	a.
DD.K =	(BL.J/ASR.J)+FCC.K	b.
FEDR.K =	DD.K/NEDD	c.
FCC.K =	(CF)(FEDR.K)	d.
PCA.JK =	(PC.K)(FCC.K)	e.
PC.K =	PC.J+(DT)(PCA.KL)	f.

* * * * *

* * * * *

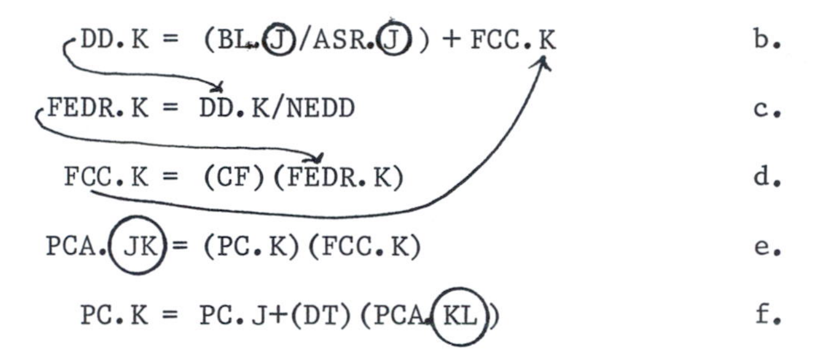
DD.K = (BL.J/ASR.J) + FCC.K
b.
FEDR.K = DD.K/NEDD
c.
FCC.K = (CF)(FEDR.K)
d.
PCA.JK = (PC.K)(FCC.K)
e.
PC.K = PC.J+(DT)(PCA.KL)
f.

Equation b is an auxiliary and should depend on levels at time K. A circular relationship exists among the three auxiliary equations b, c, and d. Rate equation e should be calculating the value for the KL interval. In the level equation f the value of the rate PCA must be from the JK interval.

W5.6 Constant and Initial Value Equations

The exercises in this section are to give practice in identifying and using the constant and initial value equations. Familiarity with the forms will make the later sections easier to read.

1. The equation

$$MHR.K = (L)(P.K)$$

is of the (level/auxiliary/rate) class. The variable P is a (level/rate). P (can/can not) be an auxiliary variable. L is a ____________.

* * * * *
* * * * *

Auxiliary, level, can, constant

(The equation is not a rate equation or the postscript of the left-side variable would be KL. It is not the standard form of a level equation. The variable P must be a level, or an auxiliary computed from levels, because only levels or auxiliaries can be inputs to auxiliaries or rates. The symbol L without a time postscript indicates a constant.)

2. AB = 7 C
BC = 8 C
CD = (AB)(DE) N
DE = (5)(BC) N

The value of CD, as an initially computed constant, is ________________.

* * * * *
* * * * *

280

(Sec.W5.6)

3. What constants appear in the following equations?

A.K =	A.J+(DT)(R.JK-S.JK)	L
A =	16	N
T =	(C)(D)	N
C =	14	C
D =	(8)(M)	N
R.KL =	$\frac{(V.K)(27)}{(D)(L)}$	R

---

* * * * *

* * * * *

T, C, D, 8, M, 27, L

(A is the initial value of the level equation. 8 and 27 are constants given as numerical values rather than symbolically. DT is also a constant but is a characteristic of the computation procedure and not of the system.)

4. The equation:

L.K = L.J+(DT)(M.JK)　　　L

has an initial value of 30. Write the initial value equation.

---

* * * * *

* * * * *

L = 30　　　N

(Sec.W5.6)

5. The graph of a level variable is:

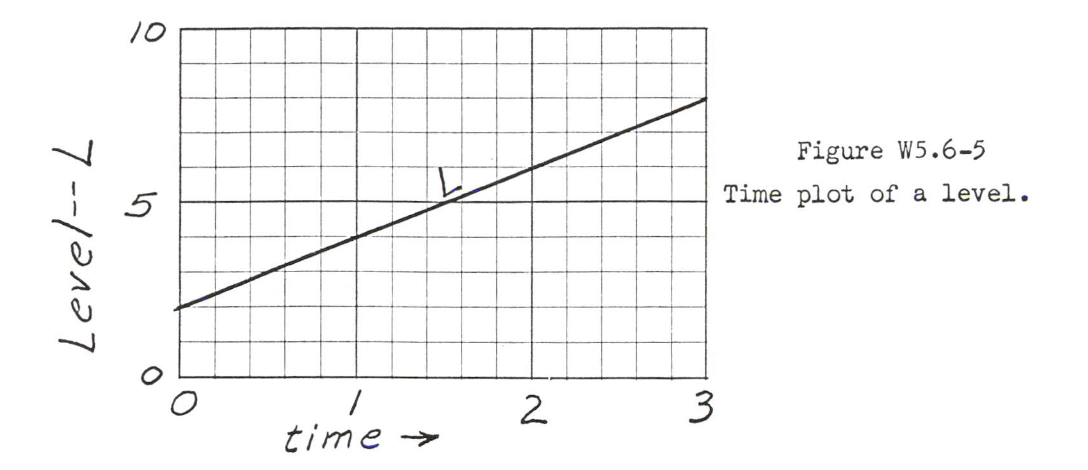

Figure W5.6-5
Time plot of a level.

The equations are

L.K = L.J+(DT)(R.JK)   L
R.KL = C   R

Write the equations for the initial value and the constant.

____

____

* * * * *

* * * * *

L = 2   N

C = 2   C

(The level starts at a value of 2 at time = 0. The level rises at the rate of two units per unit of time as shown by the slope of the line.)

6. The equations of part of a system are:

P.K = P.J+(DT)(Q.JK-R.JK)	L
P = 9	N
Q.KL = 15	R
R.JK = B	R
B = (6)(C)	N
C = 3	C

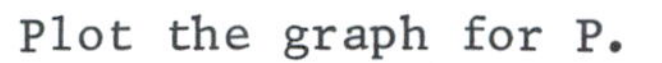
Plot the graph for P.

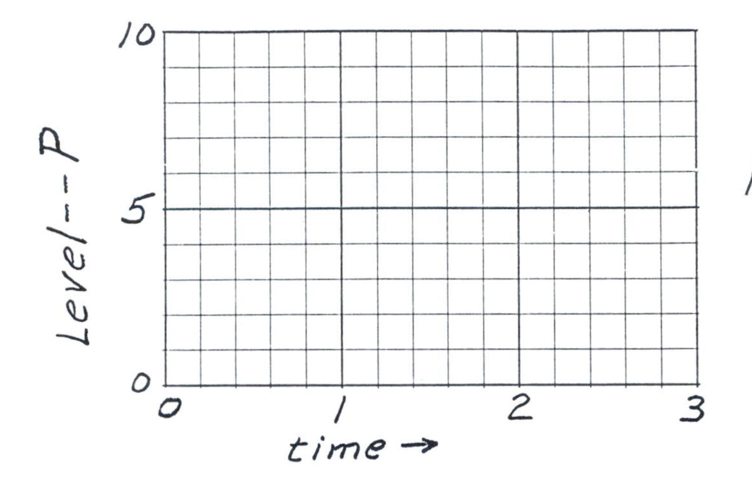

Figure W5.6-6

* * * * *
* * * * *

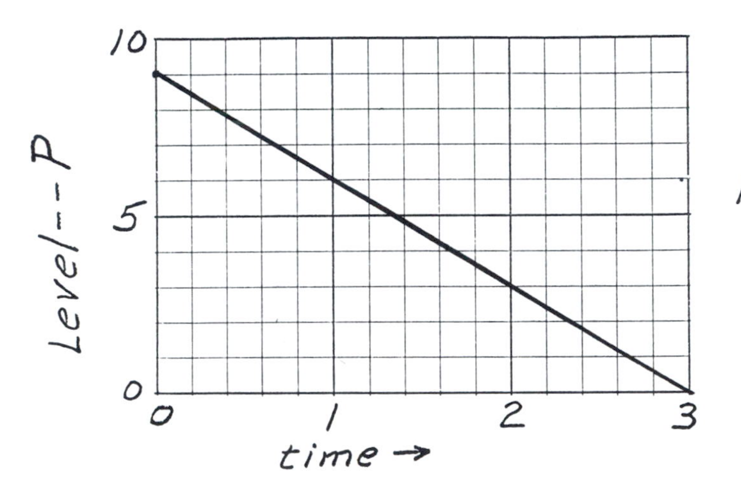

Figure W5.6-6A

(Sec.W5.6)

7. Check those of the following equations that require initial values.

   a. RM.KL = L.K/D

   b. R.K = R.J+(DT)(RM.JK)

   c. S.K = S.J+(DT)(RR.JK-RM.JK)

   d. RP.KL = (M.K)(BL.K)

   e. M.K = A.K+B.K

* * * * *

* * * * *

b, c (These are the only level equations.)

CHAPTER W6

MODELS--MISCELLANEOUS

W6.1 Dimensions

1. Using Principle 6.1-1, find and correct the error in the following rate equation

$$OB.KL = \frac{S.K}{SE.K}$$

OB--Orders booked (units/month)
S--Salesmen (men)
SE--Sales effectiveness (units/man-month)

OB.KL = ______________________________

* * * * *

* * * * *

OB.KL = (S.K)(SE.K)

In the erroneous form above,

$$\text{units/month} = \frac{\text{men}}{\text{units/man-mo}} = \frac{\text{men(men)(mo)}}{\text{units}}$$

with inconsistent units of measure on the two sides of the equation.

2. Correct the error line in the following

IS.K = B.K/SS

IS--Indicated salesmen (men)
B--Budget (dollars/month)
SS--Salesmen salary (dollar/man)

____

* * * * *

* * * * *

SS--Salesman salary (dollars/man-month)

3. Suppose we are modeling the supply sector for a commodity and wish to state that the rate of change in supply capability depends on the difference between equilibrium supply and the existing actual supply capability as follows. Fill in the missing dimensions of measure.

SCC.KL = (1/TASC)(ES.K-SC.K)

SCC--Supply capability change (units/year/year)
TASC--Time to adjust supply capability (____________)
ES--Equilibrium supply (units/year)
SC--Supply capability (____________________)

* * * * *

* * * * *

years

units/year

4. In commodity market stabilization, the transactions for the stabilization pool might have the following form. Fill in the missing dimensions.

SBT.KL = (PD.K)(PIT)(CFT.K)

SBT--Stabilization transactions (units/year)
PD--Price discrepancy (dollars/unit)
PIT--Price influence on transaction

(____________________________)

CFT--Coefficient of transaction (dimensionless)

* * * * *

* * * * *

$\frac{\text{units/year}}{\text{dollar/unit}}$ or units/year/dollar/unit

(Sec.W6.1)

5. What does dimensional analysis suggest is wrong with the following equation?

CRP.KL = P.K-AP.K

CRP--Change rate in price (dollars/month)
P--Price (dollars)
AP--Average price (dollars)

____

____

* * * * *

* * * * *

The dimensions do not balance because dollars/month = dollars. If the change rate in price depends on the difference between price and average price, then a time term is necessary to indicate how fast price changes.

$$CRP.KL = \frac{1}{TCP}(P.K-AP.K)$$

TCP--Time to change price (months)

W6.2 Solution Interval

1. The following equations could represent the time lag in managerial awareness of the changing delivery situation for a company:

$$CMADD.KL = \frac{1}{TCMA}(DD.K-MADD.K) \qquad R$$

$$TCMA = 8 \qquad C$$

$$MADD.K = MADD.J + (DT)(CMADD.JK) \qquad L$$

CMADD--Change in managerial awareness of delivery delay (weeks/week)
TCMA--Time to change managerial awareness (weeks)
DD--Delivery delay (weeks)
MADD--Managerial awareness of delivery delay (weeks)

What solution interval is the longest that could be recommended?

______________________.

* * * * *

* * * * *

4 weeks (Half the time required for the rate CMADD to correct the discrepancy between DD and MADD.)

2. The equation for charge on an electrical capacitor being discharged by a resistor is given by these equations that describe a first-order feedback loop:

$$Q.K = Q.J + (DT)(-I.JK) \qquad L$$

$$I.KL = \frac{Q.K}{(R)(C)} \qquad R$$

$$R = 10 \qquad C$$

$$C = 5 \qquad C$$

Q--Charge (coulombs)
I--Current (coulombs/second)
R--Resistant (volts/coulomb/second)
C--Farads (coulombs/volt)

What fraction of the charge on the capacitor is being dissipated per second? ______________________

* * * * *

* * * * *

$$\frac{1}{50} = \frac{1}{(R)(C)}$$

3. In Frame 2 above, how long would the initial current, if it persisted, require to discharge the capacitor to zero charge?

______________________.

* * * * *

* * * * *

50 seconds

4. In Frame 2 above, what is the time constant, or delay in the capacitor-resistor system? ______________________

* * * * *

* * * * *

50 seconds. (The time constant of a system is the time the present rate of activity would require to bring the system to equilibrium.)

5. In Frame 2 above, if the electrical circuit is the fastest-acting part of a larger system, what range of solution intervals might be considered for simulating the behavior of the system? ______________________________.

* * * * *

* * * * *

10 to 25 seconds

6. A company's reputation in the marketplace for product quality lags behind its actual performance. For durable goods, the lag is related to the normal life of the product. Suppose the following represents the market lag in becoming aware of the quality produced by a refrigerator manufacturer:

$$COQ.KL = \frac{1}{TOQ}(Q.K-OQ.K) \qquad R$$

$$TOQ = 7 \qquad C$$

$$OQ.K = OQ.J + (DT)(COQ.JK) \qquad L$$

COQ--Change in observed quality (quality units/year)
TOQ--Time to observe quality (years)
Q--Quality presently produced (quality units)
OQ--Observed quality (quality units)

Which would be the most appropriate solution interval DT for these equations? 1 week, 1 year, 2 years, 10 years

* * * * *

* * * * *

2 years

7. Suppose the equations in Frame 6 above existed in the same system with the following equations:

```
     PO.K = PO.J + (DT)(POG.JK-POP.JK)        L
   POP.KL = PO.K/DOP                          R
      DOP = 0.5
```

PO	--Potential orders	(units)
POG	--Potential orders generated	(units/year)
POP	--Potential orders placed	(units/year)
DOP	--Delay in placing orders	(years)

The order-placing flow POP can deplete potential orders with what time constant? ______________________.

For the combined equations of Frames 6 and 7 a maximum solution interval might be ____________________.

* * * * *

* * * * *

0.5 year

0.25 year (dependent on the shortest delay, which is represented by the equations in this frame.)

W6.3 Fine Structure of Computation to be Ignored

This section is not directly related to the corresponding text section but reviews principles that appeared in earlier sections.

1. A system is completely described by these equations:

```
     L.K = L.J + (DT)(R.JK)        L
    R.KL = W.K(X.K)(Z)             R
```

This is an (open/feedback) system. ________________.

* * * * *

* * * * *

open (If this is a complete description of the system, there is no connection from the level L back to the rate R. R affects L but L does not affect R.)

(Sec.W6.3)

2. The following equations are to be taken as the complete description of a system:

IR.KL =	(P)(ASR.K)	R
ACTR.K =	ACTR.J + (DT)(IR.JK-IP.JK)	L
IP.KL =	ACTR.K/PD	R

IR--Invoices rendered	(dollars/month)
P--Price	(dollars/unit)
ASR--Average sales rate	(units/month)
ACTR--Accounts receivable	(dollars)
IP--Invoices paid	(dollars/month)
PD--Payment delay	(months)

Is there a feedback loop in this system? ______________.

* * * * *

* * * * *

yes (ACTR to IP to ACTR)

3. The components of a rate equation or policy are:

____________________ ________________________

____________________ ________________________

* * * * *

* * * * *

goal, observed condition, discrepancy, action

4. The levels fully describe the present ____________ of a system. To begin a simulation computation one must have the initial conditions consisting of the values of all _________________.

* * * * *

* * * * *

condition

levels

5. A first-order, negative-feedback loop (can/can not) show oscillatory behavior.

* * * * *

* * * * *

can not (The first-order, negative loop is only capable of simple exponential response toward a goal.)

6. A first-order, positive-feedback loop (can/can not) show oscillatory behavior.

* * * * *

* * * * *

can not (The first-order, positive loop only shows exponential departure from a "goal" or reference level.)

7. The second-order, negative-feedback loop (can/can not) show oscillatory behavior.

* * * * *

* * * * *

can (The second-order negative loop is the simplest feedback structure that can oscillate around the equilibrium goal.)

8. The second-order, negative feedback loop (can/can not) show a non-oscillatory, exponential approach to its goal.

* * * * *

* * * * *

can (Depending on parameter values, the more complex systems can and often do show behavior characteristic of simpler systems.)

9. A positive feedback loop (can/can not) be converted to a negative loop by system-induced changes in the "loop gain" or amplification.

* * * * *

* * * * *

can (In Figure 2.5d, the positive loop that generates growth converts to a neutral or negative loop as the sales effectiveness falls.)

(Sec.W6.3)

10. People enter a room at the uniform rate of 10 people per minute for 20 minutes and then immediately begin to depart at the same rate for 20 minutes. Plot on the following graph the rate of people moving and the level of people in the room.

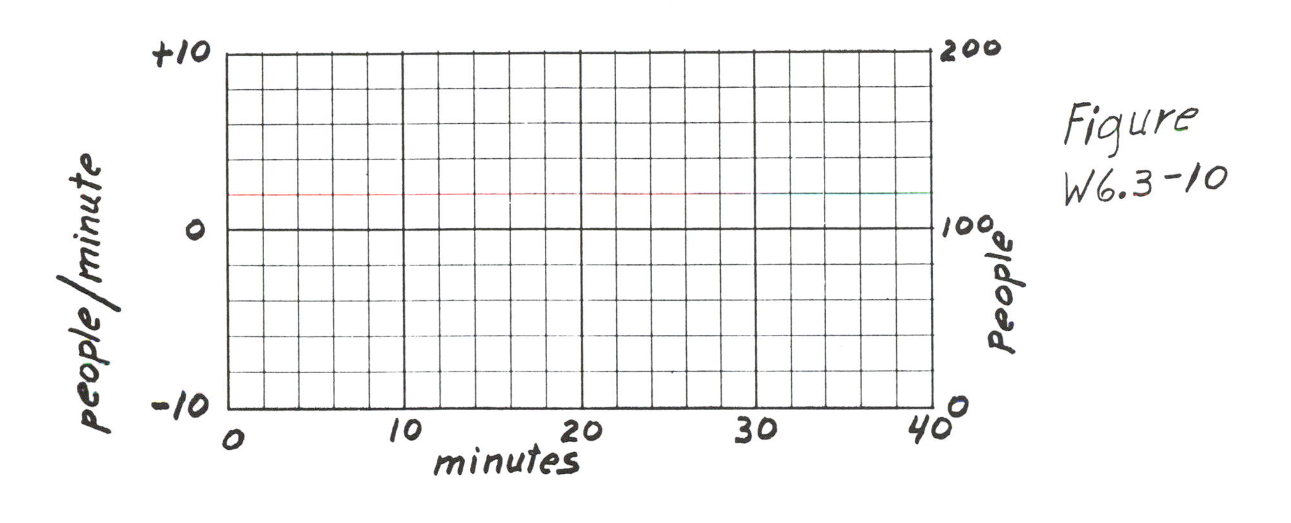

Figure W6.3-10

* * * * *

* * * * *

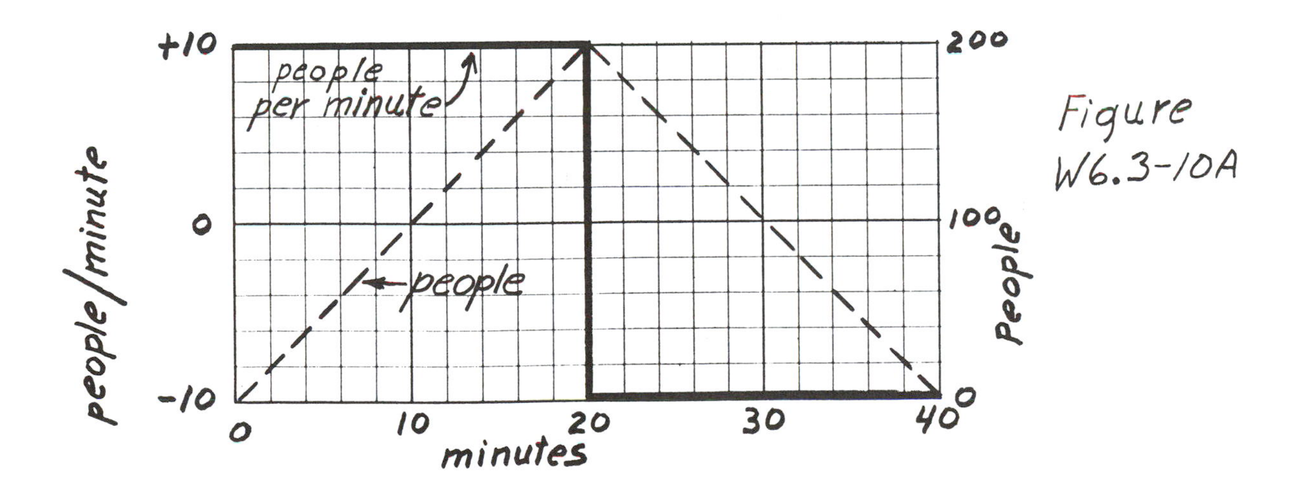

Figure W6.3-10A

Note how the magnitude of the rate determines the slope of the level. Note how, as long as the rate is positive, the level continues to increase. Observe how the process of integration (accumulating the rate of flow into the level) produces a curve of different shape from that of the rate curve. The level (integrated) curve changes less abruptly. Where the rate has a discontinuity and changes from positive to negative, the level has no discontinuity in magnitude but only a break in slope.

11. People enter a room at a uniformly increasing rate for 10 minutes, starting at zero and reaching 10 people per minute. The rate then decreases uniformly for another 20 minutes until 10 people per minute are leaving, at which time the outflow decreases uniformly to zero in another ten minutes. Plot the flow rate on this graph. Also sketch the curve showing the number of people in the room.

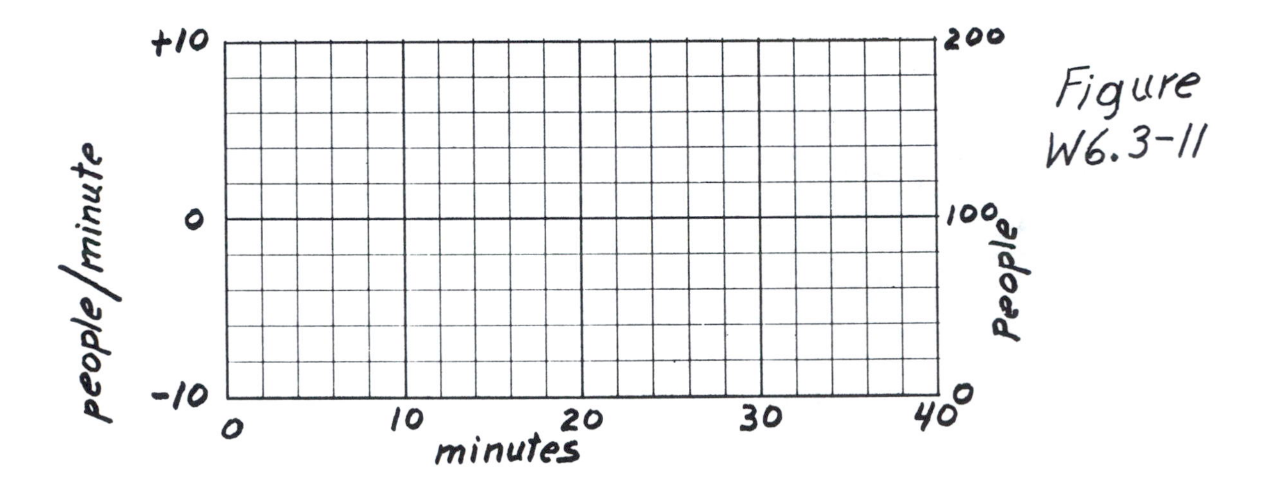

Figure W6.3-11

* * * * *

* * * * *

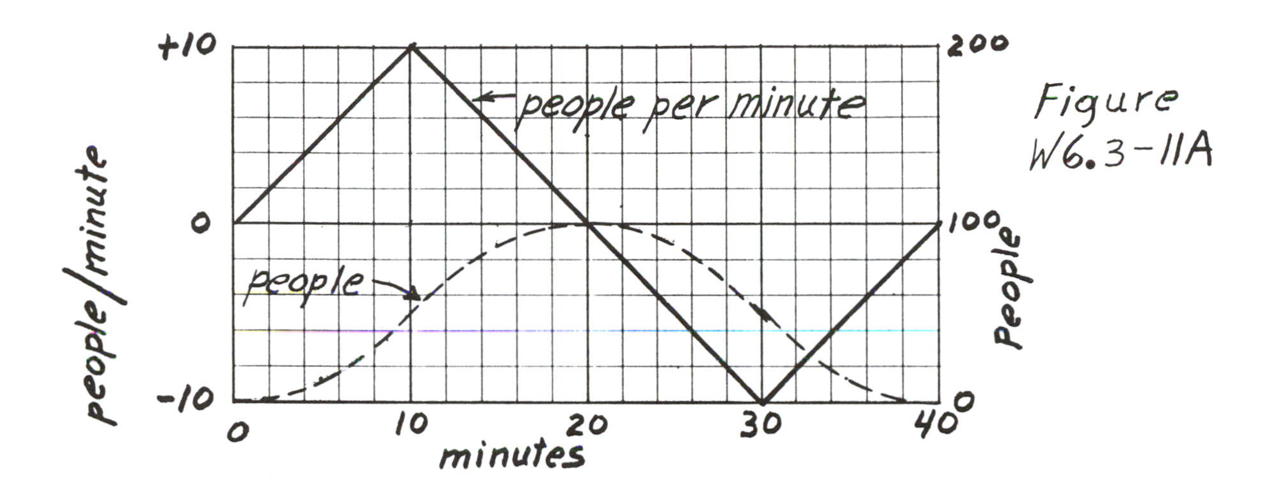

Figure W6.3-11A

Again see how the slope of the level (people) curve is equal to the height of the rate curve. Note the change in shape produced by integration. Observe how integration smooths the abrupt changes in the rate curves. Here the rate line is continuous in amplitude and has a break in its slope. The corresponding level curve has no break in slope. Integration produces a lag; the peak in the level curve occurs later than the peak in the rate.

W6.4 Differential Equations--a Digression

This section deals further with the process of integration.

1. A theatre contains 200 people. They leave at the rate of 10% of those remaining per minute. Sketch the number of people in the theatre versus time, showing any graphical aids that will help in producing the proper shape of curve. How many people remain at the end of 10 minutes? ________________.

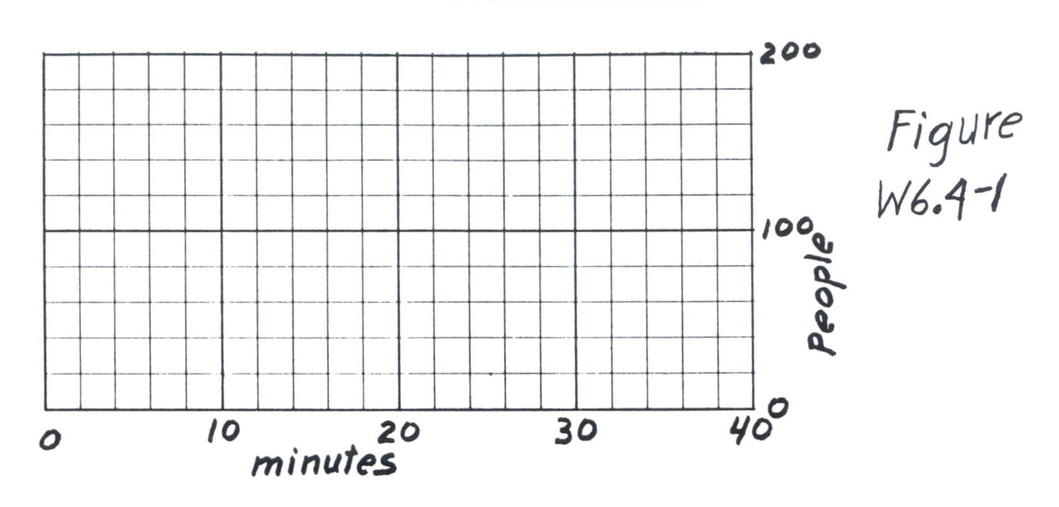

Figure W6.4-1

* * * * *
* * * * *

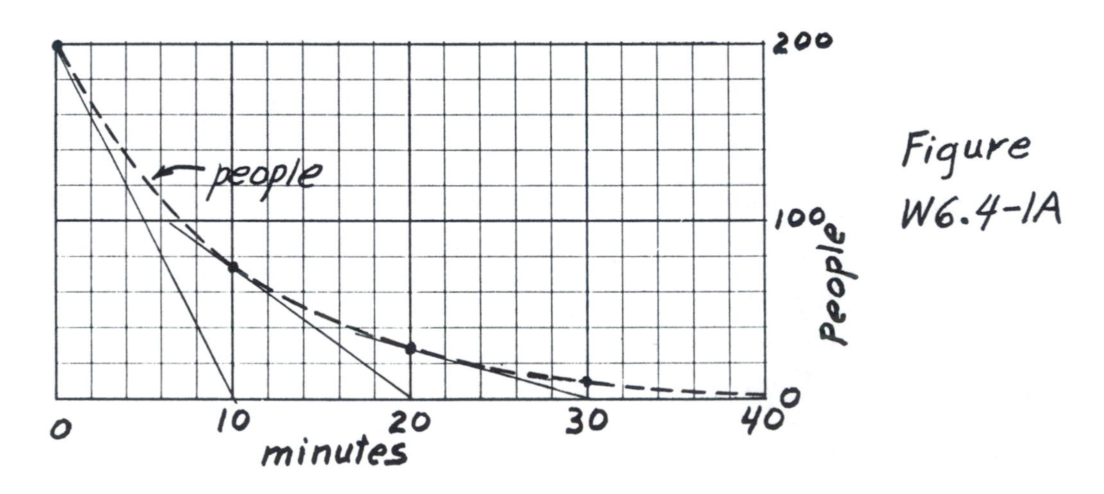

Figure W6.4-1A

74 people

(Note that, at any time, the slope is such that it reaches zero people in 10 minutes. But in any 10 minutes the number actually falls to 0.37 of the starting value. See text Figure 3.3a. The number of people

at 10 minutes is 200(.37) = 74.
At 20 minutes it is 74(.37) = 200(.37)(.37) = 27.
At 30 minutes it is 27(.37) = 200(.37)(.37)(.37) = 10.)

(Sec.W6.4)

2. In the following figure, the net inventory change is a repeating triangular fluctuation adding and subtracting from inventory with an amplitude of 20 units per month and a period of 2 months. Starting with zero inventory, add the curve for the level of inventory.

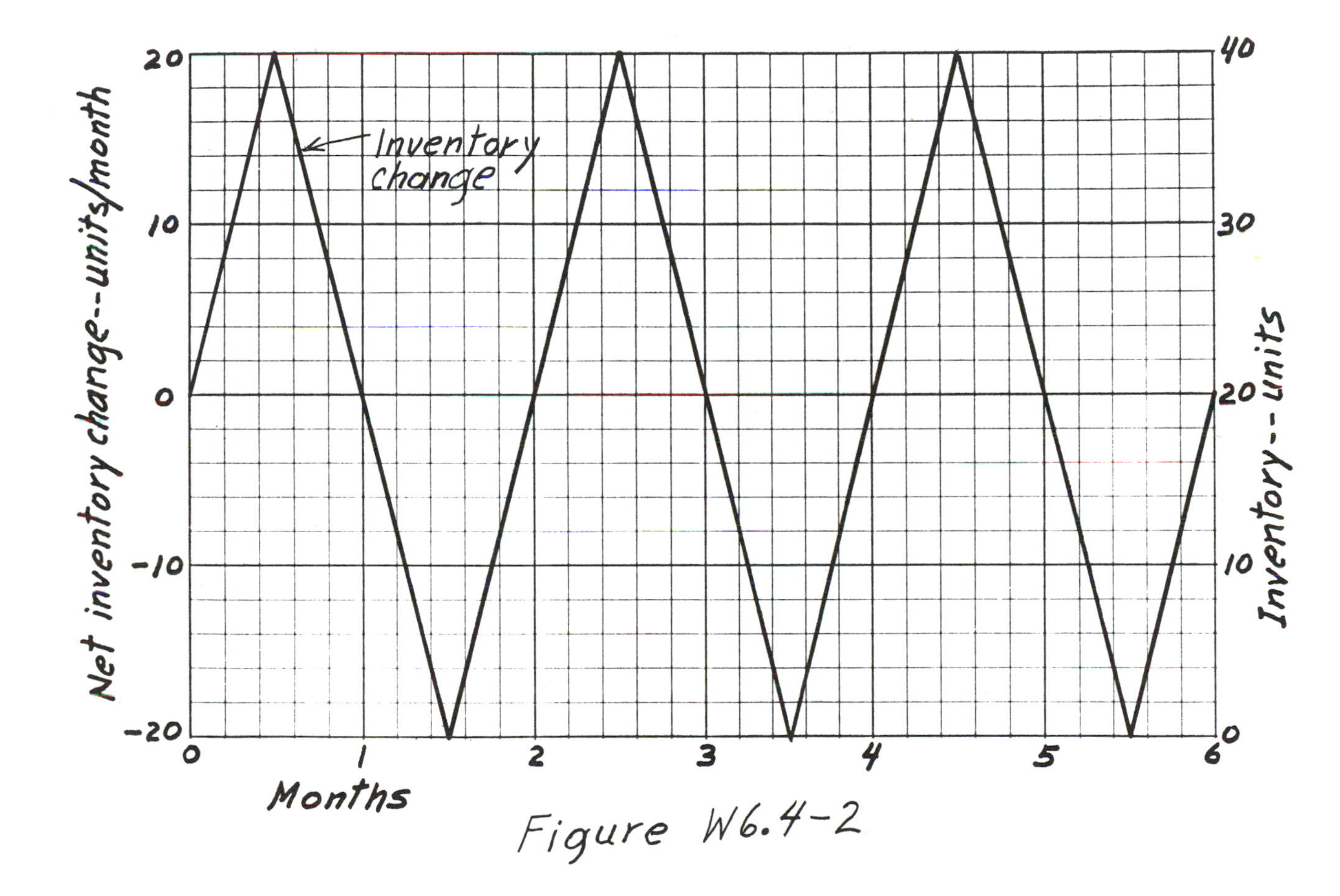

Figure W6.4-2

* * * * *

* * * * *

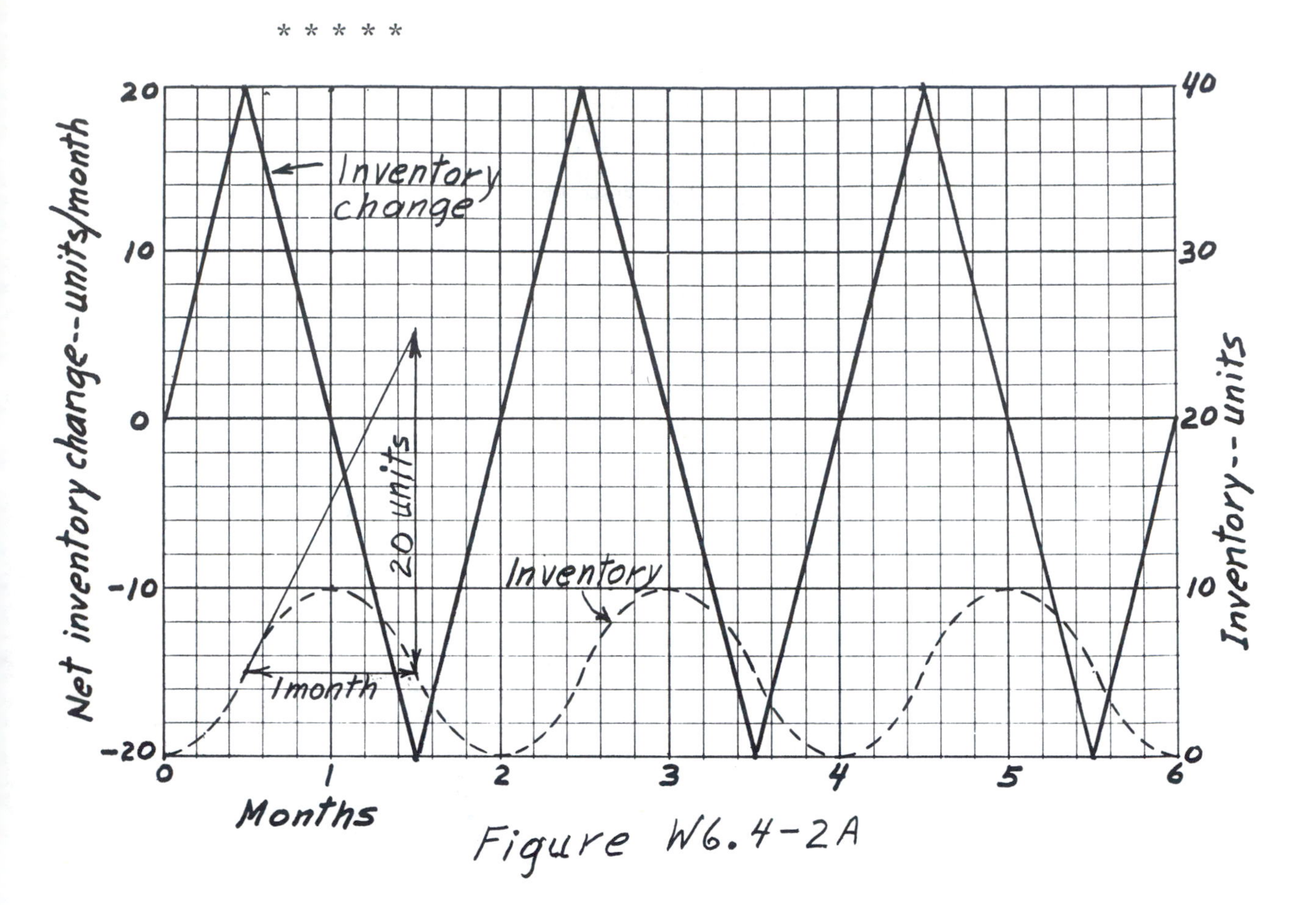

Figure W6.4-2A

Observe how the change in inventory is equal to the area under the flow rate curve. 20 units/month x 1 mo x $\frac{1}{2}$ = 10 units which is the rise in inventory during the one month that inventory is increasing. Note that the slope of the inventory curve is equal to the height of the flow rate curve. The steepest section of the inventory curve occurs when inventory change rate is a maximum or minimum. The inventory curve is horizontal when the rate curve is zero. Notice the tangent line showing the maximum slope of the inventory curve, at this point the slope of the curve is 20 units per month, equal to the height of the inventory change rate curve.

(Sec.W6.4)

3. The net change in inventory again fluctuates by plus and minus 20 units/month but this time the period is 4 months rather than 2 months. Sketch the inventory curve, again starting from zero.

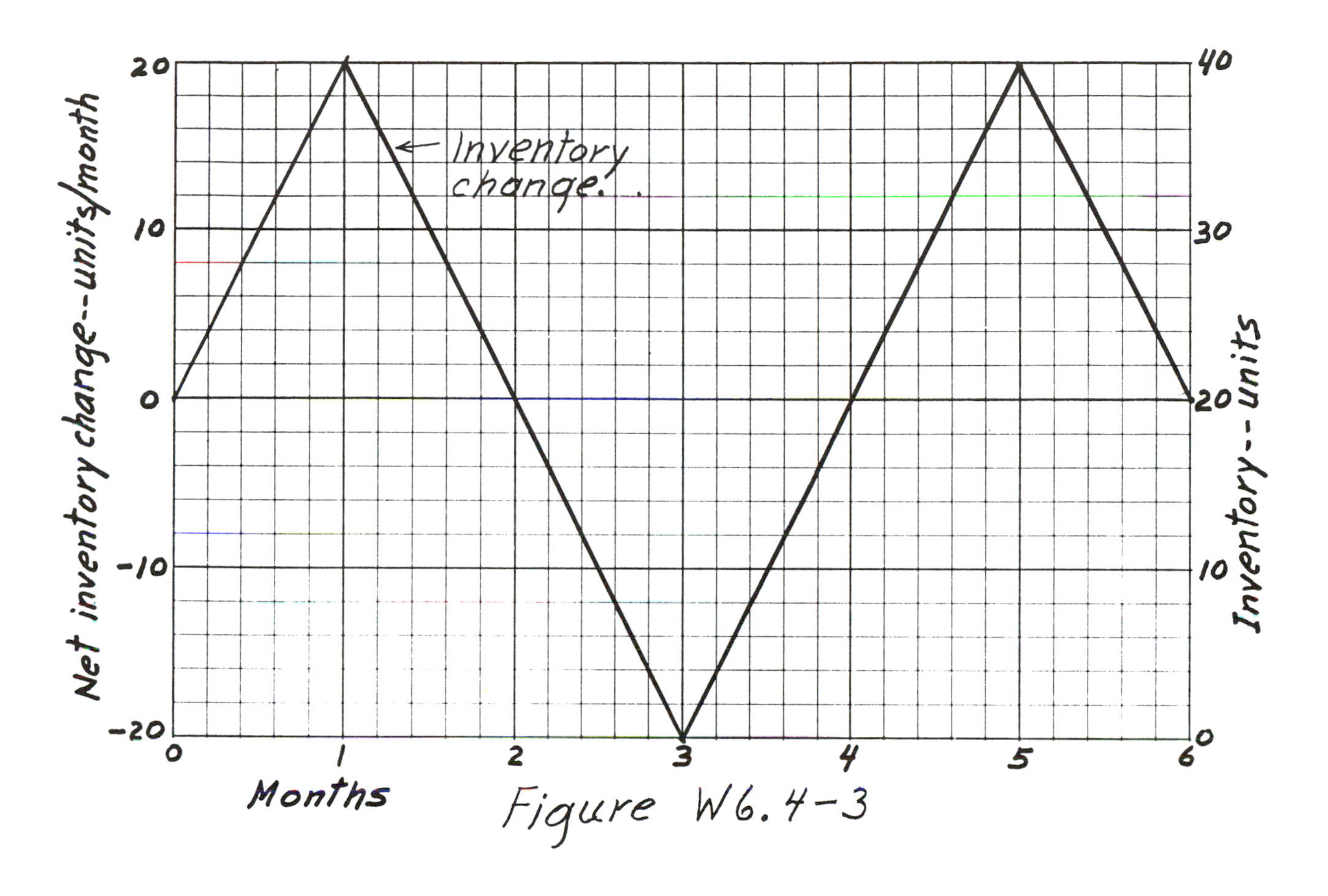

Figure W6.4-3

How does the area under a half-period of the rate curve compare with that in Figure 6.4-2A? _______________. The period of the rate fluctuation is _____________ times that in the previous frame. In this figure, is the period of the level fluctuation the same as that of the rate? ______________. How does the amplitude of the rate here compare with that in the previous figure? ________________. How does the amplitude of the level compare with that in the previous figure? _________________.

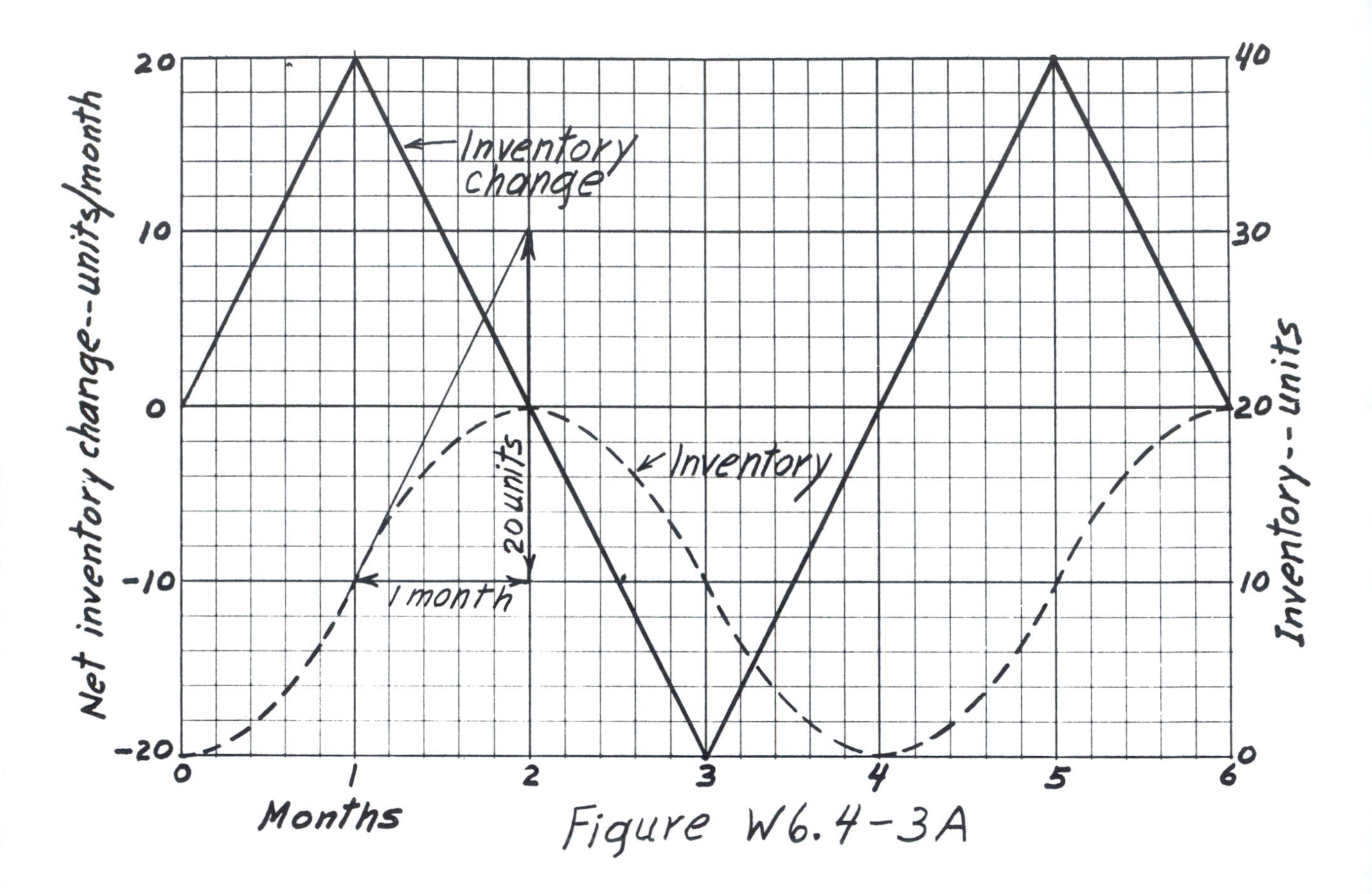

Figure W6.4-3A

double, two, yes, same, double

(With a rate curve of the same amplitude but twice the period, the area under the rate curve is twice as much as in the previous frame. This causes the level to fluctuate with double the previous amplitude. Note that at the end of one month the rate is at a peak, the slope of the level curve equals the rate and is at its steepest and has a slope of 20 units per month.)

4. Here the period of the net change in inventory again doubles to 8 months while the amplitude remains the same. Sketch the inventory curve starting from zero.

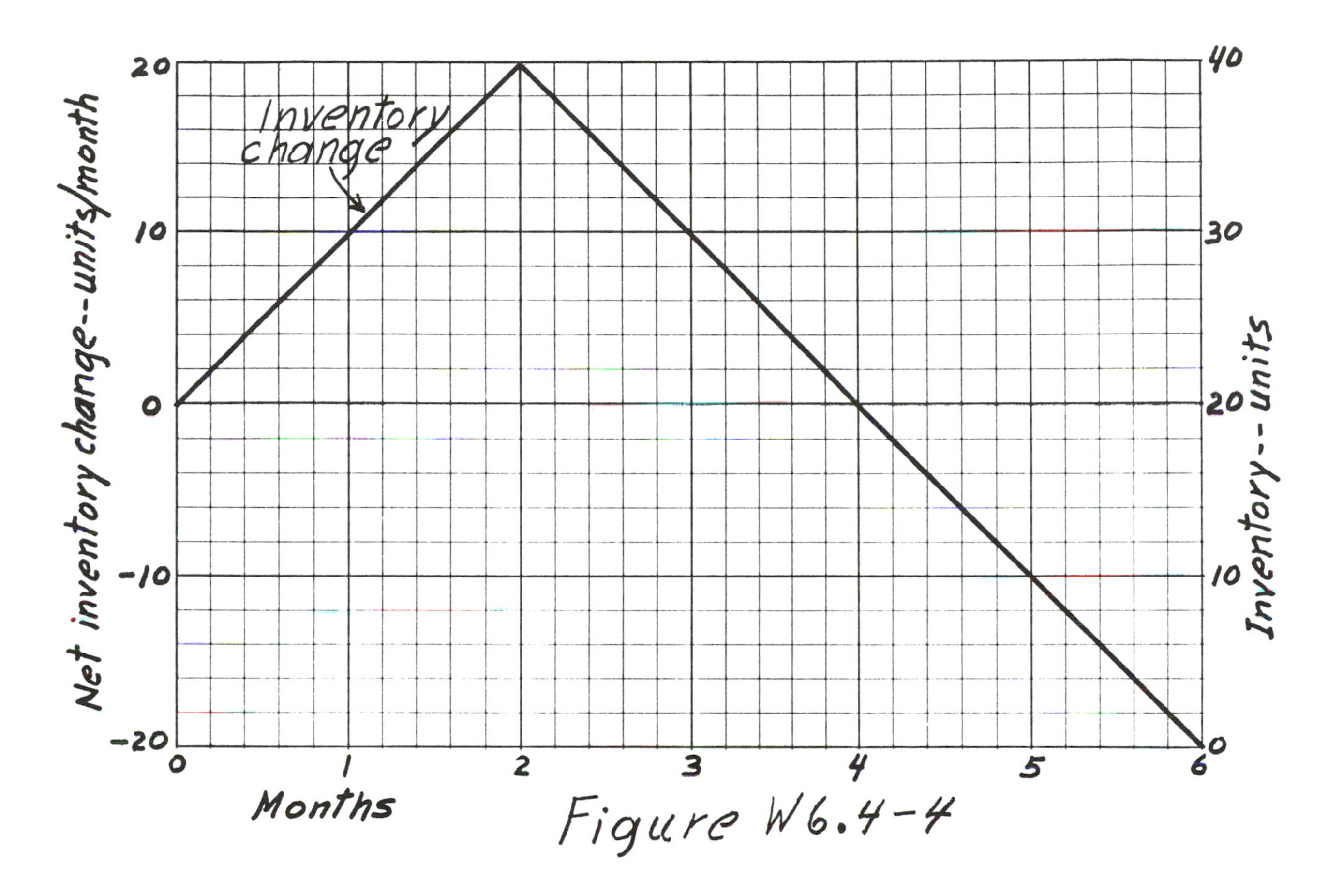

Figure W6.4-4

The maximum inventory is __________ units which is __________ times that in Figure W6.4-3A and is __________ times that in Figure W6.4-2A. The maximum slope of the inventory curve is __________ units per month and is __________ times that in Figure W6.4-3A and is __________ times that in Figure W6.4-2A. For rate fluctuations of the same amplitude and different periods, the integrated rate (level) has an amplitude proportional to the ______________.

* * * * *

* * * * *

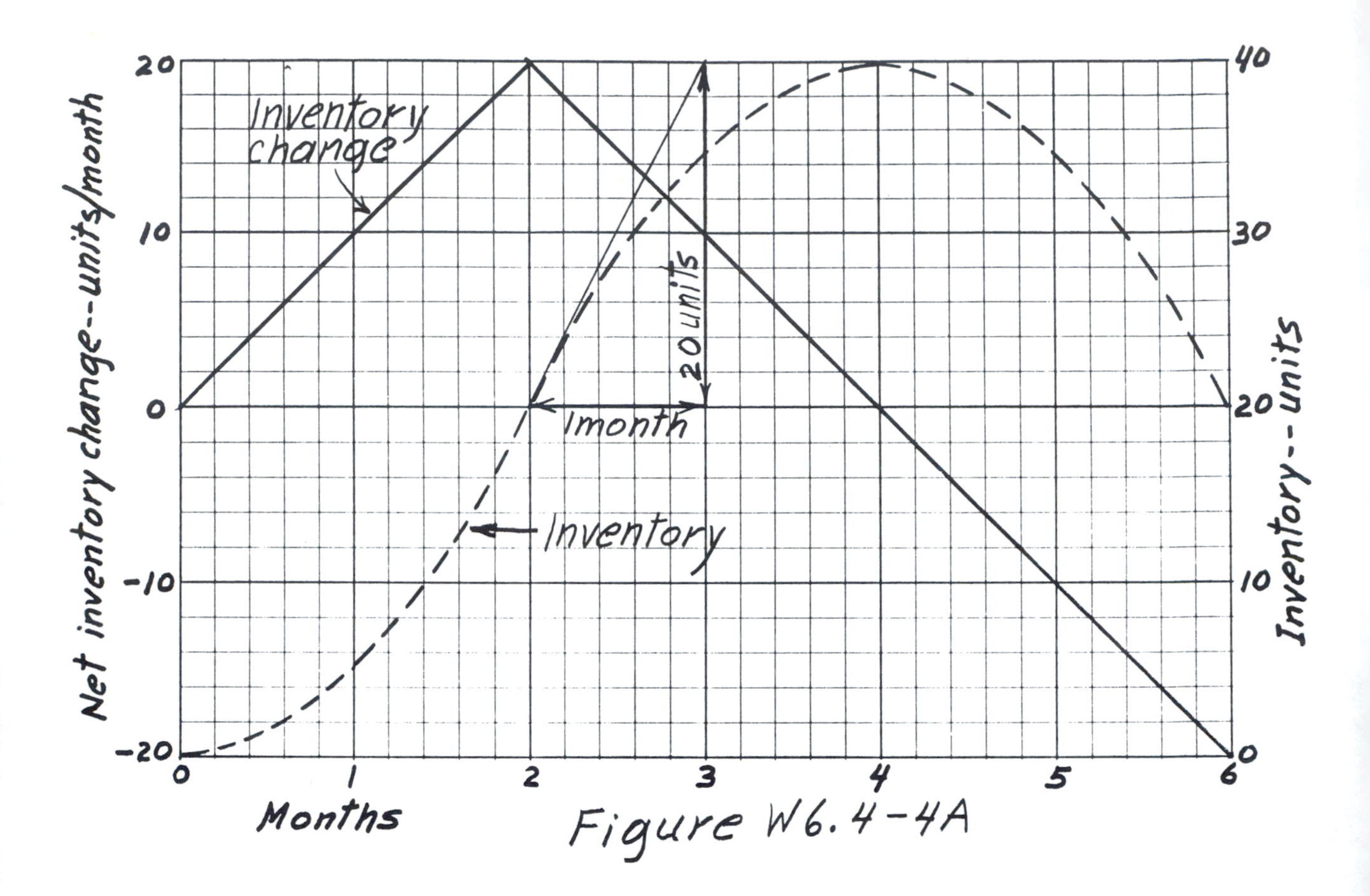

Figure W6.4-4A

40, 2, 4,

20, 1, 1

period

(Note that the amplitude of the rate determines the slope of the level, so for the same peak rate the level has the same maximum slope, regardless of the period. But the amplitude of the level curve increases in proportion to the period of the oscillation.)

CHAPTER W7

FLOW DIAGRAMS

The following questions are for Chapter 7 as a whole. Read all of text Chapter 7 first.

1. Draw only the flow diagram symbol for this equation for men where MH and ML are controlled by rate Equations W7-2 and W7-3.

$$M.K = M.J + (DT)(MH.JK - ML.JK) \qquad \text{Eq. W7-1, L}$$

* * * * *

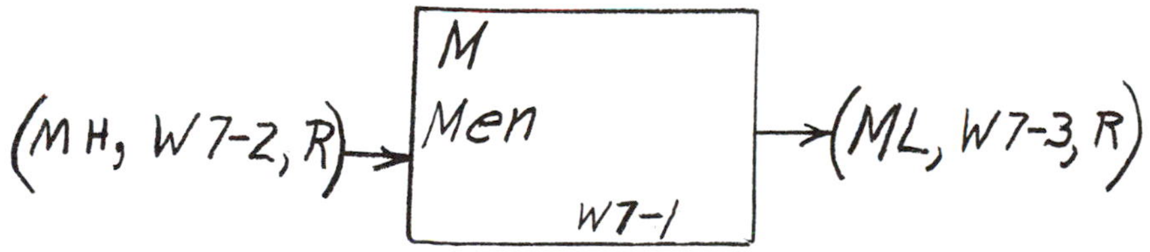

* * * * *

(MH, W7-2, R) → M Men W7-1 → (ML, W7-3, R)

Figure W7-1

2. Velocity V measured in feet per second is increased by engine acceleration EA and decreased by brake deceleration BD, both measured in feet per second per second. Show the flow diagram symbol for system equation 43 giving the value of V.

* * * * *
* * * * *

(EA) - - → V Velocity 43 - - - - → (BD)

Figure W7-2

3. Show the flow diagram symbol for the following equation for raw material.

RM.KL = FL.K/TRL Eq. W7-3, R

* * * * *

* * * * *

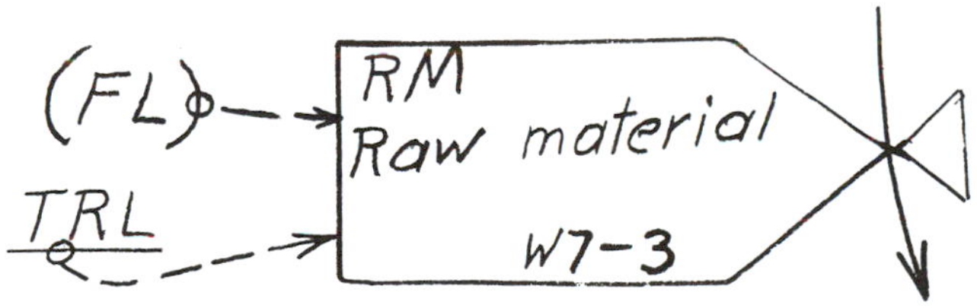

Figure W7-3

4. Show the symbols for velocity V in Equation 8 which is integrated to produce position P in Equation 9.

* * * * *

* * * * *

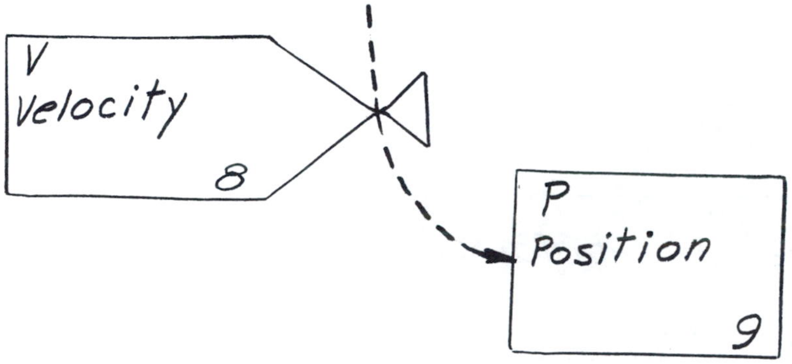

Figure W7-4

Note: If V is being integrated, it must be a rate and P must be a level.

5. Draw the symbol to represent the following equation for labor availability.

LA.K = LB.K + LC.K   Eq. 7-5, A

* * * * *

* * * * *

(LB)

LA
Labor
availability
7-5

(LC)

Figure W7-5

6. Draw the flow diagram for the following equations that define delivery delay condition, production capacity ordering, and production capacity. The manufacture of production capacity lies outside the model.

$$DDC.K = (DDRC.K/DDOG.K) - DDB \qquad 18,\_\_$$

$$PCO.KL = (DDC.K)(M)(PC.K) \qquad 19,\_\_$$

$$PC.K = PC.J + (DT)(PCO.JK) \qquad 20,\_\_$$

* * * * *

* * * * *

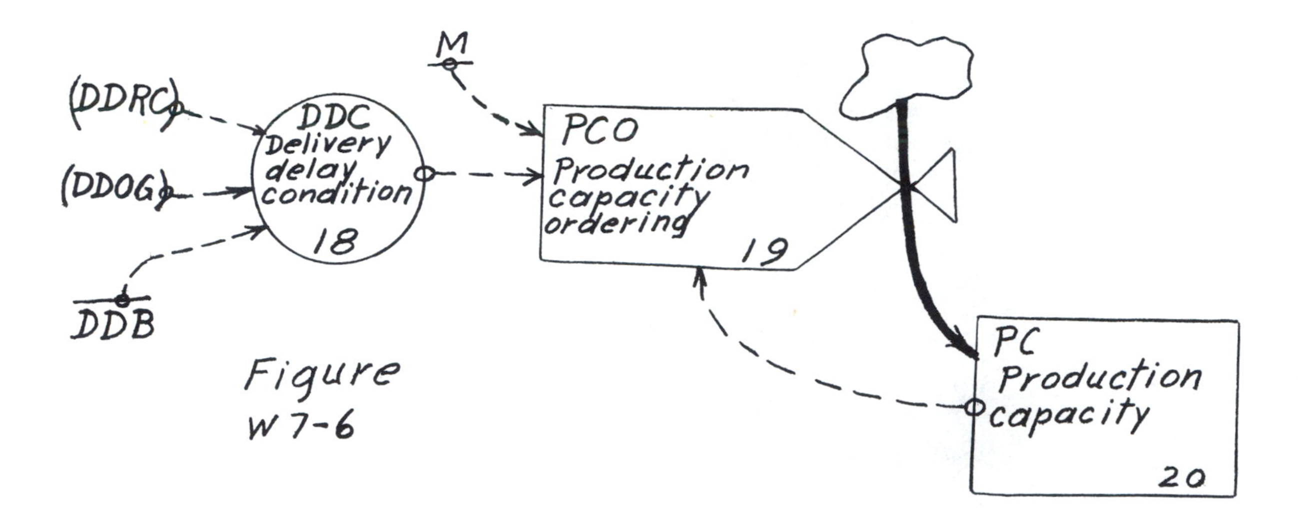

Figure W7-6

Note that the rate PCO is integrated to produce the level PC and that information about the level PC is one of the controlling inputs to PCO. The equation types should be evident from the forms of the equations.

7. Draw the flow diagram for the following equations:

MT.K = MT.J + (DT)(MH.JK - MCT.JK)	Eq. 1,__
MCT.KL = MT.K/TT.K	Eq. 2,__
TT.K = (STT)(TEC.K)	Eq. 3,__
ME.K = ME.J + (DT)(MCT.JK - ML.JK)	Eq. 4,__

MT--Men in training (men)
MH--Men being hired (men/month)
MCT--Men completing training (men/month)
TT--Training time (months)
STT--Standard training time (months)
TEC--Training effectiveness coefficient (dimensionless)
ME--Men experienced (men)
ML--Men leaving (men/month)

* * * * *

* * * * *

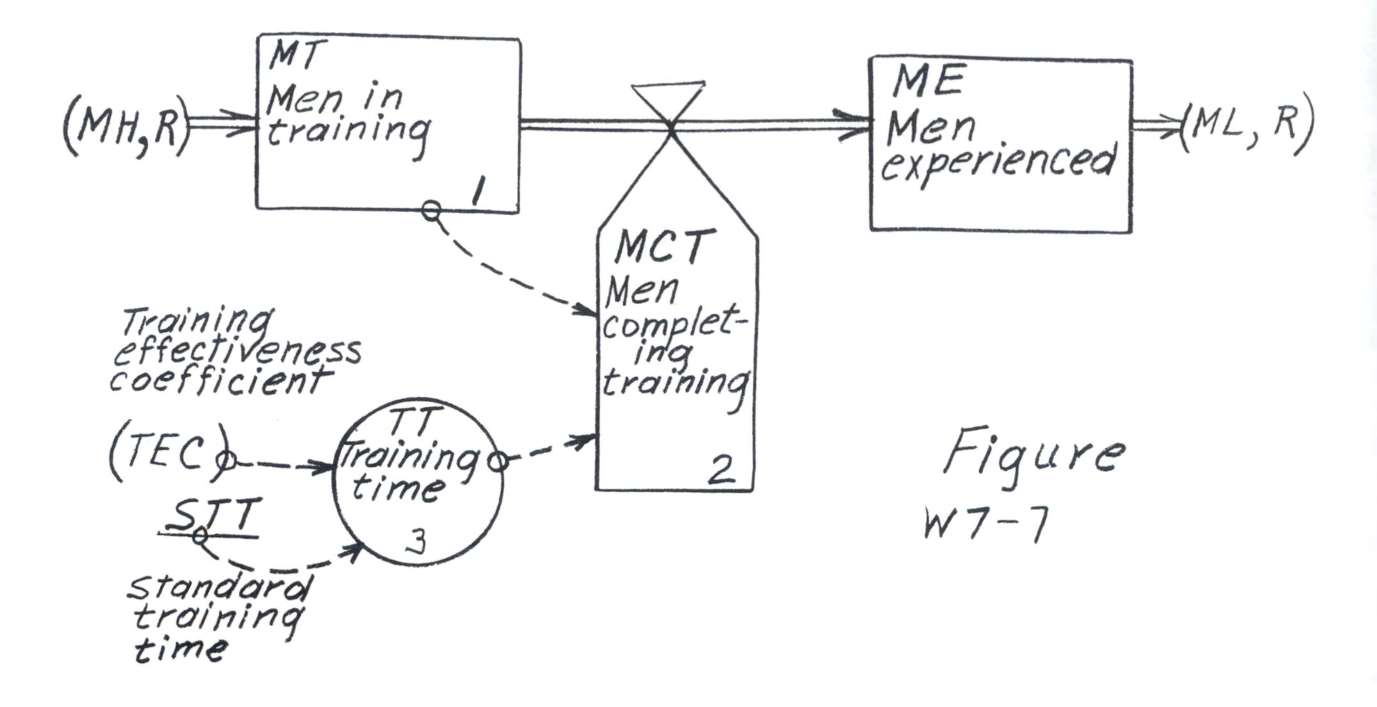

Figure W7-7

8. Draw the flow diagram for bank balance BB as Equation 48, cash payments CP as Equation 49, and accounts receivable AR as Equation 50. Bank balance is increased by the rate of cash payments. Cash payments are 1/DAR of accounts receivable where DAR is the delay in accounts receivable.

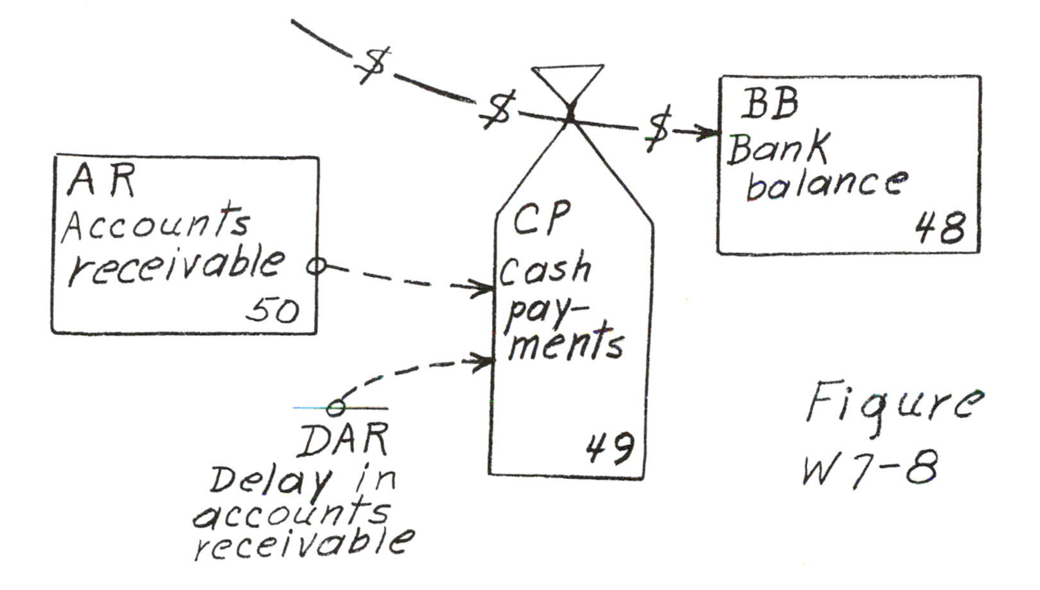
$
$
$
BB
Bank balance
48
AR
Accounts receivable
50
CP
cash pay-ments
49
DAR
Delay in accounts receivable
Figure W7-8

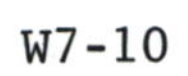

CHAPTER W8

DYNAMO COMPILER

W8.1 Functions Without Integration

1. In the following equation

   W.K = TABLE(TW,XY.K,4,12,1)

   The variable being defined is ______________.

   The function designation is ______________.

   The name of the table is ______________.

   The input variable to the table is ______________.

   The first entry in the table is for an XY value of ______________.

   The last entry in the table is for an XY value of ______________.

   The entries in the table are for values of XY

   that are ______________ apart.

   There should be ______________ entries in the table.

   * * * * *

   * * * * *

   W, TABLE, TW, XY, 4, 12, 1 unit, 9

2. If for the preceding frame the table is

   TW = 1/1.5/2/2.5/3/4/5/6/8

   plot the table and mark scales and axes on the following graph.

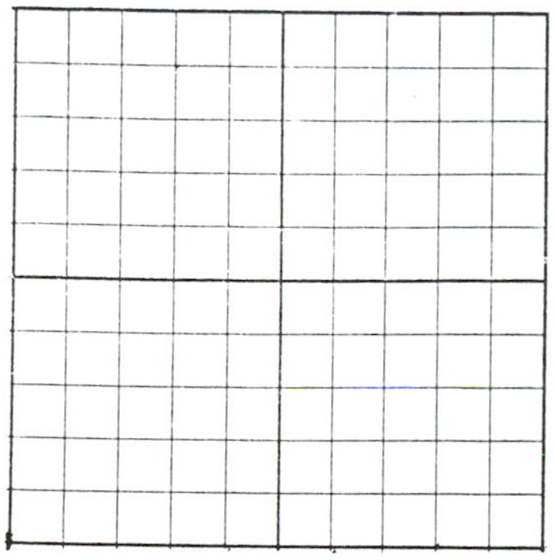

Figure W8.1-2

* * * * *

* * * * *

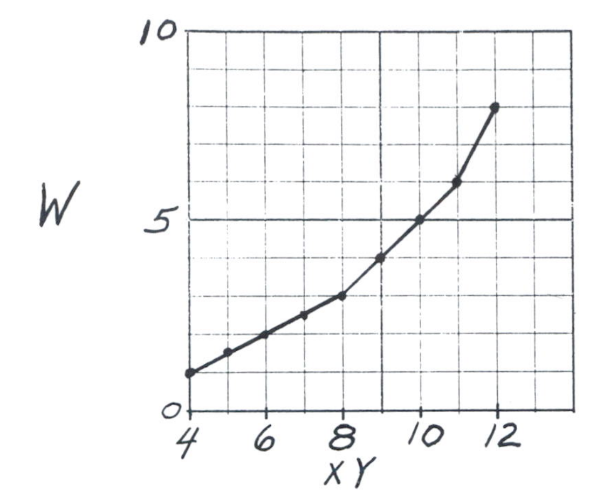

Figure W8.1-2A

3. Given the following relationship for an auxiliary variable, write the complete table function designation.

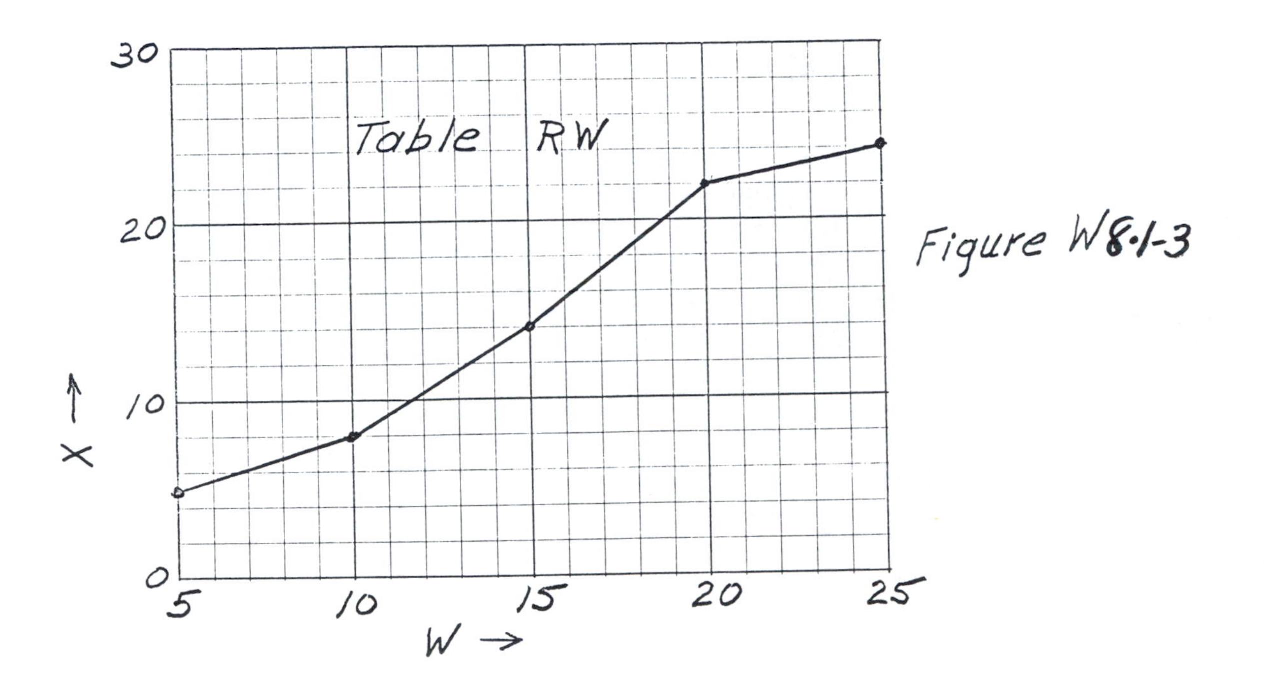

Figure W8.1-3

(Sec.W8.1)

Answer:

---

* * * * *

* * * * *

X.K = TABLE(RW,W.K,5,25,5)
RW = 5/8/14/22/24

4. Refer to text, Chapter 2. Write the TABLE function to represent the relationship in Figure 2.5b.

---

* * * * *

* * * * *

DR.K = TABLE(TDR,BL.K,0,100000,10000)
TDR = 0/5000/10000/13500/16000/17500/18500/19000/19500/19800/20000

5. Refer to text Figure 2.5c. Write the table function for sales effectiveness SE that will allow delivery delay recognized DDR to exceed 6 months, and which has table entries for each month of delivery delay.

---

* * * * *

* * * * *

SE.K = TABHL(TSE,DDR.K,0,6,1)
TSE = 400/390/350/290/210/150/100
(Note that TABHL, for high-low extension, allows delivery delay recognized to exceed 6 months while holding the last value of sales effectiveness equal to 100.)

6. Using the STEP function, write an equation for an auxiliary variable S that changes from 17 to 43 units at time 63.

---

* * * * *

* * * * *

S.K = 17+STEP(26,63)

(The initial value is 17 to which is added 26 after time 63.)

7. Using the STEP function, write an equation for Q which has the value 6 from time 0 until 18, then the value -7 until time 26 followed by the value 25 thereafter.

____

* * * * *

* * * * *

Q.K = 6+STEP(-13,18)+STEP(32,26)

8. Write the equation for an auxiliary variable R which is zero until time 8 and then slopes downward at 0.5 units per time period.

____

____

* * * * *

* * * * *

R.K = RAMP(-0.5,8)

9. What parameter is needed to specify a SIN function? __________.

* * * * *

* * * * *

the period

10. Write the equation for RM which is a sine fluctuation with amplitude 15 and period 40 time units.

____

* * * * *

* * * * *

RM.K = 15*SIN(6.283*TIME.K/40)

or

RM.K = (15)*SIN((6.283)(TIME.K)/40)

(Sec.W8.1)

11. Write the equation for variable Q having a normal noise distribution with mean value of 8 and a standard deviation of 3.

____

* * * * *

* * * * *

Q.K = NORMRN(8,3)

12. What is wrong with the following level and initial value equations?

```
P.K = P.J + (DT)(Q.JK-R.JK)     L
  P = S                         N
S.K = S.J + (DT)(Q.JK-V.JK)     L
  S = P                         N
```

____

____

* * * * *

* * * * *

The initial values of S and P depend on one another so that they are not computable from numerical constants.

W8.2 <u>Functions Containing Integration</u>

1. Text Equations 2.5-11 and 2.5-12 describe an information delay. Express the same using a DYNAMO function.

____

____

* * * * *

* * * * *

DDR.K = DLINF1(DDI.K,TDDR)

TDDR = 6

2. Write the statement for shipments received SR which is a 4-week, third-order, delayed version of shipments sent SS.

____

* * * * *

* * * * *

SR.KL = DELAY3(SS.JK,4)

3. In the following equation, how long is the delay? ______________.

```
P.KL = DELAY3(Q.JK,M)      R
   M = (3)(T)              N
   T = 4                   C
```

* * * * *

* * * * *

12 units of time

4. Examine text Figure 8.2d. The step changes by _______________ units. The delay time of the third-order delay is _____________. The output of the third-order delay rises by 0.25 unit after ____________ time units following the step. The output of the third-order delay has covered 0.75 of its total rise at _______________ time units after the step.

* * * * *

* * * * *

1, 15, 9.5, 18

5. In text Figure 8.2d, estimate the area lying between the step curve and the DELAY3 curve. _________________. If the units of measure of the STEP and DELAY3 curves are "thousands of automobiles per day" and time is measured in days, what units of measure should be associated with the area between the curves? ________________________________.

* * * * *

* * * * *

15, thousands of automobiles

(The area represents the automobiles in transit in the delay--the ones put in which have not emerged. It is given by the product of the step increase, one thousand/day, multiplied by the delay, 15 days.)

(Sec.W8.2)

6. In text Figure 8.2g, does the area between the RAMP and DELAY3 curves up to any point in time represent the units in-transit in the delay? ________________.

   * * * * *

   * * * * *

   yes

7. In text Figure 8.2d for the step input do the units in transit grow without limit? ___________________. In text Figure 8.2g for the ramp? ________________.

   * * * * *

   * * * * *

   no (they approach 15 units increase corresponding to the delay of 15 days multiplied by the flow rate change of 1 unit/day)

   yes (the flow rate is increasing continuously so that the units in transit likewise continue to increase. At any time the input rate is greater than the output rate and the difference in rate is accumulating in the delay.)

8. In text Figure 8.2g, if the axes are refrigerators per day and days, how rapidly are the units in transit in the delay increasing after the equilibrium condition is established wherein the output of DELAY3 has the same slope as the RAMP?

   ____

   * * * * *

   * * * * *

   20 refrigerators per day (This is the ramp slope of one refrigerator/day/day multiplied by the delay of 20 days.)

9. Examine in text Figure 8.2h the curves for SIN and DELAY3 where Delay = 4. The SIN period is ________ months assuming the time scale is in months. Observed in the right half of the figure after the beginning transient has been dissipated, the period of DELAY3 is __________ months.

* * * * *

* * * * *

20, 20 (In a "driven" system as here where the SIN is the input that drives the DELAY3 function, the steady-state outputs will always have the same period as the input.)

10. In text Figure 8.2h, what is the phase shift between SIN and DELAY3 for Delay = 4, that is, DELAY3 lags by ___________ months or _________________ of the period.

* * * * *

* * * * *

about 4, about 1/5 (For delays that are short compared to the period, the lag will be about equal to the delay.)

11. In text Figure 8.2h, examine the curves for SIN and DELAY3 for Delay = 20. DELAY3 lags by _________________ months or _______________ of a period.

* * * * *

* * * * *

about 12, about 3/5 (The lag can not exceed 3/4 period--1/4 period for each of the three stages or first-order sections of the delay.)

12. In text Figure 8.2h, when a third-order delay has a delay of one-fifth the period of an imposed sinusoid, what is the ratio of the amplitude of the delay output to the delay input? ___________________. What is the ratio when the delay is equal to the period? ____________________.

* * * * *

* * * * *

about 28/30 = 14/15 = .93

about 3/30 or 1/10 = .1

(Note: The longer delay absorbs the input fluctuation and only a small fraction of the amplitude reaches the output.)

(Sec.W8.2)

13. With a sinusoidal input, the output of an exponential delay is attenuated (more/less) as the delay becomes longer. The phase shift becomes (more/less) as the delay becomes longer. In determining the attenuation and phase shift, the critical relationship is the ratio of the delay to the ___________ of the sine input.

* * * * *

* * * * *

more, more, period

CHAPTER W9

INFORMATION LINKS

1. What is wrong with the system structure in the following diagram?

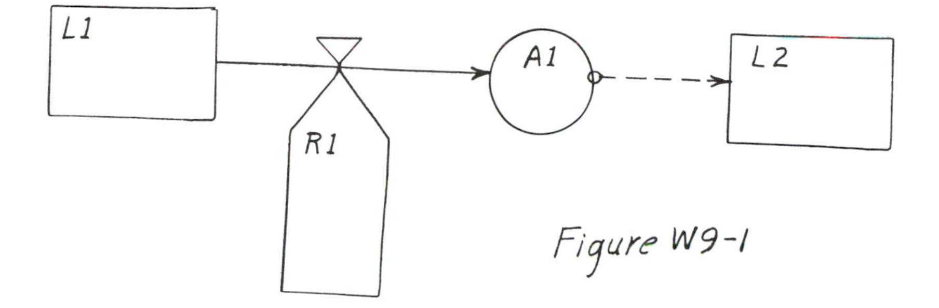

Figure W9-1

* * * * *

* * * * *

The flow R1 from level L1 must go to another level or to a sink outside the system. The auxiliary A1 can only have an information link input from a level.

2. In the production of automobiles one must use labor, money, and materials. Is there a defect in the following fragment of a flow diagram for the process?

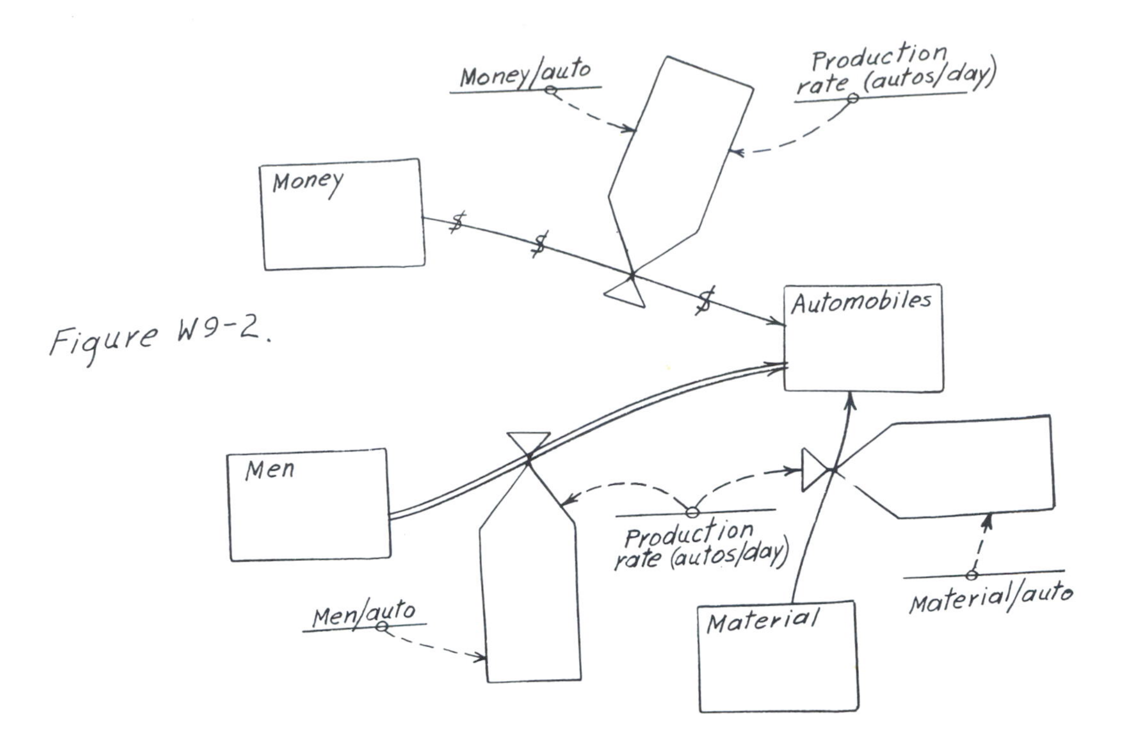

Figure W9-2.

* * * * *

* * * * *

Yes, automobiles are constructed from material; it is appropriate, if units are properly defined, to show material from the material level flowing into the automobile level. However, automobiles are not physically constructed from money or men. The money and men are not transferred out of their level to become a part of the finished automobile.

3. Show here how the men level might be related to the material level and the automobile level.

* * * * *

* * * * *

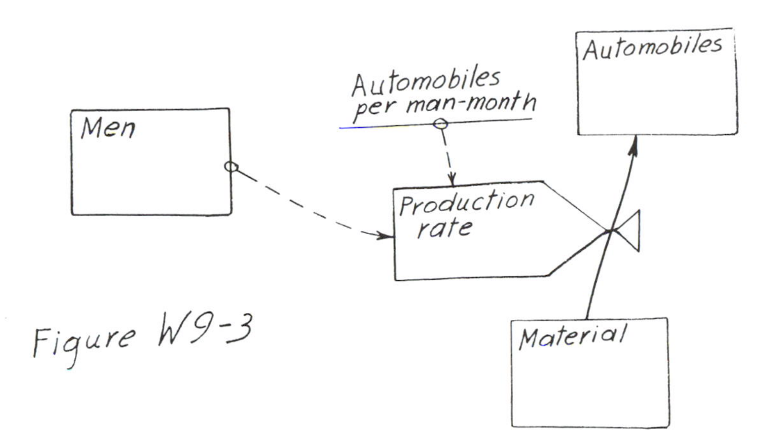

Figure W9-3

Note that the number of men control production, not by physically expending themselves but by acting through time to convert material to automobiles.

4. Show in a section of a flow diagram how money might be properly related to men in the system of Frame 2.

* * * * *

* * * * *

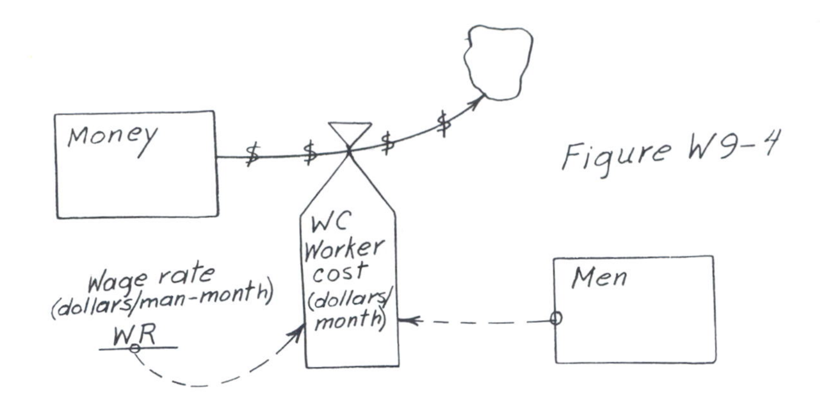

Figure W9-4

Note that the men imply a continuing expenditure of money for wages. The money flow depletes the money level, it does not flow into any of the other levels of Figure W9.2. As shown here, money for wages flows out of the system which is appropriate unless the study involves the purchasing power of the workers.

5. Suppose that an organization wants to believe that it is effective in satisfying its customers, and as a result, favorable reports from customers are given more weight than unfavorable customer reports. Assume we are modeling the rate of resource flow to quality generation. Show in a flow diagram how the bias could be inserted and write an appropriate equation for generating the bias.

* * * * *

* * * * *

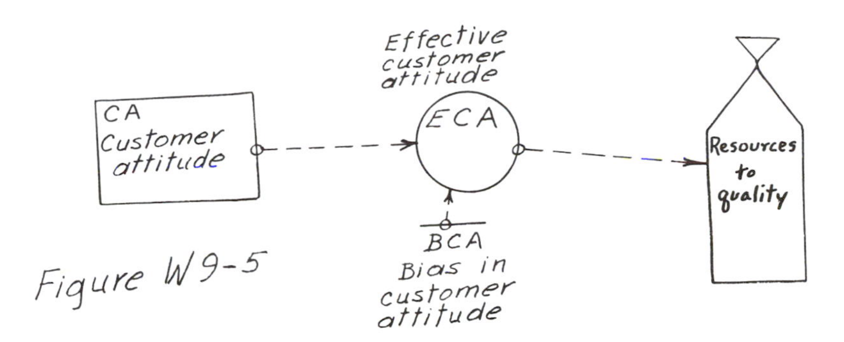

Figure W9-5

ECA.K = CA.K + BCA

ECA--Effective customer attitude (attitude units)

CA--Customer attitude (attitude units)

BCA--Bias in customer attitude (attitude units)

6. In the preceding frame, what effect will the bias have on the quality? Will the bias BCA be positive or negative?

* * * * *

* * * * *

The bias will make customer satisfaction appear greater than it is, and lead to a smaller allocation of resources to quality and to a lower quality than if satisfaction were transmitted without bias to the decision process. BCA will be positive.

7. In a time of rising sales and increasing order backlog, management may be cautious about increasing employment and production capacity for fear that the uptrend in business is only temporary. Show in a section of flow diagram how this delay might be interposed between backlog BL and the personnel hiring rate PH. Write an appropriate equation for the delayed influence of backlog.

* * * * *

* * * * *

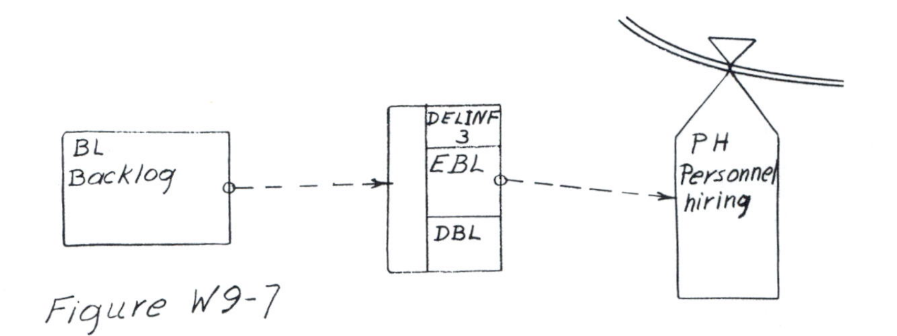

Figure W9-7

EBL.K = DLINF3(BL.K,DBL)

EBL--Effective backlog (units)

DLINF3--Operator indicating third-order information delay (See Chapter 8)

BL--Backlog (units)

DBL--Delay in backlog information (weeks)

(Of course, one might have chosen the first-order delay DLINF1 and could have drawn the diagram with the expanded detail for the delay as in text Figures 8.2e or 8.2c.)

8. Suppose that the mistakes in counting inventory and in record keeping can be represented by a mean error of zero, a root-mean-square amplitude of 3 percent of the inventory, and a long-period cutoff of 12 months. Using a function from Chapter 8, show the section of the flow diagram and the related equations that must be interposed between inventory and the reordering decision.

* * * * *

* * * * *

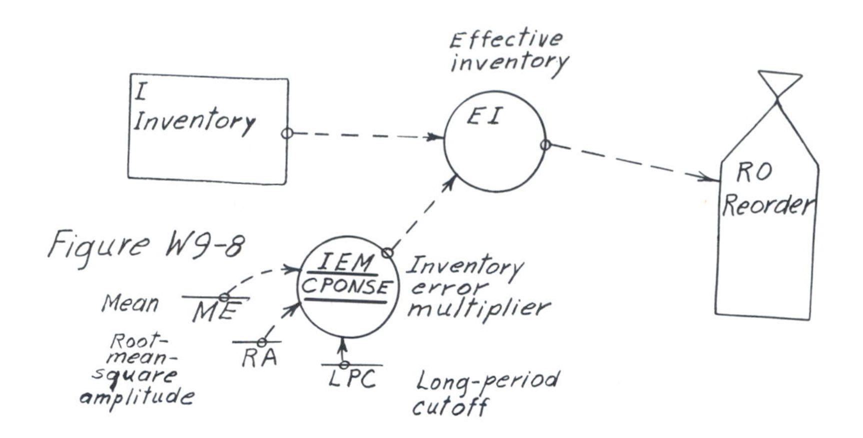

EI.K =	(I.K)(IEM.K)	A
IEM.K =	CPONSE(ME,RA,LPC)	A
ME =	1.0	C
RA =	.03	C
LPC =	12	C

EI--Effective inventory (units)

I--Inventory (units)

IEM--Inventory error multiplier (dimensionless)

CPONSE--Operator indicating constant power per octave noise generator.

ME--Mean value of noise (dimensionless)

RA--Root-mean-square amplitude (dimensionless)

LPC--Long-period cutoff (months)

(Note that although the mean value of <u>noise</u> is stated as zero, we must use a value of 1.0 if the noise term is entered as a multiplier. If the noise were added, then noise with a mean of zero could be generated, multiplied by the inventory, and the product then added to the information link from inventory. This would require a third equation.)

9. Assume the availability of shipping orders SO (machine tools/month) as an auxiliary variable. With the price P of 500 dollars per tool available, show the levels for tools T, accounts payable AP, capital equipment account CEA, and the flow diagram relationships between these variables. As tools are received, physical tools increase, accounts payable increase (actually a negative asset), and the capital equipment account increases (representing the dollar value of the tools). Write the related equations for all rates and levels.

* * * * *
* * * * *

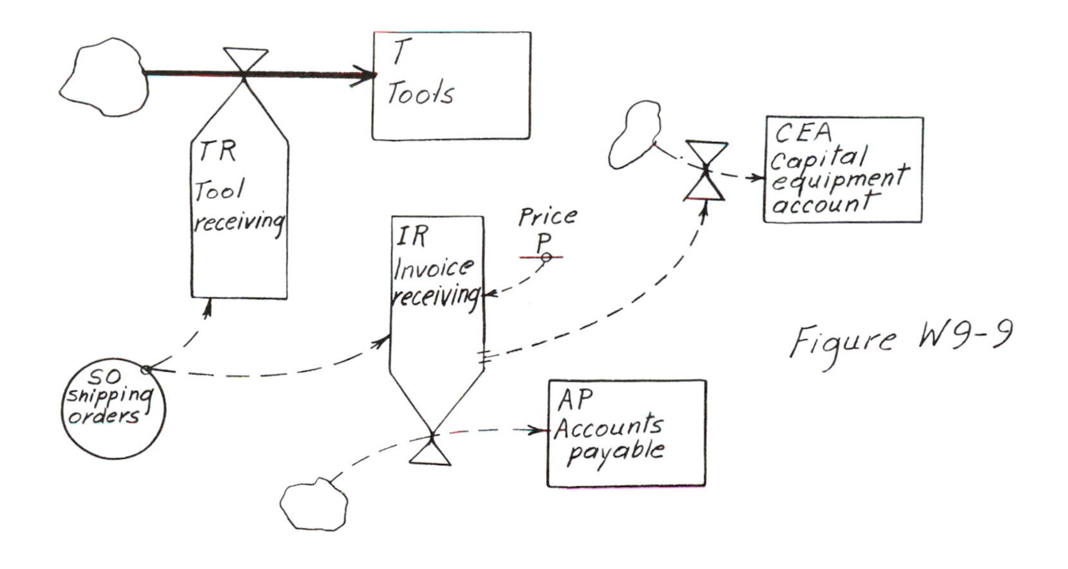

Figure W9-9

TR.KL =	SO.K	R
T.K =	T.J + (DT)(TR.JK)	L
T =	initial value needed	N
IR.KL =	(SO.K)(P)	R
P =	500	C
AP.K =	AP.J + (DT)(IR.JK)	L
AP =	initial value needed	N
CEA.K =	CEA.J + (DT)(IR.JK)	L
CEA =	initial value needed	N

TR--Tool receiving (tools/month)
SO--Shipping orders (tools/month)
T--Tools (tools)
IR--Invoice receiving (dollars/month)
P--Price (dollars/tool)
AP--Accounts payable (dollars)
CEA--Capital equipment account (dollars)

(Note that accounts payable and the capital equipment account are both in the information system and shown by dotted flow lines. Neither of the accounts are either money or physical equipment. Because the flow to accounts payable and to the capital equipment account are simultaneous bookkeeping entries, we can use the same rate IR as the input to each.)

10. Suppose that for accounts payable the average delay in paying DP is 1.5 months, that is, 1/1.5 of the accounts payable are paid per month. Show the flow diagram connecting accounts payable AP, bank balance BB,

and related rates. Write the equations for these rates and levels, including any related variables from the preceding frame.

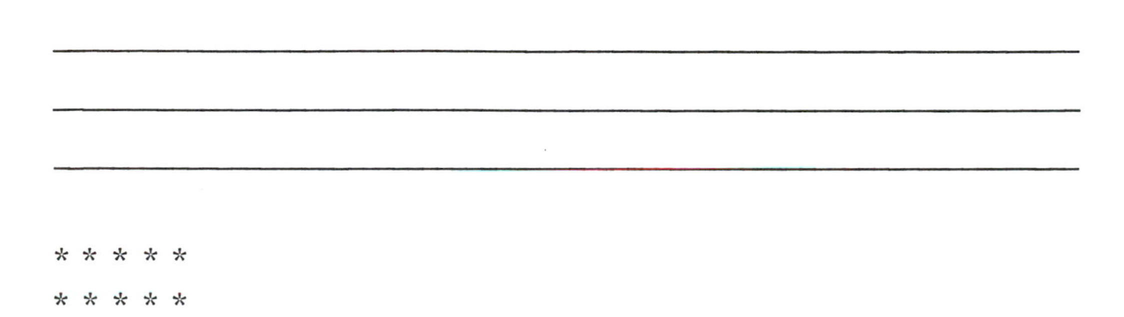

* * * * *

* * * * *

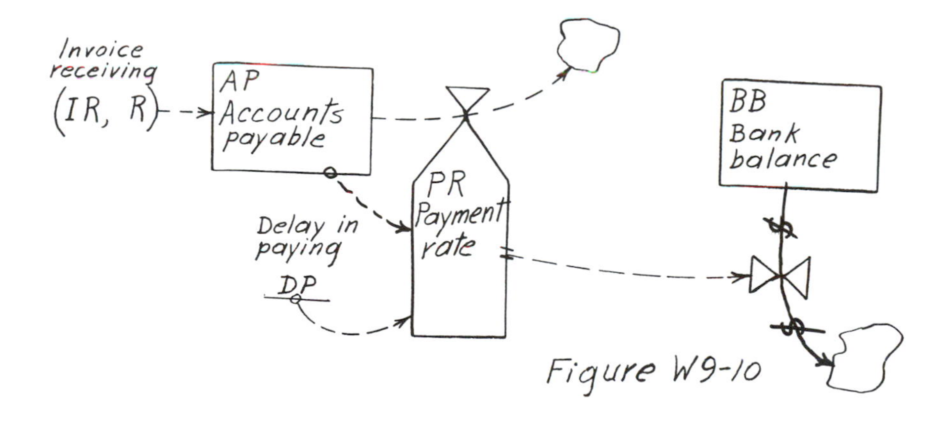

Figure W9-10

PR.KL = AP.K/DP	R
DP = 1.5	C
AP.K = AP.J + (DT)(IR.JK − PR.JK)	L
AP = initial value needed	N
BB.K = BB.J + (DT)(−PR.JK)	L
BB = initial value needed	N

PR--Payment rate (dollars/month)

AP--Accounts payable (dollars)

DP--Delay in paying (months)

IR--Invoice receiving (dollars/month)

BB--Bank balance (dollars)

11. Assume that the average life of tools LT is ten years, taken to mean that the tool discard rate TDR is one-tenth per year of the remaining tools. Assume the accounting practice to have a tool amortization rate TAR based on a 20% annual amortization AA and on the remaining unamortized balance of the capital equipment account CEA, and to ignore the scrap value of discarded tools as well as any residual value in the capital equipment account of tools that have been discarded. Amortization reduces the capital equipment account and at the same time is entered as a cost into average profit and loss APL which is computed here as an exponential average of all related profit and loss flow rates. The time to average profit and loss TAPL

is 12 months. Draw the flow diagram and write the equations, building on and being consistent with Frames 9 and 10. Ignore other rates which would contribute to average profit and loss.

* * * * *
* * * * *

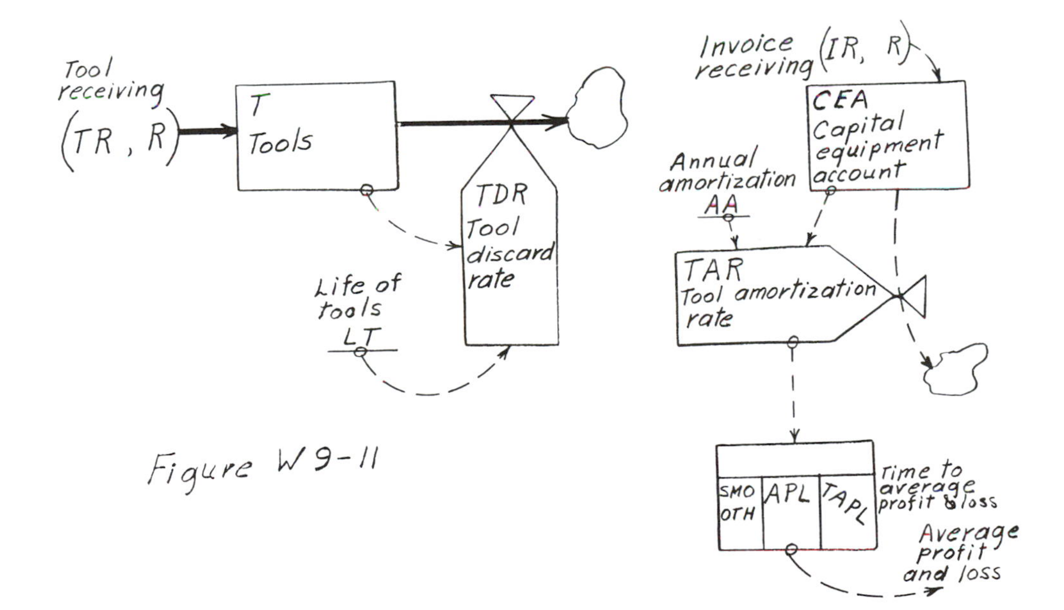

Figure W9-11

TDR.KL	=	T.K/LT	R
LT	=	120	C
T.K	=	T.J + (DT)(TR.JK − TDR.JK)	L
T	=	initial value needed	N
TAR.KL	=	CEA.K/AA	R
AA	=	60	C
CEA.K	=	CEA.J + (DT)(IR.JK − TAR.JK)	L
CEA	=	initial value needed	N
APL.K	=	SMOOTH(TAR.JK,TAPL)	A
TAPL	=	12	C

TDR--Tool discard rate (tools/month)

T--Tools (tools)

LT--Life of tools (months)

TR--Tool receiving rate (tools/month)

TAR--Tool amortization rate (dollars/month)

CEA--Capital equipment account (dollars)

AA--Annual amortization (months)

IR--Invoice receiving (dollars/month)

APL--Average profit and loss (dollars/month)

SMOOTH--Operator indicating the averaging of a rate (see text Chapter 8)

TAPL--Time to average profit and loss (months)

(Note that the actual tool life and the accounting life can be very different. The level of tools T represents the physical equipment and would be used in determining physical production rate. The financial representation in the capital equipment account CEA is an information level that responds to the accounting practices of the firm.)

CHAPTER W10

INTEGRATION

W10.1 Integrating a Constant

1. The integral of a power of the independent variable is of the form:

$$\int At^{n}dt = A\int t^{n}dt = A\left(\frac{1}{n+1}\right)t^{n+1}$$

Write the integrals of the following:

$$\int (AB)t^{4}dt =$$

$$M\int t^{B}dt =$$

If H = (C)(F) , C = 2 , F = 3 ,

$$C\int t^{H}dt =$$

* * * * *

* * * * *

$$\int (AB)t^{4}dt = (AB)\left(\frac{1}{5}\right)t^{5}$$

$$M\int t^{B}dt = (M)\left(\frac{1}{B+1}\right)t^{B+1}$$

$$C\int t^{H}dt = (2)\left(\frac{1}{7}\right)t^{7} = \frac{2}{7}t^{7}$$

2. If a fragment of a system is represented by the following flow diagram,

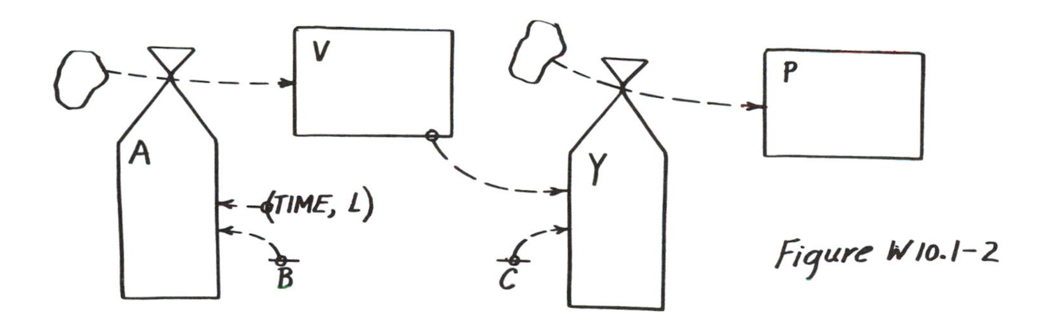

Figure W10.1-2

and if the equations of the system include the following,

A.KL = (B)(TIME.K)

Y.KL = V.K/C

B = 6

C = 3

TIME--Time (seconds)

write the expressions in continuous calculus notation that describe V and P as a function of time, assuming initial values of zero.

V =

P =

* * * * *

* * * * *

Where (TIME) = t

$$V = B\int t dt = \frac{B}{2}t^2 = 3(\text{TIME})^2$$

$$P = \frac{1}{C}\int Y dt = \frac{1}{3}3t^2 dt = \frac{3}{3}\left(\frac{1}{3}\right)t^3 = \frac{1}{3}t^3 = \frac{1}{3}(\text{TIME})^3$$

3. From calculus, the general form of the integral of an exponential is

$$\int Ae^{Bt}dt = A\frac{1}{B}e^{Bt}$$

The process of integration does not change the exponential except for a multiplier term. Complete the following

$\int e^{7t}dt =$ ______________________

$\int e^{t/4}dt =$ ______________________

If C = (A)(B) , B = 7D , A = 5 , D = 1/5

$\int Ae^{Ct}dt =$ ______________________

$\int e^{t/T}dt =$ ______________________

* * * * *

* * * * *

$$\int e^{7t}dt = \left(\frac{1}{7}\right)e^{7t}$$

$$\int e^{t/4}dt = 4e^{t/4}$$

$$\int Ae^{Ct}dt = (5)\left(\frac{1}{7}\right)e^{7t}$$

$$\int e^{t/T}dt = Te^{t/T}$$

4. In the following flow diagram for a first-order feedback loop,

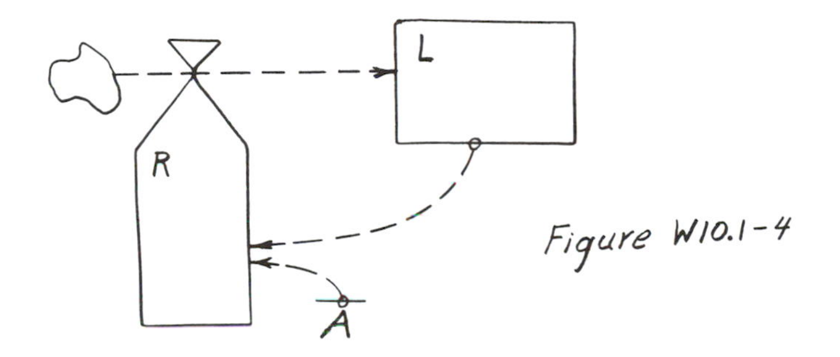

Figure W10.1-4

the equations are

R.KL =	L.K/A	R
L.K =	L.J + (DT)(R.JK)	L
L =	8	N

Suppose we guess that the time-shape of L can be given in continuous calculus notation by

$$L = 8t^2$$

Could this be correct? When t = 0 what is L?

____

____

* * * * *

* * * * *

No. The solution says that L = 0 at t = 0 but the initial condition of L is 8.

5. For the system of Frame 4, suppose we guess that

$$L = 8 + t^2$$

Could this be correct based on conditions at t = 0? ______________

With this assumed value of L, what is R?

R = ______________________________

* * * * *

* * * * *

Yes.

$$R = \frac{8}{A} + \frac{1}{A}t^2$$

6. With the value of R from Frame 5, and with Frame 4 showing that L is the integral of R,

L = ______________________________

= ______________________________

= ______________________________

* * * * *

* * * * *

$$L = 8 + \int_o^t R\,dt$$

$$= 8 + \left[\frac{8}{A}t + \left(\frac{1}{A}\right)\left(\frac{1}{3}\right)t^3\right]_o^t$$

$$= 8 + \frac{8}{A}t + \frac{1}{A}\left(\frac{1}{3}\right)t^3$$

7. If the assumed solution for L in Frame 5 is to be correct, it must be identical to the statement for L in Frame 6. Is there any value of A which will make the two expressions for L identical for all values of t? ________________________

Can the assumption for L in Frame 5 be correct? ________________

* * * * *

* * * * *

No, no.

W10.2 Integration Creates the Exponential

1. A system is described by:

   M.KL = (2)(P.K)/4 R

   P.K = P.J + (DT)(M.JK) L

   P = 5 N

   DT = 5 C

   This is a (positive/negative) loop. ________________

   The time constant is ______________.

   The equilibrium value is ________________.

   Is the value for solution interval reasonable? _______________.

   * * * * *

   * * * * *

   positive

   $2 = \frac{1}{2/4}$

   zero (In the rate equation, any value of P above zero causes positive growth, any value of P less than zero causes negative growth.)

   no (According to Principle 6.2-1 it must be less than 1, perhaps 0.4.)

2. In the following system,

   SA.K = SA.J + (DT)(LB.JK) L

   SA = 20 N

   LB.KL = (10 — SA.K)/6 R

   Is the loop positive or negative? ______________________

   What is the equilibrium value? __________________________

   What is the time constant of response? ___________________

   What value has the rate LB at t = 0? ____________________

   What value of DT should be used? ________________________

* * * * *

* * * * *

negative (SA enters the rate equation with a negative sign.)

10 (Note that the rate declines to zero when SA = 10.)

6 (1/6 is the multiplier of the level SA in the rate equation.)

$-\frac{10}{6}$

About 1 time unit for DT

3. On the following graph, sketch the value of S vs. time for the following system.

B.KL =	(5 − S.K)/10	R
S.K =	S.J + (DT)(B.JK)	L
S =	15	N

Show the time constant clearly on the graph.

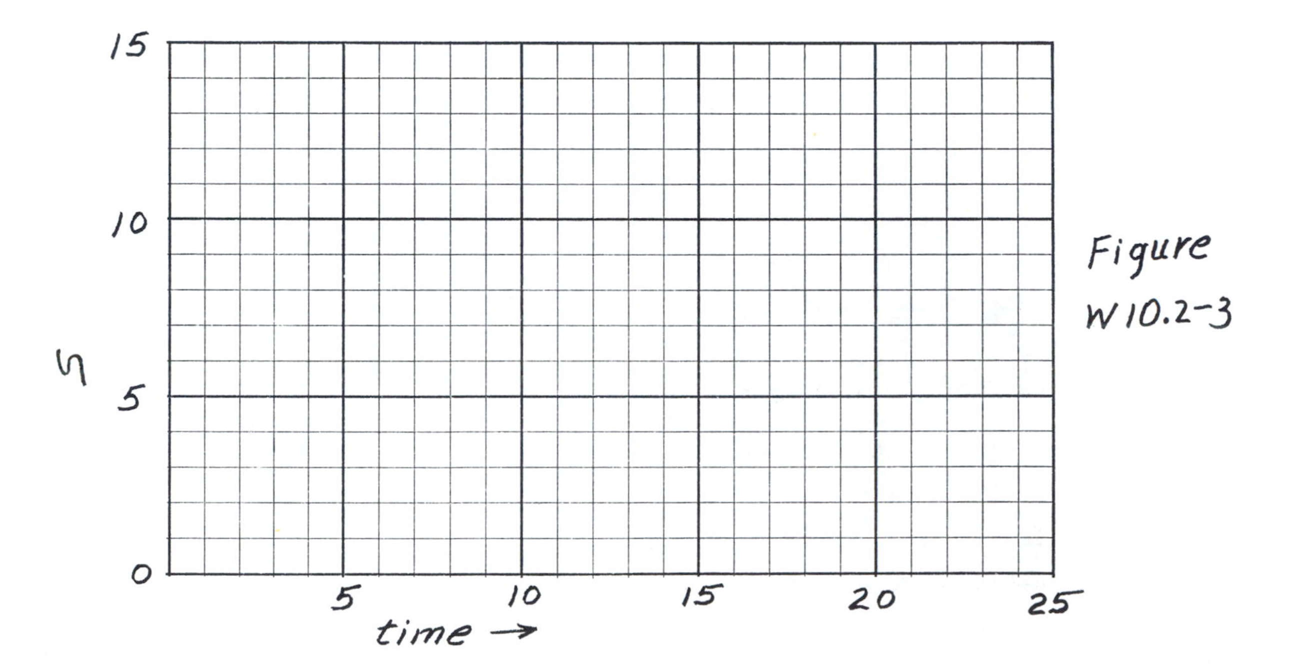

Figure W10.2-3

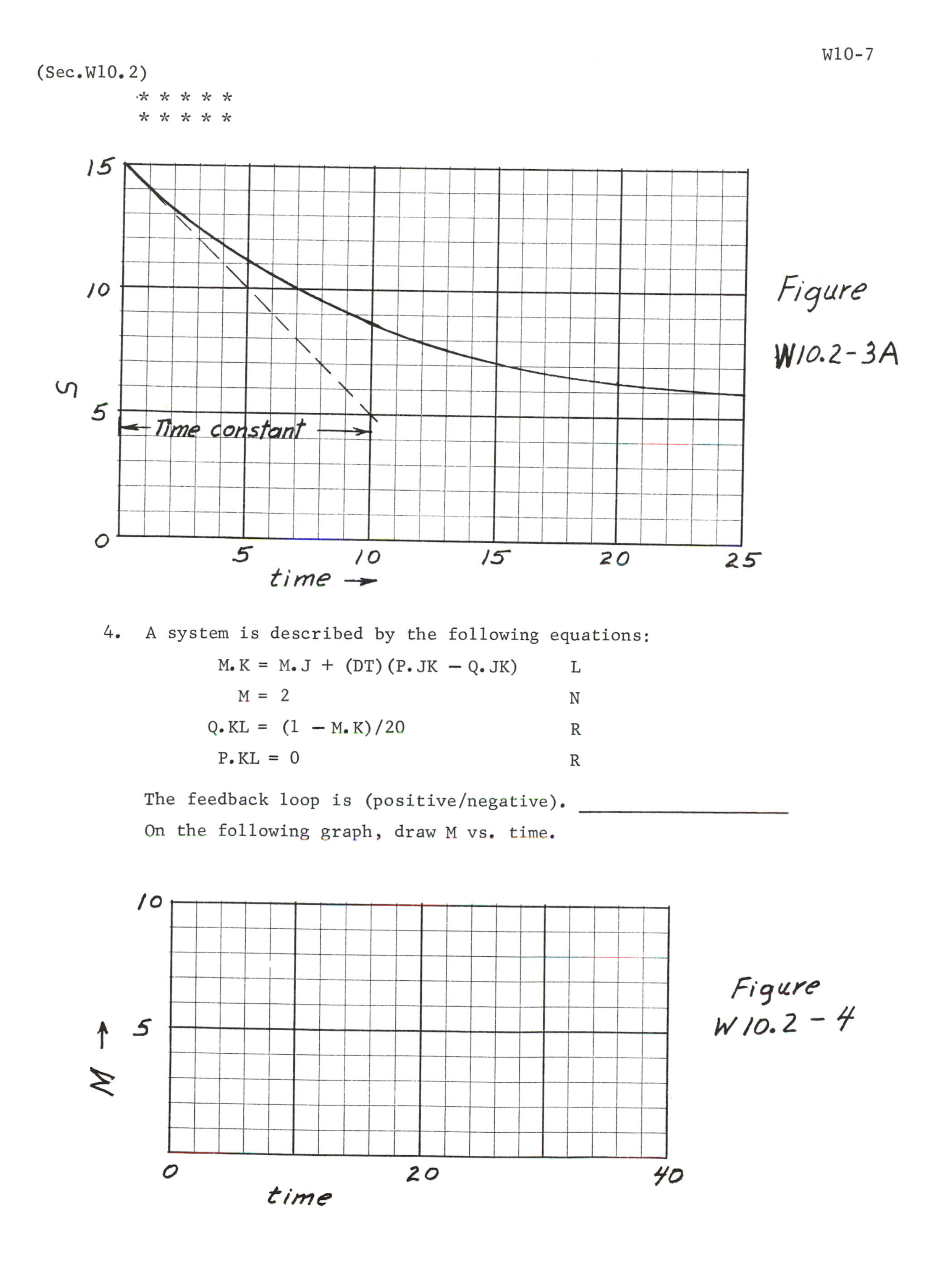

(Sec.W10.2)

* * * * *
* * * * *

Figure W10.2-3A

4. A system is described by the following equations:

M.K = M.J + (DT)(P.JK – Q.JK)	L
M = 2	N
Q.KL = (1 – M.K)/20	R
P.KL = 0	R

The feedback loop is (positive/negative). ____________

On the following graph, draw M vs. time.

Figure W10.2-4

* * * * *

* * * * *

Positive (Note that the negative rate enters the level equation with a second negative sign.)

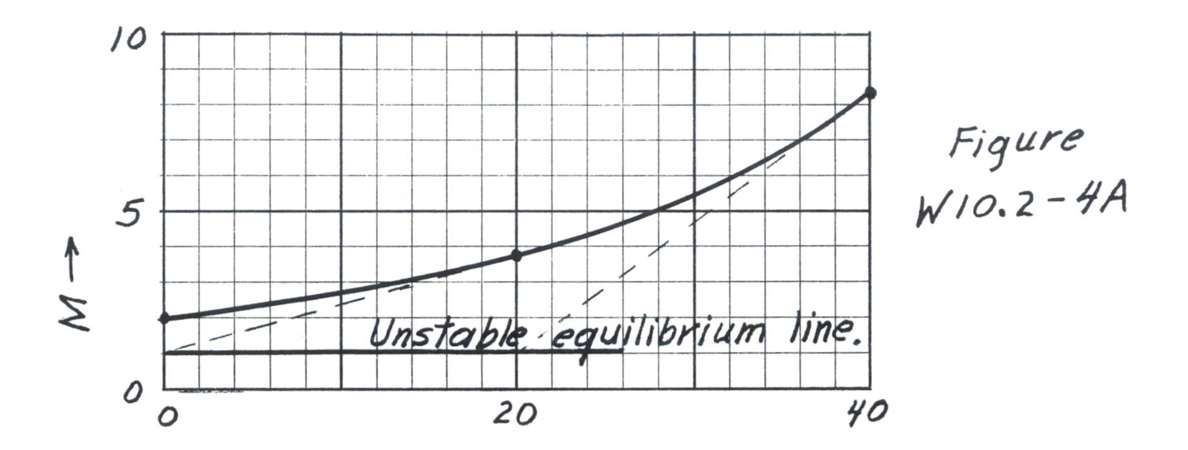

Note that the initial value lies above the "goal" value of 1 which for the positive feedback loop is the unstable equilibrium value. If the initial value of M were less than 1, the value of M would move ever more rapidly in the negative direction. In one time constant, 20 time units, the deviation from equilibrium increases by a factor e. At time 20

$$M = (1)(2.718) + 1 = 2.718 + 1 = 3.718$$

where the "1" that multiplies 2.718 is the amount that M starts above equilibrium and the "+1" term is the equilibrium value. At time 40, $M = (2.718)(2.718) + 1 = 7.39 + 1 = 8.39$

5. Assume that the growth of a consulting firm, which is operating in a field where the demand exceeds the supply, is limited by the rate at which new members of the firm can be found and trained. Suppose that the firm has 15 men and that on the average each member of the firm finds and trains a new man each 12 months. Write the DYNAMO format equations for this system. A complete set of equations should include the solution interval.

* * * * *

* * * * *

$$M.K = M.J + (DT)(HR.JK) \quad L$$

$$M = 15$$

$$HR.KL = M.K/12 \quad R$$

$$DT = 2 \quad (\text{or } 3) \quad C$$

6. What is the system time constant in Frame 5? ____________
   How many men will the firm have at the end of one year? ____________
   Two years? ____________

* * * * *

* * * * *

12 months, 40, 109

7. Suppose that acquiring professional staff for the consulting firm operates as a two-stage process. Men TM are added to the training-and-recruiting office at a rate TMR proportional to the professional working staff PS of the firm. (PS again starts at 15 men.) One man per year is added to TM for each ten men on the professional staff. The professional staff expansion rate PSR is five men per year for each training office man TM. Draw the flow diagram of the system.

* * * * *

* * * * *

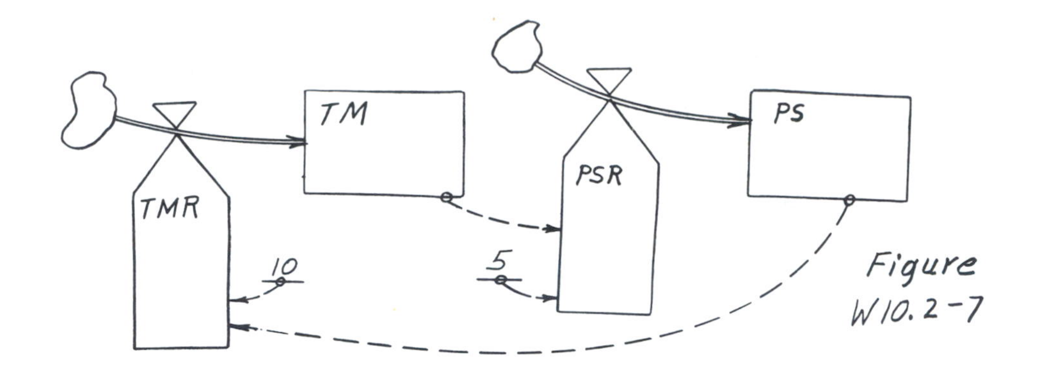

Figure W10.2-7

8. Write the equations for the system in Frame 7.

* * * * *

* * * * *

TMR.KL	=	PS.K/10	R
TM.K	=	TM.J + (DT)(TMR.JK)	L
TM	=	Initial value	N
PSR.KL	=	(TM.K)(5)	R
PS.K	=	PS.J + (DT)(PSR.JK)	L
PS	=	15	N
DT	=	.1	C

(The solution interval is here taken as half the "time constant" of 1/5 in the equation for PSR. The value of 1/5 would be the time constant of a simple first-order loop that was closed through a level directly back to PSR. Actually as could be seen from the actual loop time constant in Frame 9, the interval might be longer.)

9. What is the growth time constant of the system in Frames 7 and 8?

____

* * * * *

* * * * *

$T = \sqrt[2]{\left(\frac{1}{5}\right)(10)} = \sqrt{2} = 1.4$ years.

(As given in the description of the time constant of growth in Principle 10.2-3)

10. In Frame 9, about how long is required for the professional staff to double? ________________________

* * * * *

* * * * *

Approx. 1 yr. (The doubling time is approximately 0.7 of a time constant. Doubling time = 0.7 x 1.4)

11. In Frame 8 what should be the initial value of TM to be consistent with the exponential growth rate? ____________________________

* * * * *

* * * * *

$TM = 2.1 \text{ men} = (15)\sqrt{\frac{1/5}{10}} = 15\sqrt{\frac{1}{50}} = \text{approx. } \frac{15}{7}$

(Equation 10.2-18 gives the relationship which must exist at any time if unperturbed exponential growth is in progress.)

12. Suppose the effectiveness of the training staff TM were less so that they add only 2.5 professional staff per year per man in the training-and-recruiting office. What would be the growth time constant? _______________________.

What time would be required to double the staff? __________________.

* * * * *

* * * * *

2 years, 1.4 years.

13. For the conditions of Frame 12, what is the relationship between PS and TM during exponential growth? ____________________________.

* * * * *

* * * * *

$$TM = PS\sqrt{\frac{1/2.5}{10}} = PS\sqrt{\frac{1}{25}} = \frac{PS}{5}$$

(Note that TM is 1/5 of PS rather than $\frac{1}{7}$ as before. Growth takes 40% longer and the cost of the training-and-recruiting office is almost 50% greater than before.)

W10.3 Integration Creates the Sinusoid

1. For a pendulum swinging through a small angle, the equations of motion are often stated as

$$A = \frac{G}{L}(-P)$$

$$V = \int A\,dt$$

$$P = \int V\,dt$$

A--Acceleration of the pendulum (feet/sec^2)

G--Acceleration of gravity, constant = 32.2 (feet/sec^2)

L--Length of pendulum (feet)

P--Position of pendulum (feet)

V--Velocity of pendulum (feet/sec)

Rewrite these equations in DYNAMO compiler format adding anything necessary for a complete system description for a pendulum which is 8.05 feet long, started from rest at an initial position displacement of one foot.

* * * * *

* * * * *

```
 A.KL = (G/L)(— P.K)          R
    G = 32.2                  C
    L = 8.05                  C
  V.K = V.J + (DT)(A.JK)      L
    V = 0                     N
PC.KL = V.K                   R
  P.K = P.J + (DT)(PC.JK)     L
    P = 1                     N
```

PC--Position change (feet/sec)

(Note that the rate which changes the position is given a separate name from velocity, implying a dimensionless unity multiplier.)

2. Sketch the flow diagram for the system of Frame 1.

* * * * *

* * * * *

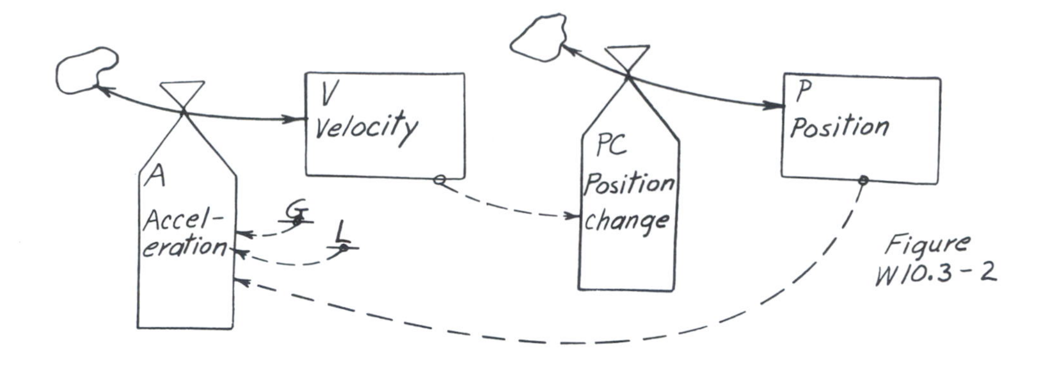

Figure W10.3-2

3. Using Principle 10.3-1, what will be the period for a complete swing and return of the pendulum? ____________________

* * * * *

* * * * *

$$P = 2\pi\sqrt{\left(\frac{L}{G}\right)(1)} = 2\pi\sqrt{\frac{8.05}{32.2}} = \pi = 3.14 \text{ sec.}$$

4. Suppose the hiring rate of workers is proportional to the excess of backlog over 2,000 units. When the backlog is 2400 units, workers are hired at 5 men per month. Each man manufactures 2 units per month. Orders flow in at a uniform rate of 250 units/month. Shipments are made and the backlog is decreased as the items are manufactured. The initial backlog is 2000 units and initial employment is 70 men. Draw the flow diagram.

* * * * *

* * * * *

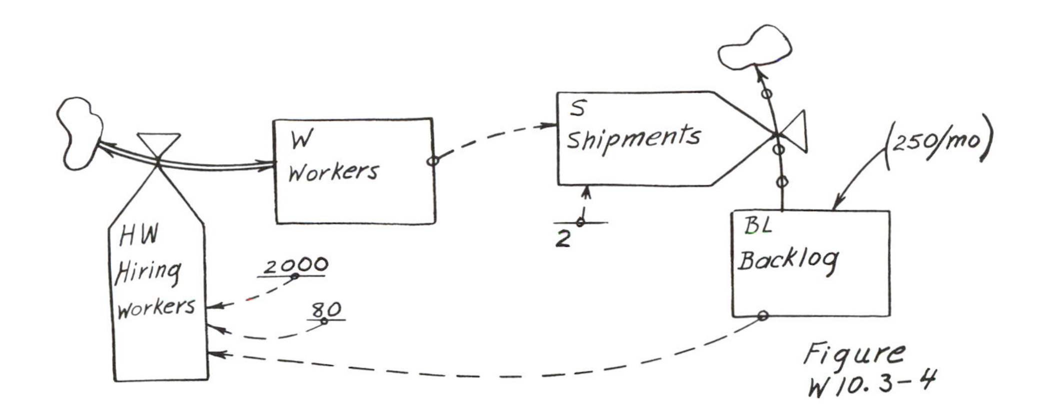

Figure W10.3-4

Note that the hiring rate is $\frac{1}{80}$ $\frac{\text{man/mo}}{\text{unit}}$ times the backlog discrepancy.

$$5 = \frac{1}{80}\ (2400 - 2000)$$

5. The system in Frame 4 is a (positive/negative) feedback loop.

   Why?____

   ____

   ____

* * * * *

* * * * *

Negative. Rising backlog increases hiring which increases workers which increases shipments which <u>reduces</u> backlog. There is one reversal of sense around the loop.

6. Write the equations for the system of Frame 4.

* * * * *

* * * * *

$$HW.KL = \frac{1}{80}(BL.K - 2000) \quad R$$

$$W.K = W.J + (DT)(HW.JK) \quad L$$

$$W = 70 \quad N$$

$$S.KL = (W.K)(2) \quad R$$

$$BL.K = BL.J + (DT)(-S.JK) \quad L$$

$$BL = 2000 \quad N$$

7. Given the system described by the equations in Frame 6, will fluctuation of employment occur? ______________________

Why? ______________________

____

____

* * * * *

* * * * *

Yes. The system is a second-order negative-feedback loop with simple integrations so it will show sustained oscillation if disturbed. The initial employment is less than the equilibrium level so backlog will begin to build up.

8. In the preceding system, what will be the period of oscillation in employment?

* * * * *

* * * * *

$$P = 2\pi\sqrt{(A1)(A2)} = 2\pi\sqrt{(80)\left(\frac{1}{2}\right)} = 2\pi\sqrt{40}$$

$$= \text{about 40 months}$$

9. Suppose in an agricultural commodity market that prices rise and fall with world inventories I without time lag. Inventories are normally a one-year supply. Change in production capacity CPC is proportional to price, being zero when inventories are normal and being increased 1 percent of equilibrium production capacity EPC per month for each 10 percent fall from equilibrium inventory EI. Production capacity PC starts at normal; inventory is 20 percent below normal. The commodity is produced at a rate PR equal to the production capacity. Assume consumption is constant. Draw the flow diagram. Write the rate equation for CPC.

* * * * *

* * * * *

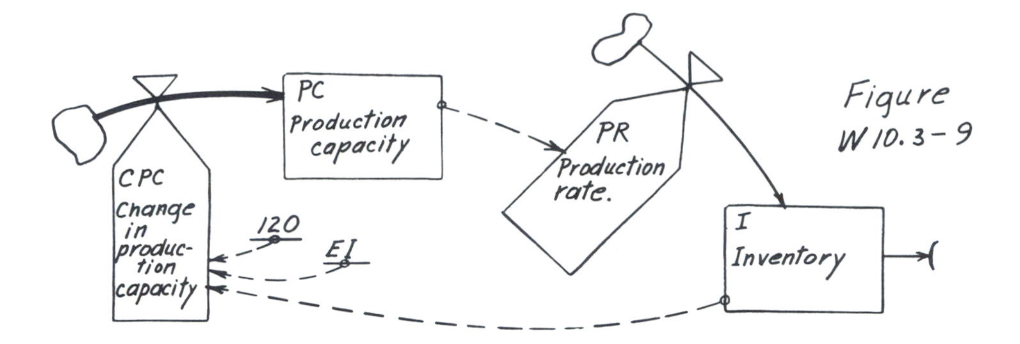

$$CPC.KL = \frac{1}{120}\,(EI - I.K)$$

Note that equilibrium production rate is 1/12 per month of equilibrium inventory. The statement of the problem gives EPC = EI/12 and

$$.01\left(\frac{EI}{12}\right) = \frac{1}{C}\,(.1)(EI)$$

$$\frac{.01}{12} = \frac{.1}{C}$$

$$C = 120\;\left(1/mo^2\right)$$

10. Sketch inventory versus time and label the scales.

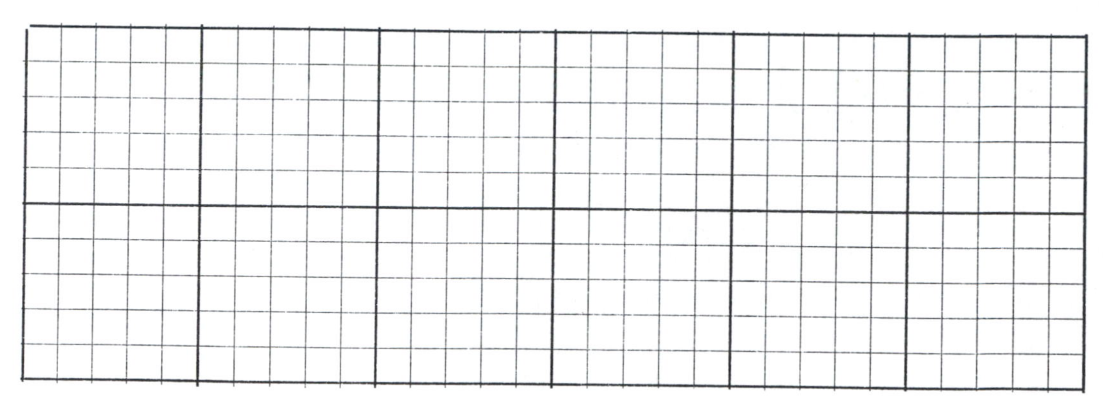

Figure W10.3-10

* * * * *

* * * * *

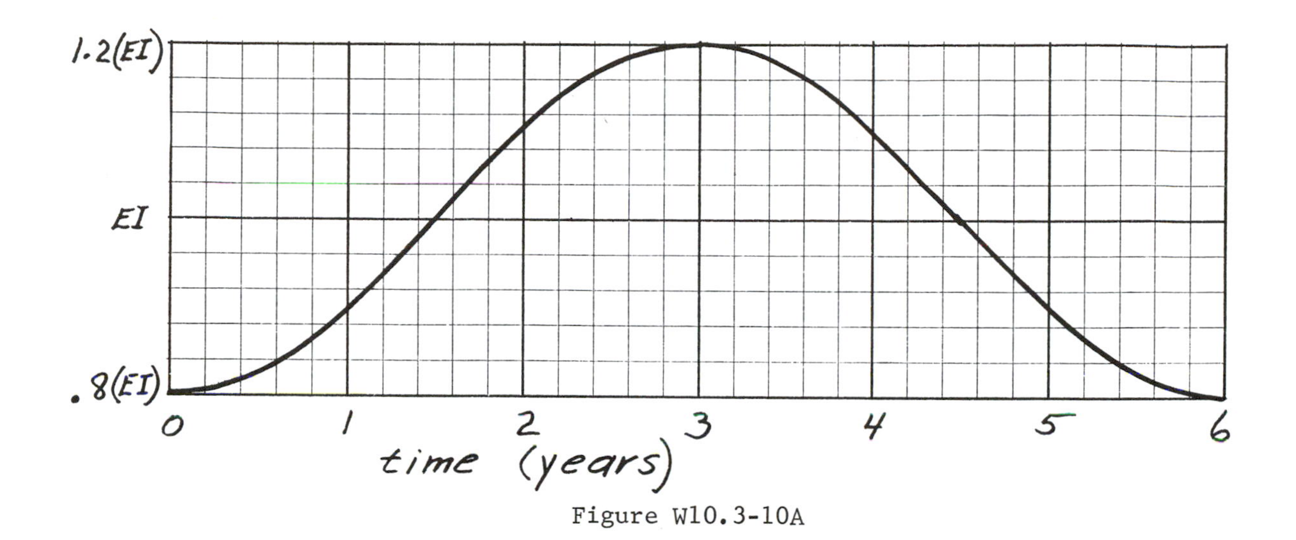

Figure W10.3-10A

$$\text{Period} = 2\pi\sqrt{(120)(1)} = 2\pi\sqrt{120}$$

$$= \text{approx. } (6.3)(11) = \text{approx. } 70 \text{ mo.}$$

$$= \text{approx. } 6 \text{ years}$$

W10-20